Sludge Tracker

SLUDGE TRACKER

Toxicus ad Infinitum

The Adverse Impact of Land-Disposed

Toxic Sewage Sludge

on Human & Environmental Health

Richard C. Honour, PhD
&
Michelle Horkings-Brigham

HORUS PUBLISHING, INC.

SLUDGE TRACKER

TOXICUS AD INFINITUM – The Adverse Impact of Land-Disposed *Toxic Sewage Sludge* on Human & Environmental Health

Cover designed by Melissa Peizer

ISBN: 978-0-9652621-5-6

HORUS PUBLISHING, INC.
P.O Box 1127
McKenna, WA 98558, U.S.A.
https://www.sludgetracker.com
info@sludgetracker.com

Disclaimer: All events presented in this book are real; however, all conversations are recorded "as remembered" and have been edited for storytelling purposes. Names have been substituted to protect identities, except where public figure comments exist in the government record. Please note: The authors do not intend to cause harm to any individual mentioned or referred to herein, or otherwise noted or mentioned. Please contact the publisher and we will correct any misinformation presented.

Key words and terms: *Toxic Sewage Sludge*, sludgers, biosolids, wastewater effluents, wastewater treatment plant (WWTP), publicly owned treatment works (POTW), physical inerts, manufactured inerts, landfill leachates, legacy landfill leachates, global warming, climate change, climate crisis, climate emergency, ecocide, environment, environmental degradation, contamination, contaminants, pollution, pollutants, toxin, toxicant, toxics, public health, pathogen, infectious disease, oncogen, carcinogen, cancer.

Dear Reader,

This creative work and effort of investigative reporting represents the authors' actual, real-life, personal experiences as an after-the-fact narrative about the wanton and reckless practice of land-disposed *Toxic Sewage Sludge* and the observed adverse impacts on people and the environment. The work is not presented as being the outcome of rigorous scientific endeavor, but is offered solely to inform and empower, for free and open public discussion and disclosure assure public awareness and participation.

Land disposal of *Toxic Sewage Sludge* is hardly a simple matter of nuisance odors or unsightliness, but rather, *Toxic Sewage Sludge* is toxic solid waste. *Toxic Sewage Sludge* disposed in our forests and sprayed on our lands used to grow food is an ongoing practice of substantial concern because it kills living things on an immediate and/or long-term basis, without regard for the essential nature of our air, food, soil, and water. What then could be the solution to this growing *Toxicus ad Infinitum* (i.e., Toxic to Infinity) crisis?

The word "toxin," as used herein, generally refers to poisonous chemicals, materials or substances produced by living things, while the word "toxicant" generally refers to poisons generated or derived from non-living entities. Both are abundant in *Toxic Sewage Sludge*. Collectively, toxins and toxicants are referred to as "toxics," meaning any substance or material that contains a poison that may be capable of causing debilitating harm, disease or death.

Being toxic or poisonous is not a matter of dose, it is a matter of inciting interference with, or the impairment of, any biological function at any cellular, tissue, organ or organism level, at any concentration, either singly or in combination with any other chemical, chemicals or substances.

For this reason and many more, a discussion of the dangers of *Toxic Sewage Sludge* disposal is not only critical, but must include a workable solution, while revealing the consequences if we continue along our current path of "pollution as policy" and "commerce versus the better interests of human and environmental health."

Resources that support the viewpoints of the authors are included at the back of this book, arranged by chapter for ease of reference, and added for your further investigation. More information can be found at www.sludgetracker.com.

This book is dedicated to those many good souls,
past and present, who strive to make this world a better,
healthier, and more enlightened experience for all.

TABLE OF CONTENTS

PRAEFATIO

BON APPÉTIT!

The Adverse Impact of Land-Disposed *Toxic Sewage Sludge*

You know it's darkest before the dawn,
this thought keeps me moving on.
If we could heed these early warnings,
the time is now quite early morning.

Pete Seeger, *Quite Early Morning*, 1975.[1]

By the very food we eat, we are self-exposed to toxic wastes on a daily basis, if we intend to eat at all. No matter if we are vegans, vegetarians, carnivores or persons of mixed diet, with few exceptions, we are all exposed throughout our lives to toxic wastes, the inciting agents of nearly all chronic diseases. Consuming foods that have been free from exposure to fertilizers, pesticides or *Toxic Sewage Sludge* can be a life extender, if such foods can be found. The fact that we are experiencing a global explosion of cancer and other chronic diseases, far exceeding the rate of population growth, bears witness to our ever-increasing exposure to toxic wastes.[2]

Through inquiries of directly affected citizens, analysis of peer-reviewed scientific literature, and on-site field investigations of ground-zero *Toxic Sewage Sludge* dump sites in forests and on farmlands, we have learned that land-disposed *Toxic Sewage Sludge* results in a broad spectrum of adverse human and environmental health impacts. Importantly, we have learned that the term "organic" does not and must not be associated with any form or source of *Toxic Sewage Sludge* or its resulting commercial products. While "organic" may be relevant when associated with soils, compost or foods, it must never be used recklessly when the core matrix of the matter is associated with any form or source of toxic waste, and *Toxic Sewage Sludge* is toxic waste.

During passage or processing of raw sewage through a wastewater treatment plant, the actual "organic" content of raw sewage is digested and removed in large part by the microbial degradation of organic matter, a field of science in and of itself. The remaining residual *Toxic Sewage Sludge* product includes additional partially digested or undigested organic matter, plus the more toxic components of sewage, including toxic metals, and hundreds of thousands of chemicals and biochemicals also containing carbon, but that do not pass any criterion or test as being "organic." By these definitions and our findings, it becomes clear that you cannot be a champion of land-disposed *Toxic Sewage Sludge* and also be an advocate for human and environmental health.

In chemistry, the chemical definition of organic is based solely on chemical structure. With very few exceptions a chemical is classified as organic if it contains at least one carbon atom, regardless of its source or activity. The agricultural definition of organic differs from the strictly chemical use of the term and is focused on product labeling. For a food or feed product to be labeled as "organic" there are agronomic and animal husbandry practices that must be followed. According to the U.S. Department of Agriculture (USDA), organic meat, poultry, eggs, and dairy products come from animals that are given no antibiotics or growth hormones. Organic food must be produced without the use of pesticides, fertilizers made with synthetic ingredients, or sewage sludge. USDA's Organic Rule applies not only to food but also to any and all agricultural products. Of particular note: the USDA publishes guidelines for making the kind of compost that is permitted in organic production, and sewage sludge is excluded.[3]

However, the Clean Water Act (CWA), Clean Air Act (CAA), Endangered Species Act (ESA), Safe Drinking Water Act (SDWA), and Food Safety Modernization Act (FSMA) all allow for the land and water disposal of *Toxic Sewage Sludge* as a business opportunity in the name of "organic," "beneficial reuse," "soil amendment," or the like. The outcome is that people and the environment suffer, as witnessed by increasing rates of human cancer and other chronic diseases, as well as by the continuing decline of salmon, orcas, and other indicator species, all as a direct result of failed infrastructure and legislation. It is the land disposal of *Toxic Sewage Sludge* and the resulting leachates therefrom, plus toxic wastewater effluents from wastewater treatment plants, plus leachates and runoff from landfills and land-based agriculture, that poison our living environment.

Globally, there are two main categories of agriculture focused separately on plant- and animal-based products, and that provide most nutrition for humans. The U.S. is the world's largest beef producer and the second-largest beef exporter, while milk has a farm value second only to beef among livestock products. The U.S. is also the world's largest poultry producer and the second-largest exporter of poultry, mainly as "chicken meat." As well, the U.S. is the world's second-largest pork producer. Farmers' sales of livestock—beef, dairy, poultry, and pork—account for more than half of all U.S. agricultural cash receipts, exceeding $160 billion per year.[4]

Almost all feed materials for the global beef, milk, poultry, and pork industries are comprised of cereal grains, corn, and various byproducts derived largely from corn and wheat. Notably, most cereal grain feed products used in animal agriculture in the U.S. and in much of the rest of the world are grown in soils exposed to fertilizers, pesticides, and other toxic materials, including *Toxic Sewage Sludge*. In the U.S., more than half of all plant agriculture products are cultivated in soils exposed to *Toxic Sewage Sludge*.[5]

When considering that the single greatest cash value animal food product in the world is chicken meat, whether proclaimed as organic or not, it is to be recognized that most chickens in the world are fed cereal grain products and byproducts grown in those same fertilizers, pesticides, and *Toxic Sewage Sludge*. If chicken meat is at the top of the human food chain, then what contaminants and pollutants are carried forward from agricultural soils and animal feed products into the resulting chicken meat products? The answer is known just marginally.

Bonus Appetitus!

As you are about to discover, it is all done with broad agency acknowledgement and approval, despite the vast trove of science that speaks to the contrary. In any case, the authors do not come from a place where doing nothing is an option.

RCH & MH-B

Sludge Tracker

Introduction

It's Not What You Think

Every industrial and municipal polluter in the country has a huge stake in protecting lax EPA regulations, which allow pollutants concentrated at wastewater treatment plants to be applied to every available parcel of land, from farms and forests to school playgrounds.

David L. Lewis, *Science for Sale,* 2014.[1]

Land-disposed *Toxic Sewage Sludge* and other toxic wastes are destroying the quality of our air, food, soil, and water, not just in the U.S. Pacific Northwest, the focal point of this tome, but worldwide, while filling the coffers of farmers, contractors, collaborators, and their co-conspirators.[2] Here in Washington State, farmers especially like the practice, because it enhances their profitability, since they can use this low-cost, agency-provided toxic waste product as a contrived "Fertilizer Substitute." It's much cheaper than buying commercial fertilizers. Truckers, spreaders, and farmers thrive on the application of farm-dumped *Toxic Sewage Sludge*, even though it endangers the health of us all, and that of our living environment.

Forest-disposed *Toxic Sewage Sludge* is especially gruesome, in that actual people use the same access roads and trails to gain entry for hiking, biking, camping, fishing, and hunting, as well as for so many other family- and outdoor-oriented recreational opportunities by which we are blessed in Washington State. The locals, without realizing the danger, live in it.

As one prime example, King County, located in Washington State, generates 120,000 or so wet tons of *Toxic Sewage Sludge* per year, most of which originates from wastewater treatment plants within the county itself, as well as from some surrounding regions and communities. When considering that King County represents perhaps half the population of Washington State, the rest of the state may produce yet another 100,000 wet tons or more of *Toxic Sewage Sludge* per year.

King County's *Toxic Sewage Sludge* represents the residual solids product from about 178 million gallons per day of raw sewage and other wastewaters that pass through the system. Nearly all of that *Toxic Sewage Sludge* product is destined by way of double dump trucks for the forests and farms of Washington's Cascade Mountains and for the critical agricultural lands to the east. One individual has been tracking and monitoring this state- and county-sponsored public health debacle for decades, much to the dismay of county and state agencies, as well as of their *Toxic Sewage Sludge* dumping minions.

Hauling, dumping, spreading, and thereafter ignoring the adverse consequences of land-disposed *Toxic Sewage Sludge* in Washington State is a multimillion-dollar, cash-generating industry. The suspected adverse consequences include air, food, soil, and water contamination and pollution, with the continuing demise of our state's once pristine ambiance as a recognized outcome. As well, the concepts of environmental degradation and ecocide appear to have no measure of concern here at all in the presence of the deep-rooted *Toxic Sewage Sludge*-dumping trade.

Toxic Sewage Sludge from King County's wastewater treatment plants (WWTPs) and publicly owned treatment works (POTWs) represents about half of Washington State's *Toxic Sewage Sludge*, thereby representing likewise about half of the state's population. As Washington's population grows, so too does its need for modern and functional infrastructure to process and manage its greater-than-proportional generation of wastes, including gaseous, liquid, and solid forms of waste alike. The rate of increase in waste generation is actually exceeding the rate of population growth because of expanding commerce and consumerism, and while this may be good news for commerce, the resulting adverse consequences on human and environmental health are both ignored and disastrous.

Lagging behind is the basic infrastructure needed to properly manage the ever-growing deluge of overly desired, constantly produced, often non-essential products and the resulting cumulative waste streams generated thereby, as both solid and liquid wastes, plus the toxic gaseous outpourings therefrom. The primary victims of this out-of-control generation of wastes are our air, food, soil, and water, plus all of the condemned and dependent life forms in the way or associated therewith, which includes us, due to our seeming inability to safely manage the disposal of our toxic wastes.

One logical solution to this dilemma is curtailment of the production of excess or non-essential products beyond bare necessity, but who then would decide the criteria that define "essential?" As well, our global culture and economy have become highly addicted to making, transporting, selling, and buying things, so modifying consumer behaviors is also quite unlikely. Instead, would it not be more prudent to consider how best to safely and efficiently manage and dispose of today's excessive and oftentimes dangerous waste products?

As a society we must awaken to the fact that it is unhealthy to live in our own wastes, much less in the cumulative, collected, and transported wastes of others, no matter the cover story. The ill-conceived practice within the United States and elsewhere of the land disposal of *Toxic Sewage Sludge* (aka "sludge," "sewage sludge" or "biosolids") represents collectively an environmental crisis, a public health crisis, and a probable contributing source of the exigent global climate crisis. Writing for *Bulletin of the Atomic Scientists* in March, 2022, Robert Socolow noted: "Climate change is a manifestation of prosperity. Other things being equal, climate change will arrive more slowly if greenhouse gas emissions from human activity plunge to a small fraction of their current value and stay down."[3] It's all a matter of infrastructure, the total failure of which reflects the growing health crisis here in Washington State, as one blatant example, and globally.

An important lesson, before we begin, is that there is no such thing as "biosolids," and there are no such things as: "Class A, Class B or Exceptional Quality (EQ) Biosolids."[4] These are artificial social constructs and marketing terms created by agencies and the *Toxic Sewage Sludge*-disposal industry. Collectively, they seek no more than to achieve the lowest cost, most convenient disposal of our ever-growing waste streams. Public health appears to be of little concern or consideration, much less with regard to environmental health or sustainability. *Toxic Sewage Sludge* is toxic waste; it is NOT "biosolids," whatever that contrived marketing term may be intended to mean or imply.

Also, and to clarify, *Toxic Sewage Sludge* is not a "Valued Resource," a "Fertilizer Substitute" or a "Soil Amendment." Land-disposed *Toxic Sewage Sludge* does not represent "Beneficial Reuse" or "Residuals Management," and furthermore, land-disposed *Toxic Sewage Sludge* does not "Enhance Wildlife Habitat" or "Embody Resource Conservation," and it is not "Resource Recovery." These terms are agency monikers intended to deceive the public and others. *Toxic Sewage Sludge* is toxic waste, end of story!

None of the public's concerns about open toxic waste disposal appear to be of any genuine interest to our state and county agencies, even though many of their waste disposal practices and the consequences therefrom are in direct or at least marginal violation of, or in contradiction to, the Clean Water Act (CWA), the Safe Drinking Water Act (SDWA), the Endangered Species Act (ESA), and the Resource Conservation and Recovery Act (RCRA). A key shortcoming of these Federal Acts is that they all include carve-outs that are intended to protect and preserve the land disposal of *Toxic Sewage Sludge* and the water disposal of toxic wastewater effluents from our WWTPs and POTWs, no matter the adverse human and environmental health consequences. Commerce rules, while public health and environmental quality are compromised, and it is time for that to change. If the decline and demise of Washington State's major waterways, such as and including the Columbia and Snake Rivers, Lake Washington, Puget Sound, the greater Salish Sea, and the near-countless other in-feeding waterways and tributaries that eventually make their way to the Pacific Ocean, are not enough to make you squirm after reading this narrative, then we all need help.

The Washington State Departments of Health, Natural Resources, and Ecology have applauded the low cost and more convenient practice of disposing of *Toxic Sewage Sludge* in our forests, simply because the forests of the Cascade Mountains are so much closer to WA's most populated areas, which can be preferable to making the near-daily, hundreds of miles longer hauls to Central Washington to dump polluted wastes on soils that provide for our vast food resources. One only has to consider the obvious adverse effect of greenhouse gas emissions on human health and the environment by the very fact that thousands of double dump trucks from King County trek over the Cascade Mountains belching endless toxic exhaust gases just to deliver their deadly loads of *Toxic Sewage Sludge* to Central Washington's farmlands, as documented herein.[5]

If contamination and pollution of our air, food, soil, and water resources are in fact matters of human and environmental health, and thereby a noted threat to national security, how can they pull this off? And, if the greenhouse gases emitted from the fossil fuel guzzling, *Toxic Sewage Sludge*-hauling double dump trucks, and generated by processing raw sewage in wastewater treatment plants and publicly-owned treatment works, plus from the aerobic and anaerobic decomposition of land-disposed toxic sludges and water-disposed toxic wastewater effluents of and by themselves are not enough to make you wonder about the peculiar stories told by our municipal, county, and state infrastructure managers, then we all need to take a deeper look at the vast and multiple sources of agency-generated contaminants, pollutants, and greenhouse gases produced by those said mechanisms, facilities, and practices, and the adverse consequences therefrom.

Monitoring these appalling agency practices has been a dedicated campaign, decades-long, by a self-identified sludge tracker whose story and discoveries are chronicled in these pages as a warning to us all. The adventure told is a result of up-close and personal encounters with tens of thousands of children with cancer, none of whom had ever incited their own disease. It was always someone else before them who caused their cancers, and it was and remains the willful and reckless open disposal of toxic wastes into our living environment that causes the pain, and just as often the death, of many of cancer's victims, as well.

Sludge tracking, in this personal instance, has involved hiking and mountain biking into the forests of Washington State's Cascade Mountains or driving through the high-country prairies and farmlands of Eastern and Central Washington, to investigate the consequences of agencies' convenient approaches to toxic waste disposal. By discussing the monitoring of regularly scheduled toxic waste disposal practices, as supported and encouraged by elected officials and the agencies they manage, this work intends to reveal and state directly that they know very well what they do, and the adverse human and environmental health consequences thereof, all the while living in the self-imposed darkness of excuse and denial. There are safe and effective means of toxic waste disposal or repurposing, as are engaged in some areas of the U.S. and in other countries, but they are ignored for reasons so incomprehensible that they defy even our wildest imagination.

Land disposal of *Toxic Sewage Sludge* is a contentious business, addressed by agencies as a burden, nuisance, or revenue opportunity, rather than as a call for care and consideration. In any case, land disposal of toxic wastes in any form is costly to the public, both in treasury and as a matter of human health, including diminishing the quality of a vibrant forest experience, clean food production, or the health of the greater environment.

Land disposal of *Toxic Sewage Sludge* is a public health crisis, not a subtle detail or one of those environmental issues that may be secreted in forests or on prairies or farmlands, much less in composts, hazardous waste operations, soil amendments, potting mixes, or the like. No matter if such products are labelled "Organic" or otherwise, it is land-disposed toxic waste, well disguised beyond public view.

Part I

Sludge Tracking Tales

Prologue

Not On My Street!

The most alarming of all of Man's assaults upon the environment is the contamination of air, earth, rivers and sea with dangerous and even lethal materials.

Rachel Carson, *Silent Spring,* 1962.[1]

Yelm, Washington, is a curious city located on a prairie above the Nisqually River. Nisqually's waters stem from the still pristine glaciers of Mount Rainier. For eons, salmon have journeyed from the Pacific Ocean to spawn along the upper reaches of this magnificent river. Blood and sacrifice have been shed by Tribal members protecting fishing rights, and many individuals have worked for decades to enhance safe passage for the area's beloved Sockeye, Coho, and other salmon. My name is Michelle Horkings-Brigham and I live on a ridge above the river.

One quiet day, in 2017, a friend called with an alarming message. "You're kidding me," I said to the person on the phone, "I just defeated a massive, illicit pot operation and now this! On my road!"

"Didn't you read the sign?" he asked patiently.

"What sign?" I yelled louder than I should have to my sympathetic colleague.

"There's a sign that says the Department of Ecology intends to spread sewage at the end of your street," he stated matter-of-factly.

"No. I haven't seen it. How could Yelm allow sludge to be trucked in from all over the state and dumped HERE? Didn't we stop this once before already? Where's the sign?" I gruffly demanded.

"At the corner," he quickly ended the conversation.

"I can't believe this!" I wanted to scream but instead thanked him for the unpleasant news as I hung up the phone.

Hit with that same sinking feeling as when marijuana growers tried to establish an industrial-sized enterprise next to my home, I was overcome with despair. "This just can't be true," I mumbled to myself. "We don't have the resources or energy for another battle." Life seemed in that moment so ridiculous and unfair. "Why us? . . . Again?"

I knew that years prior, Yelm had fought to prevent the spraying of King County sludge on a logged property not far from city limits. I wasn't around at the time, but had heard stories of the busloads of angry citizens who traveled to Seattle to protest the idea. The media caught on and the people won. Yet, as witnessed many times and recorded herein, the sludgers inevitably found another location. They, the legislators, and the Washington State Department of Ecology (DOE) are responsible for such decisions and they simply took the problem to someone else's back yard.

This had never really hit home to me until I was threatened with the same. Nor had it done so to anyone who, back in those years, had adamantly refused to allow residential, hospital, industrial, and all manner of unknown toxic waste from all over the state to be dumped outside of our then sleepy town. If it had, we would have done something to correct this blight in government oversight. What I also hadn't considered, until personally confronted, is just how dangerous this toxic mix can be, nor how little political incentive there had been to improve what can only be called today a public and environmental health crisis. All are impacted, yet few are aware of this gross negligence within our supposedly modern society.

When my friend informed me this issue was once again rearing its ugly head in Yelm, I thought I knew about sludge. But I didn't. I had no real concept of the havoc and suffering the land disposal of sewage sludge can bring. Only as my personal drama progressed, did the magnitude of this global concern awaken my sleeping brain. As a result of my now firsthand experience, I came to understand how, when we are not directly threatened by dangerous chemicals, poisoned water, filthy air, or the negative impacts of a changing climate—nature in revolt—we tend to live inside our own bubble. I had been asleep to a problem that is so severe I could not believe my own ignorance. Every moment of every day someone flushes, hospitals remove waste, industries of all shapes and sizes discharge chemical pollutants. Where does all that poop and everything else unimaginable go?

Not until a government agency, whose supposed mission is to protect people and the environment, was going to allow such wastes to be dumped at my doorstep did I wake up to the threat lurking on and below the surface of the land. Unbeknownst to many, this threat invites disease, death, and economic downfall. The unseen enemy is *Toxic Sewage Sludge* and it is everywhere in our society. Even more stunning was my discovery of the motives driving this public health debacle and why there has been so little done to improve the situation. If it's everywhere, then eventually all will be impacted—even government entities who are ultimately responsible.

In her world-renowned book, *Silent Spring*, Rachel Carson revealed the dangers and repercussions of toxic chemicals recklessly applied as fertilizers and pesticides during the middle of last century. While struggling to overcome her own illness from cancer, she warned:

> Along with the possibility of the extinction of mankind by nuclear war, the central problem of our age has therefore become the contamination of man's total environment with such substances of incredible potential for harm—substances that accumulate in the tissues of plants and animals and even penetrate the germ cells to shatter or alter the very material of heredity upon which the shape of the future depends.[2]

In hindsight, her dire warning rings just as true today as back then.

Carson brought public attention to the potential demise of man and the environment by explaining in detail the fragility of our bodies and nature and the many problems associated with the spreading of deadly chemicals—cancer, reduced fertility, habitat loss, to name but a few. Despite some improvement, the problem never disappeared, regardless of the attention that has since been paid to her monumental efforts. Six decades later, the abuse continues in yet other, barely noticeable forms. I learned that here, in the United States of America, through clever marketing and human apathy, we have allowed this toxic menace to enter our forests, our waters, our air, our farmlands, our food supply, and even our own back yards.

For a long moment after my friend called, I sat staring into space, lost in victimization. I tried to convince myself of a phrase I'd heard many times at Ramtha's School of Enlightenment (RSE), situated on the outskirts of town between Yelm's ornate Buddhist Temple and the local high school: "Adversity makes you greater," RSE students are taught, when confronted with a difficult challenge.[3]

"Sure," I grumbled to no one. "Haven't I had enough adversity already?"

Quite obviously I hadn't, and I certainly wasn't feeling great, but I did know that if I cowered from this fight I would be accepting a terrible fate. My home would become worthless. Just like the intruding pot farm, the health and safety of my family would again be at stake. "At least make an effort," I mumbled, as I forced myself off the couch and went out to look for the sign.

The Department of Ecology sludge notification was barely visible at the corner. The roughly two foot by one foot placard was on a short stake almost hidden in the tall summer grass. The print was so small you could hardly read it. I had to get out of my car.

My friend was correct. Word was immediately sent out and I felt a glimmer of hope as Yelm rallied together, once again, for the fight.

The proposed *Toxic Sewage Sludge* disposal site at the end of my street is a stunning property perched above the Nisqually River. Mount Rainier towers in the distance yet appears close enough to be hugging the river's edge. The roughly one-hundred-acre site is bordered on several sides by Douglas fir and western red cedar that shelter acres of conservation easement; lands intended to remain unmarred into perpetuity. Along the river and for many acres beside the property lies a protected migration trail that provides safe passage to wandering black bear, elk, coyote, and other creatures as they move among the Bald Hills from Centralia to where the river enters Puget Sound. This beautiful acreage slopes gently toward a large, private lake where ducks and migrating geese gather in droves. Beneath the property, astoundingly, flows Yelm's aquifer recharge. Any proposal to dump sludge here, of all places, was insane. Especially when considering the following contradictory statement on the DOE's own website: "The goal of establishing critical aquifer recharge areas (CARA) is to protect a community's drinking water by preventing pollution and maintaining supply. We help cities and counties comply with CARA requirements that are required by the state Growth Management Act. Washington State residents get at least 60 percent of their drinking water from groundwater. Most people in rural areas get their drinking water from groundwater."[4]

Many, including local citizens, business owners, Tribal members, government officials, scientists, attorneys, the Nisqually River Council, and all three county commissioners, opposed the DOE's shortsighted view. They wrote letters, made phone calls, held meetings, gave public presentations, knocked on doors, signed petitions, and, in my case, formed a neighborhood activist group to counter the proposal.

We called our group "Preserve the Commons," a name chosen because sludge spreading would endanger not just those living nearby, such as myself, but the entire town.[5] Land shared by a community was called "The Commons" in early European times. In this instance, it wasn't the land the people shared, it was what ran beneath—some of the cleanest, sweetest water to be found. Flowing from the mountain and filtered by glacial outwash, Yelm's water supply is highly valued by all who live in the area.

A wealthy timberman had purchased the property a few years prior, logged some areas, then sold off half that was mostly locked into grassland conservation and could not be developed. He lived in a neighboring county. As a businessman, he was known to buy up large parcels around Yelm and elsewhere, log the timber, then divide the land for future development. Less than a mile away he had purchased another property and logged the forest without permit. That property was put on a six-year moratorium.

At the potential sludge spreading site, this fellow had agreed to allow, more than a year before the proposal was brought to the public's attention, a sludge hauling company from yet another county, Fire Mountain Farms, to spray sludge across the landscape.[6] Up to fifty semi-trucks a day would suddenly wreak havoc through our town and roar up and down my quiet, dead-end street. Toxic dust and the stench of human and other wastes would fill the air creating an unbearable living environment for neighboring landowners. Water runs downhill; therefore, toxic, unknowable chemicals would flow into the lake and river potentially poisoning aquatic species, migrating creatures, and birds. The large elk herd regularly seen in these parts would be impacted. Leached into the aquifer, drinking water would become contaminated over time causing untold human suffering. Vectors would be drawn to the stench and filth inviting all kinds of potential diseases. This situation was unacceptable to the residents of Yelm.

Short notification of a town hall meeting, intended by Ecology to convince the community of their nonsensical reasoning, gave citizens little time to prepare. Yet, at least two hundred people packed the local senior center that evening. After listening for well over an hour to "expert" opinions and a Fire Mountain Farms family member, individual after individual went forward to counter their unconvincing arguments.[7] I suggested to a DOE representative that she ask for a show of hands to see how many present were opposed. She would not allow the question. Regardless, everyone raised their hand. Many in the audience had fought the previous sludge battle with King County and would not tolerate this repeated threat to their safety. Over the years, Yelm had successfully fought against sludge, a NASCAR racetrack, a plan to incinerate toxic hospital waste, and a state prison from disrupting the peace and quality of life citizens enjoy in this odd, melting pot of a city. Yelm would not be bullied by a proposal that clearly ignored public or environmental health and safety.

Perhaps the greatest significance to come out of that town hall meeting was to realize that the sponsors of this proposal were absolutely determined to go forward with the project, despite overwhelming citizen objections. Officials had a standard response for our every concern. Even more alarming was realizing the scope of the agreement already in place between the DOE and Fire Mountain Farms. The proposed site had been awarded a Determination of Non-Significance (DNS) in 2018.[8] We were given only a few weeks to challenge this outrageous assessment.

Every day toxic human waste increases around the planet. When we flush, it has to go somewhere, was basically the argument. Land-disposed *Toxic Sewage Sludge* was said to be well treated and harmless according to expert reasoning. The DOE referred to data points and graphs and the Environmental Protection Agency's (EPA) latest research to make their case. Everyone understood, following that meeting, that this would be a David and Goliath battle—commercial interests versus public health.

Not long after, seeking any information I could find to help solve the problem, I attended a conference hosted by the Northwest Toxic Communities Coalition at the University of Washington.[9] I arrived late. A group of around forty individuals was already seated and listening to the speaker. The only empty seat was on the other side of the room next to a chap I assumed to be somewhere in his sixty's. Noisily, I plonked myself in my seat, irritated by the monster pounding at my door. I wanted answers and wasted no time. As soon as I could I raised my hand.

"How can Yelm stop biosolids being spread on our rural landscapes? What other alternatives exist for the disposal of human waste?" I emphatically wanted to know. Before my questions were even answered, this fellow sitting beside me slapped my wrist.

"It's not BIO solids," he interrupted. "There's no such thing. Never use that term again."

"What? What are you talking about?" I snapped. "They want to spread this stuff on my street. How can you say there's no such thing?" Then I continued to rattle on about what the DOE was planning with "biosolids" and I was slapped again. I was trying to sound educated on the matter by using the correct lingo, being in a university and in front of this apparently well-informed group. Instead, I was being chastised for exhibiting my ignorance regarding a trap that many before me had fallen into; the false notion of a benign product that in truth has no guarantee of safety. The prefix "bio" made one think it was harmless; just another organic or "natural" substance. In fact, as my playful neighbor explained to me, the term was given to successfully market an organic-sounding product that has little "bio" about it. Sludge might include human sewage but what about all the other chemicals and toxins mixed into that waste? That was this crazy guy's point as he continued to slap my wrist every time I made a slip and mentioned the word "biosolid."

"I'm Richard Honour," he announced, as I was getting over being repeatedly slapped in front of everyone. "Never use that term. It's sludge; *Toxic Sewage Sludge*, and that's all that it is, and it's toxic!"

He was right of course. I had fallen for the marketing. But I was suddenly more intrigued by his name. "Honor," I thought, "I sure wonder if he lives up to THAT name."

Dr. Honour's conversation and antics that day entertained my frustrated and angry mind. He threw me tidbits about his many sludge adventures as we chatted during the break. He said he was 80 years old and still rode a mountain bike, alone, for miles into the forest chasing sludgers. "What does your wife think about that?" I wanted to ask but held myself in check.

"Come on, you're not 80 and still riding a bike up mountains," I teased. I was enjoying the lighthearted conversation as we bantered around. "By the way, I was a ballerina and wrote a book about pot," thinking he couldn't beat that one.

"I worked in the Navy, went to Antarctica to do research for an expedition to Mars, then led a ten-year program to cure cancer. I did find a cure. I also know a lot about sludge," he informed me.

"This guy can't be for real," I thought, as I tucked away his card. "Why wasn't HE up there presenting if he knows so much about bio … oops, sludge!"

Back in Yelm, I reported on the conference at our next Preserve the Commons meeting. "Sounds like quite a character," was the general response when I mentioned Dr. Honour. "Maybe we should schedule a meeting to see if he's legit," was the overall consensus.

So, we did. On several occasions a group of us drove to the Poodle Dog, an old-fashioned diner in Fife, to find out if this unusual fellow really knew about sludge and if he could help us resolve our problem. Richard told me later he hated the place, but besides defeating cancer, removing sludge from the environment was his passion and he was willing to go anywhere to make his case. We met several times at the Poodle Dog before he agreed to give a presentation in Yelm. He never asked for a penny.

During those early interactions, I was incapable of understanding the extent of Dr. Honour's experience and knowledge about almost anything you might be curious to ask him. He can talk your head off for hours and is always novel, humorous, and captivating. At first I was challenged to believe his varied and somewhat outrageous personal life stories. I thought he was exaggerating or maybe even deluded. The snippets he shared sounded larger than life and so unusual it all seemed impossible for one person to experience. When he asked to read my book, I was even more surprised. I mean, and I do understand this, who really wants to read a book about *Cannabis*? Smoke it, maybe, but reading a book about the conundrum associated with the industry is not your everyday kind of thing.[10] Yet Richard was genuinely interested and did read my book. Every word. I was impressed. As it turned out, he informed me *Cannabis* byproducts are a superb fuel for the combustion of toxic waste; however, little did I know at the time, he also wanted help with a book about an ever-increasing threat that he calls: "*Toxicus ad Infinitum.*"

Through a series of live presentations, Dr. Honour helped to inform our community about the hazards of land-disposed *Toxic Sewage Sludge* and what that was doing globally to harm the quality of our air, food, soil, and water—in other words, its negative impact on human and environmental health.[11] He had ample data and photographs, collected over years from his own research, to prove the point. The local mayor and a county commissioner made time to attend whenever he presented at the Yelm Community Center. They too had not grasped the severity of the issue. The mayor wanted a solution to Yelm's wastewater treatment crisis and was garnering information to make an informed decision before leaving office, so we even arranged a private meeting.

Yet, despite all our efforts, nothing changed with the DOE's intentions to land-dispose *Toxic Sewage Sludge* on my street. Finally, Preserve the Commons decided to raise funds to print banners. Quietly, under the light of a crescent moon hanging like a jewel above the horizon, large banners were erected again and again on the street corner after being stripped down and removed by mid-morning. Eventually, public pressure swayed the outcome. Not because the DOE came to their senses, or the sludge hauler had a change of heart, but because the landowner withdrew his consent to have sludge sprayed on his land.[12] His reputation was clearly at stake in the community.

Following Yelm's success, Dr. Honour and I began to collaborate with the intent of bringing awareness not only to the issue, but to the solution he had found. Our meetings were upgraded from the Poodle Dog to a noisy corner at the Olive Garden, halfway between his home and mine. Over brunch I would ask questions and record everything he told me. And the more I listened, the more I realized the significance of such a modern-day, unsung hero's story. Between meetings he would write vignettes, provide updates of his escapades over the phone, and send emails with additional thoughts and photos before heading off on his mountain bike to locate yet another sludge dump in the Snoqualmie forests. In the process of putting the book together, this adventurous, sludge-tracking rebel turned eighty-two.

At the beginning of our first recorded session, I said to Richard: "Your life seems so outlandish. Who are you? No one will take you seriously if you can't convince us your adventures are for real."

Without ruffling a feather, Richard dove into this personal narrative, and I became fascinated by his quest to cure cancer and save humanity from themselves. Relating Dr. Honour's journey is important because so many threads come together to explain why this humble genius was able to discover something no one had previously thought of. In compiling this "scientific adventure story," forthcoming chapters are relayed through his first-person account. None could convey better what Dr. Honour must tell the world than through his own colorful expression.

During that first meeting at the Olive Garden, when everyone was unsure where the world was headed, any remaining doubts I had about the sincerity of Richard's story fell away. I was intrigued as he described the events of his remarkable life. It now made sense. Although Dr. Honour's expertise on sludge was clear to everyone attending his presentations, questions had still lingered in my mind: "Had he really been to Antarctica to do government research for Mars? Could he really have found a novel, immunotherapeutic agent for the treatment of triple negative breast cancer? Has he truly discovered a solution to transforming toxic waste into renewable clean energy?"

As Richard related his adventures, I understood why he was able to make a discovery that we believe will not only transform the sludge industry, but all our lives as well. He is a man of science, thorough and persistent, who leaves no stone unturned.

In reflection, I find it interesting that Coronavirus (COVID-19), an infectious disease caused by the SARS-CoV-2 virus, was raging across the world as these chapters emerged. Suddenly, individuals in positions of influence were ready to listen to what Dr. Honour had long been saying: "*Toxic Sewage Sludge* dumped on the landscape is of no benefit to society." Just maybe, by mid-2021, policymakers were beginning to realize that our air, food, soil, and water are being contaminated to such a point, through an antiquated system of toxic waste disposal in this country, that our immune systems are being weakened, thereby causing greater susceptibility to diseases like COVID and its variants. Sadly, it may have taken a global pandemic to bring an awareness that we cannot sustain ourselves in a lifestyle poisoned by our own waste.

Land disposal of *Toxic Sewage Sludge* is a threat to everyone. Although unsuspected by the everyday citizen, commerce-driven greed and the personal need for power are overriding the public's health and potential for longevity. Resources are being mismanaged, polluted, and non-replenished as required for environmental sustainability and the quality of human life. Clearly, if we don't find a solution to the disposal or repurposing of our waste it will destroy us all eventually. More whales and dolphins will beach themselves as a wakeup call to our polluted waters. More salmon will disappear from our rivers.

Dr. Honour's multifaceted experience is legit, I am now certain. And I think all will agree, after reaching this book's conclusion, that he has made important discoveries while most assuredly spending a varied and fruitful life living up to his honorable name.

Chapter 1

Dr. Honour

All through life, be sure to put your feet in the right place, and then stand firm.

Abraham Lincoln, 16th President of the United States, 1862.[1]

Richard Honour, that's me. I was born in Roselle, New Jersey, next to the city of Elizabeth, right across from the port of New York. From my rooftop, as a kid, I could see the Statue of Liberty and the Empire State Building. I was just another kid of the city, the ports, the sewers, and the garbage dumps. I lived there until I finished high school in 1958.

My mother was from Waterford, New York. Her parents were descendants of generations of farmers dating back to when they came to America from Holland in the early sixteen-hundreds. My grandfather was a potato farmer, the Justice of the Peace, the Keeper of the Locks on the nearby Hudson River and Erie Canal, and the Game Warden. He was the local cop for everything. His name was Charlie Pollock.

My mother was one of two children. When she was sixteen, she got a job serving tea in a tearoom until my father, who was thirty-two, ran away with her. Old Charlie Pollock, as one would do in those days, chased them with a shotgun but he never caught them.

My father was born in 1896. He was a dirt farmer from Greenville, South Carolina, where the family inhabited the remains of an old mansion that had been burned-out at the end of the Civil War. They were slavers and everything was destroyed. The only way to survive was to get out of the South, so my father and his brothers went to New York City on a job hunt. It was hard times. That was during the early depression when no one had a steady job.

My father joined the Army and was sent to Texas, where he contracted tuberculosis and was released from service. Not long after, he became a ditch digger for the first sewer system in New York City that drained Manhattan through Battery Park into New York Harbor. My father dug trenches in Manhattan and made sewer pipes out of brick and rock so that each home could discharge its own sewage.

Prior to that all human waste went into the streets. There were no sewers at the time, just gutters along the edges of the streets that drained all sewage from the city. Manhattan has a gentle slope toward Battery Park and the water, and with Battery Park located at the southern tip of New York City, all waters that drain therefrom enter New York Bay, and on to Governors Island, Ellis Island, Liberty Island, and beyond. More than 12 million immigrants passed through Ellis Island from 1892 to 1954, and we seem to have greeted them all. Welcome to "Sewage City."

My father's family came to the Carolinas and Georgia to escape religious persecution in England where the Church of Rome was trying to take over. Christianity in its local form was not acceptable, so the zealots got on ships and headed for America. They came ashore in the Carolinas in the early sixteen-hundreds.

The family name was Honour, back then. The original name was Honoré because of their origin in the northwest region of Brittany, France, in the city of Brest. They escaped from there to Devon, England, via a northward crossing of the English Channel. In 1623 they emigrated to North America by ship, and that was it, a typical American story. Both sets of parents originating from different parts of Europe, then arriving in North America at about the same time, all because they were Christians who had to get out of town.

In 1959, I too had to get out of town. I could no longer tolerate New York or New Jersey. From when I was twelve to about sixteen, I had worked in various shops that included a bakery and an auto repair shop. I had played track and football in school and was even a New Jersey state champion pole vaulter in 1958. But New Jersey, New York, and their ports were no more to me than dirty, filthy, sewer drainage, crime-ridden places, and I had other ideas. I couldn't get out of there fast enough, but I couldn't quite figure out how to get out.

New Jersey, where I lived, was kind of like the burbs. Many kids saw their parents get on trains in the morning to go to New York City. That's where people worked. All too often, from kindergarten through high school, and most especially in the early years of elementary school, you never knew who would appear in your classroom the next day. In the 1940's, you never knew who was going to be in class when you showed up for school, because people from around the world were abandoning their homelands to come to America. Nearly every day ships came into the harbors bringing Jews who had escaped persecution, and Germans who were leaving their country. People arrived from Poland, Spain, France, Italy, and elsewhere. Those were downtrodden people who escaped to America, and it went on for decades, as it does still to this day.

War-torn Europe just wasn't rebuilding fast enough. No one could live there. So, where I went to school was really a melting pot of the oppressed world. That also included people from Russia, China, Korea, Armenia, Latin America, Scandinavia, the Caribbean Islands, and more, all escaping oppression, and who were totally destitute, but who had been taken in by the United States.

Social studies classes in school became less about how the world was configured and more about where these new kids came from, and how and why they landed here. The English language was a mystery to them, yet it must be learned. Their parents typically remained at home, shut in, fearful of how they would be received, still in severe stress.

As mentioned, from my rooftop I could see the Statue of Liberty. It had a lot of meaning back then, most especially for the people who had just arrived. We learned a lot about cultures from around the world simply by being right where we were. Seeing fresh pasta drying on the fence out front, for example, was a bit alien to us at first, but we got used to it. And yet for me, New Jersey still was not an acceptable place to live. Therefore, upon graduation, I walked into the Navy recruiter's office in Elizabeth and jokingly asked in a shy way, "Do you have any room for me? I want to get out of here."

The recruiter responded, "Yeah. We're starting a few new wars around the world. Where would you like to go and what would you like to do?"

He was joking back so I answered, "Yeah, well, I wanna get out of New Jersey. How far can I get out of here?"

"How about San Diego, California?"

"Good idea."

"What do you want to do?"

"Well, let's see. I wanna be an electronics guy on a submarine and a diver. How about that?"

"All of those things?"

"Yeah, all of those things."

"Ok, it's a deal. Can you leave on Monday?"

"Yes! Good! Let's go!"

So, I did. I was nineteen, and on my way out of town, deal closed. The following Monday morning, very early, I went to New York City and found the Navy Induction Center. I fell in line with a hundred or so other can't-wait-to-get-out-of-here teenagers, and off we went on a fully arm-guarded train to Chicago. There was not to be any getting off along the way. They didn't want anyone having a sudden change of mind or heart.

Once in Chicago, we were shipped off to the U.S. Naval Training Center, Great Lakes, Illinois. After that, I was sent to New London, Connecticut, to submarine school. From there, I flew on a 707 to San Diego to attend a series of electronics schools at Point Loma, then to several diving schools to become an electronic surveillance specialist and diver on a submarine, just as I had requested. Upon completion, I was sent a few blocks away to the USS Catfish, SS-339, that looked more like a black porcupine with tumors than a submarine with all its various antennas and domed things sticking out of the hull. The odd thing was, it didn't have a visible name or number, or any other means of identification. There were just some thugs standing at the gangway, fully armed, seemingly awaiting my arrival.

"Get aboard!" one of them commanded. And that was it. We shipped out right away and snooped around the Pacific in way too many places, in retrospect, from the Arctic to south of the equator in multiple and various directions, always on alert.

Who knew we were at war in 1960, and with whom, and about what? Not us, for sure! Worse yet, we always had these skinny, shaggy, nearly post-teenagers on board who seldom washed and sported thick glasses and beards. They wore faded jeans and old T-shirts as their dress code. They had an amazing ability to not speak for months. We soon learned they were the actual managers of all things important who pulled the reins, decided the course, and ran the boat (submarines are referred to as boats, not ships).

The Cold War was blooming, not looming, and the United States Navy had made a serious mistake. We didn't get on to it soon enough, so we were way behind. We didn't have enough sailors, we didn't have enough ships, we didn't have enough submarines, we didn't have enough specialists, or anyone to make them go. That's why we were grabbed off the streets as teenagers, and that's when I learned, far after the fact, that when I walked in the door of the recruiter's office and asked, "Do you have any room for me? I want to get out of here." I'm sure he was thinking, behind his snarky, smirky smile, "Go ahead, make my day."

"Good! Let's go!"

Unbeknownst to me, I was fresh meat and that's what they needed. The United States Government was paying high bonuses for old time submariners from the Korean War to come back and run some reconfigured submarines, along with some naive teenagers. Grandpas and their surrogate grandsons made things go.

We worked the Pacific Rim from Wrangle Island in the Arctic off northern Siberia, down to warmer Asian waters of the South China Sea and elsewhere, always engaging in learn-as-you-go cold water and warm water surveillance work. Surveillance of what?

Wrangle Island, our first deployment, is where the Russians built a pop-up missile base. If you look at the globe from space, it's a direct shot over the top to Grand Coulee Dam in Washington State and Hoover Dam in Arizona and Nevada, thereby presenting an imminent threat, well poised to eliminate our key power sources of the day. We were in a Cold War and my job was to photograph things and record signals emanating from remote and unusual sources and places, and to employ various other technologies to assure the peace.

Then away we went to the border of North Korea and China, and on south to the perimeters of Southeast Asian nations of interest. Next, a straight shot east across the Pacific to the coasts of Latin America, then back to Siberia and the Kamchatka Peninsula, moving along the perimeters of their Navy bases. Our restored and noisy WWII submarine could be tracked easily by anyone, anytime, anywhere, so we clearly understood the well-defined rules of non-engagement.

Our opponents made some serious mistakes. All waters and lands within or adjacent to the convergence of the Sea of Okhotsk, the Bering Sea, and the Chukchi Sea were referred to as, "Up North." Curiously, there they employed undersea cables for their "secret" communications. However, little did they know there are strange people who escape from submarines into dark and cold waters to place recording devices on such cables. All was recorded, then we would ship off the precious recordings, never to know what may have been learned, or not.

Each underwater surveillance excursion was followed by a long stretch in a warm shower to recover and help alleviate the headaches incited by exposure to frigid waters for far too long. Many days aboard Catfish were spent in Arctic sleeping bags, bunny-hopping from one place to another, not wanting to get chilled further.

From Up North we worked our way south again, engaging the China, North Korea, North Vietnam, and South Vietnam borders as surveillance targets, including Hainan Island in the South China Sea, plus certain other islands of the nearby South Pacific along the way. We were in a serious, albeit occult war in Southeast Asia, involving several areas to the far north and west, of which few in the U.S. could locate on a map. What were we looking for?

On one particular foray, and as a very bad and nearly endgame example, we entered the outflow of a river that drained from the border of two Asian nations. The job was to record the acoustical output from a far upstream hydroelectric plant. The techies back home could then work backward to determine the electrical energy output from the power plant, or so they said.

In the dark of night, we pointed the bow toward the river flow and cruised at "All Ahead, Slow Turns" into the upstream course as far as the water depth would allow—not far, as it turned out. Someone (most likely a newbie Naval Academy Midshipman) had miscalculated the tides, and thereby the currently prevailing water depth. With the first screech of the hull scraping on a sandbar, we had to slowly reverse to get out of there. Dawn was emerging, and so were the local fisherman setting their nets for the morning catch—so much for remote mission planning and intelligence for nighttime adventures while submerged in unknown waters! In the shock of, "All Back, One Third," we then entangled massive amounts of fishing nets and lines in the propellers and had to leave stat, no matter the obstacles.

"Get me the hell out of here!" bellowed the Skipper, who had very little sense of humor in an emergency, of which we seemed to have oh so many.

Once in deeper waters, the diver, "moi," had to go overboard with bolt cutters, hacksaw, and knife to free the props from the spaghetti-like mass of fishing gear, amidst the shouting, scurrying, and arm waving of enraged fisherman who ventured over to see the spectacle.

It was light now. I signaled to them in international code with my middle finger, but they seemed non-conversant in the common language. They knew not who we were or from whence we originated, simply because we were painted black, with some last-minute, lighter grey splotches of finger paint here and there as an attempt to achieve functional camouflage. We looked like any ordinary black and grey submarine disguised as your typical sea-going porcupine; just in the wrong place—now caught in daylight. We definitely blew that one!

The majority of my time was spent on the USS Catfish in various oceans, harbors, and rivers. From my personal perspective, the warm waters to the south were just perfect, mostly. I have an attraction to water and have had some interesting experiences in peculiar waterways, clean and otherwise, but not with toxic chemicals at that time, at least none of which I was yet aware.

Somewhere along the way the United States entered the Laotian War, of which few Americans have ever been informed.[2] I'm a Laotian War Veteran, not a Vietnam War Veteran. After the fact, we were conferred the honor of being documented as Vietnam War Vets, even though that war was just brewing and came after us.

The Laotian War was supposed to be a secret. In the history of the United States, this was the largest air bombing war ever conducted. One objective was to blow up all the Viet Minh in Laos, Cambodia, Thailand, and North Vietnam, as they were called, before they were later named the Viet Cong. The Viet Minh sought to gain independence for Vietnam from the French Empire. The subsequent Viet Cong were South Vietnamese supporters of the communist National Liberation Front in South Vietnam during the U.S.–Vietnam War. I'm still not sure why we were there, or who won, but we did it.

Saturation bombing of Laos, etc., with over three hundred sorties a day, continued for perhaps a year, originating from certain islands in the Pacific—a toxic load dropped on those countries that is today inconceivable. One sortie could include around three hundred aircraft, each loaded with three hundred bombs, deployed for three hundred consecutive days to bomb nearly anything and everything, seen or not. The objective was to obliterate anyone and everything that could have been supportive of the unseen and nebulous enemy on the ground. It was our biggest secret war, and my part was surveillance of ship and boat traffic in adjacent waters. Our squadron's motto stated: *Detect, Attack, Destroy*. It may as well have been: *Get Cold, Get Lost, Get Warm, Run Aground, Get Tangled, Get Caught, Catch On Fire, Get Away As Fast As You Can!* But we were very good at what we did, considering the near emergency status of getting somewhere important on a refurbished, antique submarine at the right time to do something.

We began this enormous bombing campaign throughout Southeast Asia because we were of the opinion that we were facing an enormous threat based on the domino theory of the day, which is to say, if one country falls and becomes communist, then all will fall and become communist. Truth be told, we desperately needed their rubber trees, or so we were informed, because the U.S. didn't have any rubber plantations. What? Didn't we already invent plastic in the 1860s? Colleagues were actually working with people on the ground, capturing and smuggling rubber tree seedlings out of Southeast Asia into Hawaii, where they still grow today among isolated plantations of exotic industrial, medically useful, and otherwise beneficial tropical plants. We were sure we would need all that stuff someday.

Meanwhile, we were also very busy in the Philippines, Guatemala, Cuba, and elsewhere, facing down real or perceived communist threats. The serious threat of a Russian-Cuban consortium then culminated in our presence at the infamous Bay of Pigs at Playa Girón.[3] In each case we went on station, only to be sent elsewhere when it was over to chase greater perceived harm, when and as needed.

This was also the time of DDT and Agent Orange, and my first real introduction to toxic chemicals.[4] One of my friends died not so long ago from Agent Orange exposure way back then. I was around Agent Orange, but I wasn't any part of the operation. No one knew what it was. The deadly stuff was designed to deforest Southeast Asian jungles because if you killed the vegetation then the enemy couldn't hide, and we could bomb them even better.

Agent Orange was an herbicide that primarily killed plants. It was used to get rid of the jungle and expose underground tunnels and settlements. It meant nothing in the end, other than having made an environmental mess, and killed many people.

Although I loved what I did, and would have done it forever, after several years of sub-ocean faring adventures, there was a burning drive to enroll in college. The Navy was offering substantial enticements to go to additional submarine training schools in Florida and Washington, D.C., with higher pay and cash incentives, but it was June and summer college sessions were beginning in San Diego.

The Navy hires you for six years, and at four years you have the option of skipping town; however, they tell you quite bluntly how much it cost to train you, likely to make you feel guilty. Nonetheless, I went to San Diego State University and then to the University of California for a total of ten years, and never paid a penny. I studied engineering, math, and life sciences, and then specialized in microbiology and infectious diseases. The catch was, every time you went to the Dean of Admissions office, four quarters a year, and signed the forms whereby the government paid the bills, you got notices from Uncle Sam that you still owed someone something. I left the Navy in April of 1963, with two "still-owed" years to go.

After I left the Navy, I migrated back to a beer bar with some friends on New Year's Eve. It was on the waterfront in San Diego and there I met this woman. Her girlfriend brought her for an evening out. They were celebrating someone's birthday. That was on New Year's and the following December, Ruth and I got married. I was just starting school.

Ruth came from New York State not far from where my parents had met and run away. She was 22 and had just finished college when she moved to San Diego to live with her sister. We met, got married, and spent the first ten years of our life together pretty much on campus in student housing. Our two children were born there, as well, and went to the preschool experimental learning centers at the University of California and Cal Poly Pomona when they were three.

While at San Diego State University, in 1965, I started looking into wastewater treatment plants. My marine biology studies had included additional classes at Scripps Institution of Oceanography. Nearby was a huge pipe that ran raw sewage into the ocean from La Jolla at La Jolla Cove and, as a diver, I was going up into that sewage pipe, photographing the debris hanging off the rough edges and seams. It had to stop, and it was, but I was interested in the kinds of microorganisms that could live there; and there were many. Studying microorganisms from unforgiving places is a discipline known as Harsh Environment Microbiology, and the subject microbes are labelled as "Extremophiles."[5] All of that became my objective.

After we moaned and caused a lot of trouble, the city of San Diego finally removed the pipe and the pollution, and then created an improved wastewater treatment plant near the submarine base at Point Loma. A nuclear reactor was being built on the beach there as well.

While working with raw sewage from this facility, no one quite realized at the time that the microbes in the sewage, including from the adjacent medical centers and research labs, perhaps even more so than the toxic chemicals, were having an extreme, adverse effect on the marine life of this beautiful ocean park. What were those relationships, and why was untreated wastewater being released into the Pacific Ocean? After all, this was the 1960s, and we must have advanced past this life-threatening debacle by now. Not so back then, and not so now, more than 60 years later.

California eventually cleaned up its environmental disasters, now being at the top of anyone's ranked list of environmental and public health consciousness, by focusing heavily on the adverse consequences of their infrastructural operations, while more primitive states continue to sacrifice human and environmental health in favor of lowest cost and most convenient methods of concealing and disguising their toxic wastewater effluents, sewage sludges, and landfill leachates. Moving and hiding toxic wastes has become an art form, covered by codes, regulations, laws, common practices, and legislative indifference or interference, rather than practicing proper management, treatment, and elimination for the public good. Our precious air, foods, soils, and waters are actually matters of national security and, as I came to understand, it is the legislative squandering of these critical resources that drives us into disbelief, dismay, and compromised health.

Curiously, while our complaints back then in La Jolla centered on misdirected human wastes as raw sewage, today the main concern there is fecal contamination from sea lion and sea bird feces. The cove is now protected as part of a state marine reserve, while raw sewage and other wastes go to the greatly upgraded wastewater treatment plant at Point Loma, San Diego.

At the University of California, Riverside (UCR), where I ventured next, such things as sewage sludge and wastewater effluents were of interest, as those materials were being used to irrigate and fertilize many crop plants, experimentally, on the campus research farms. The major challenge was that no one on Earth knew what was in those toxic matrices, much less the adverse effects on human and environmental health. Did anyone care? After all, the proponents would proclaim (and still do), that the Chinese had been using human poop on their food crops for untold millennia. Little do those proponent folks appreciate that in long-ago times, none of our air, food, soils, and waters, or anything else to which we are now exposed, was contaminated with any of the untold hundreds of thousands of toxic chemicals and infectious agents which we find in sewage and other wastes today.

Back then, poop was poop; today, poop, aka *Toxic Sewage Sludge*, includes toxic biological and chemical wastes—domestic, medical, and industrial, plus highly toxic stormwater runoff and so many other polluting sources—such that nature's built-in processes of microbial degradation and decomposition simply do not and cannot render such materials as being less-toxic, non-toxic or in any way safe. The Environmental Protection Agency scientists tell us safety can only be determined by rigorous safety/toxicology testing, not by innuendo or personal preference, or for the purposes of commerce or convenience as is most prevalent today.

One of my professors at UC Riverside was Crellin Pauling, son of Linus Pauling, a leading scientist at the time who participated in the advent of nucleic acid (DNA) chemistry.[6] They, and so many other scientists of that period, provided the basic science that was the gateway to a broader understanding of the general microbiota and greater spectrum of life forms inhabiting once pristine environments. They and others set the stage for our ability to more fully evaluate the adverse effects of contamination and pollution by which we are all affected, and by which we unknowingly remain at risk for disease and death to this day.

When I went to the University of California in 1969, I continued studying harsh environment microbiology, mainly pathogens—meaning microbes that can infect us and cause disease, including epidemics and pandemics that could be the primary cause of death in immune suppressed people, such as in cancer patients and people subjected to immune suppressive therapies. My fascination was with microorganisms, microbiology, and infectious diseases.

Because of the nature of microorganisms and the processes of infection and parasitism, I first began studying infectious diseases in plants and agricultural crops. That moved into human infectious diseases, especially in immunocompromised people, such as cancer patients. I was looking for exotic microorganisms in places where no one had ever looked before. I had a suspicion that science at the time only knew of perhaps one thousandth of one percent of the microbes that existed, but I was wrong. We knew much less than that, which was only revealed by genomics later with the advent of whole genomic sequencing and next-generation sequencing (NGS) technology. The advent of DNA technology came upon us, and we started science all over again.

My transient and ever-elusive objective was to work for the Food and Agriculture Organization (FAO) of the United Nations, which leads international efforts to defeat hunger and improve nutrition and food security globally. Its Latin motto, *fiat panis*, translates to: "Let there be bread," and that seemed to be a worthy objective. My dissertation subject microbe, *Phytophthora parasitica*, was devastating the crops of the world back then, just as it does today.[7]

Somewhere amongst all the toxic sludge and effluent work, I had developed a new method for mass production of the sexual spores (oospores) of *Phytophthora parasitica*, one of the most devastating pathogens of food crops globally. To me, it was a matter of food security. The role of sexual spores in the pathogen's survival and pathogenicity was an unknown factor to science.

Phytophthora species are oomycetes, not fungi, and no one knew how to produce the sexual spores *in vitro*, and therefore we lacked a key research tool. Now we could mass-produce the sexual spores and thereby investigate such issues as their rapid adaptation to a broad range of crop plant hosts, their survival mechanisms in harsh conditions, and their resistance to pesticides, among other issues, all of which are based on gene-mediated recombination events in response to environmental stimuli. When I declared the required Statement of Research Focus to my Major Professor and Department Chairman at UCR, they responded in concert that they had no idea of what I spoke. Where do you go from there? Antarctica!

Not far from UC Riverside is NASA's Jet Propulsion Lab (JPL) at California Institute of Technology (Cal Tech), Pasadena. I was thirty-two when just before completing my graduate studies at UCR, and while working on the last details of my dissertation, NASA invited me to go to Antarctica to investigate harsh environment microorganisms, with a beginning mind to perhaps discovering what kind of microbes may be encountered if we ever went to Mars. A friend and colleague had done previous work with the people at

NASA, JPL, and Caltech, and told them, "You should talk to this guy." So, they called me.

JPL was within an hour's drive from Riverside, so I drove over there, and they said, "We're doing some harsh environment microbiology work in Antarctica, and don't know how to do what you're doing. Would you like to go along with us?"

I said, "Ok."

We left on a plane for Hawaii, then from Hawaii to Pago Pago, and then to Christchurch, New Zealand, and down to Antarctica. I soon realized my methods for isolating and identifying extremophiles in this vastly different environment had to be revised, for certain.

Extremophiles are microbes that usually operate only in extreme or harsh environments, such as in raw sewage, wastewater effluents or in *Toxic Sewage Sludge*, or in the Antarctic or on Mars, for example. The challenge at hand was to determine if our earthly methods could be adapted to some form of a miniaturized device that could be sent to Mars to collect and evaluate soil samples, so as to determine if we may be learning anything at all about what we were allegedly after.

What were we after? Maybe we could figure out how to collect, isolate, grow, and identify extremophilic microbes beforehand, right here, for Antarctica has some similarities to Mars, or so they said, but would they be different, similar, or the same? I was instructed by the Feds not to ask such questions, but to offer only at least some plausible answers. I was dealing with the likes of NASA, the National Science Foundation, and the State Department, so compliance, formality, and expertise were required, not offhanded remarks.

"Expertise in what?" I asked. What was being proposed had never been done, and just who were these guys, anyway, and just who was paying the $250,000 per day for the helicopters that transported us to previously untouched locations in a distant, cold wilderness?

Looking for extraterrestrial life forms was easily a twenty-five-year project, at least, or so I thought. There were lots of people they could find to do exotic projects, but who ever said extraterrestrial microbes would be anything like those here at home, even at the South Pole? Unlikely, for sure, in my mind.

While we were "On the Ice," as the Antarctic was referenced, the State Department intervened and said, "We're having a really serious public relations problem because some people of the world are starting to realize that research bases in exotic places, such as the South Pole and on the coast of Antarctica, generate trash and laboratory wastes that are being put on ice flows and sent out to dump over into the sea when they melt. That's not going over very well. So could we divert your work to figuring out a plan on how we could get trash and garbage out of there without contaminating anything?"

From that point on, I was investigating garbage and effluents in Antarctica. There were no sludges because everything flowed into the sea directly, wholly unobstructed, just as it had at home. I may have been the first person to have ever done that work. I had to figure out what kind, how much, how dangerous, where should it go, what were the alternatives, and how would we get it there, anyhow.

The alternatives were really bad: Shipping garbage out to sea and letting it disappear was not a good idea. Draining sewage into the sea was a bad idea, as well. Then they thought, "How do you move it, so as to remove it, and where should it go?" The project thereby became a waste relocation project, just as it is to this day.

My question was: "Well, it's U.S. garbage from U.S. research stations, and there are twelve treaty nations, meaning twelve countries that have the right to do clean investigations, so how are we going to get everyone else on board with a U.S. push for cleanliness?"

The State Department would take care of that. We were going to be the lead nation on getting it off the ice, and so I said, "We should ship it to America." Well, that was the dumbest suggestion anyone had ever heard because that's nearly the furthest point on Earth from McMurdo Station. But to my mind, anywhere else you sent it meant that all you were doing was moving the pollution around to another location, not eliminating it. You were not getting rid of it because logical destinations, such as India, China or Australia, were just other dumping grounds, and if they burned it, they would simply cause even more trouble—a quandary of the day, indeed, and one that still persists, 50 years later.

As things turned out, and under the guidance of others, all of that cold garbage was eventually shipped to Los Angeles to be put on a train to Spokane, Washington, there to be dumped in the landfill atop their local aquifer, way back when.[8]

The only electrical power at that station in Antarctica while I was there was a small nuclear power plant. Ten years prior, I had been stationed in San Diego on a submarine next to a model nuclear power plant that was being developed to someday end up on a submarine, or so the story went at the time. It ended up being shipped to Guantanamo Bay, Cuba, to provide autonomous power at that shady facility, in the event the Russia/Cuba alliance panned out for them. So, I knew how they worked, what they did, and what would be the likely consequences of failure. I opined to the multiple and various federal agencies involved that this nuclear power plant in Antarctica already leaked radiation. We had to be careful about a leaky nuclear reactor, because, just as on a submarine, you can't simply open the windows at those Antarctic latitudes to let the contaminants out.

"No, it doesn't leak," I was told repeatedly by the Feds.

"Yes, it does," said I.

"No, it doesn't."

"Yes, it does," and so we all went on.

Finally, it was tested, and "Yes, it did leak," so the nuclear power plant was wrapped up and shipped off, and no one knows to where, or even if it got very far offshore before it disappeared at sea, or not.

I went down to the South Pole to do harsh environment microbiology related to the Martian thing, but the diversions were many. While I had to perform the research work, yet another item was added to my ever-mutable repertoire. I was to track the activities of some Russian scientists.

Russia is a treaty nation among the twelve Antarctic treaty nations, and it somehow became my job to surveil and become familiar with their behaviors "because that's what I did in the Navy," or so they said. There must be a connection. Other graduate students from other universities did not have the same background. The whole thing about JPL, Caltech, and all the rest of that—Martians, extraterrestrial microbiology, etc.—was a big fat ruse. Remember, every time you went to the Dean's office and signed that paper to go to school for ten years on their dime, they let you know you owed them something. They kept paying the bills, so I kept being available.

When I finally graduated from UC Riverside in 1973, after coming back from the South Pole, I began working at the University of Southern California, Comprehensive Cancer Center, for the next ten years. During that time, I received more calls to do a few other things. You just don't get to say, "No."

This was a difficult time for those seeking new employment opportunities. Fresh graduates with PhDs couldn't get jobs out of college, universities were cutting back, and professors couldn't find work. However, I soon received a call on a Sunday evening. The caller suggested I contact the Dean of the Medical School at the University of Southern California, Los Angeles, at his home that same night. He provided the number, and I did as suggested. The Dean said, "Can you be at my office at the USC Medical Center at eight in the morning?"

"Yes," said I, and he gave me the address.

When I walked in the next morning, he said, "Take your coat off and sit back."

Shortly after completing my dissertation at UC Riverside, and when others were seriously unemployed, I was hired to manage an international clinical trials research program in pediatric cancer that was fully funded by the National Cancer Institute and the pharmaceutical industry. The Feds actually funded it all. I was the Executive Director of this program at the Cancer Center of the USC Medical Center from 1973 to 1983.[9]

Very little of the clinical trials research was actually done at USC. My office had about 33 clinical records analysts and data managers, plus five PhD biostatisticians. The program included 22 of the major teaching research hospitals throughout the United States and Canada, and each of those had several satellite medical centers, also specializing in pediatric oncology. Overall, there were about 120 hospitals and clinics in the program throughout the U.S. and Canada.

More than a thousand physicians in the program were pediatric oncologists and they each had data managers, clinical records analysts, oncology nurses, multiple affiliate clinics, and so on. There were several thousand people in this program, all funded by the National Cancer Institute. Established several years prior to my arrival, it was moderately dysfunctional, having lost some key staff, so I was offered the opportunity to improve things and make it work. And it worked. The research grew over the next ten years, with important results that translated into improved treatments and care for children with cancer.

We entered about 7,500 children each year into the various clinical trials programs. They were children with cancer, and they were treated with various combination therapy regimens to evaluate new drugs and drug combinations, as well as with radiation therapy, biologicals, and immunotherapies. Immuno-oncology was just becoming something of merit at that time. All of the pediatric oncologists, oncology nurses, clinical records analysts, programmers, social workers, and statisticians became my personal heroes. It was emotionally challenging work, day by day, just to go in there, and some had been in the field for decades.

While managing the pediatric oncology clinical trials research operations work at the USC Cancer Center, I was also teaching environmental regulatory issues on Reservations that included soils, soil microbiology, soil chemistry, soil physics, landfills, garbage, *Toxic Sewage Sludge*, and more, because those were the issues that represented the inciting agents of cancers in children, cumulatively over generations, as we now know. The unbroached point back then was just as it remains today, that our primary objective must be prevention of cancer, not just focusing on proposed new experimental treatments.

Concurrently, NASA/JPL and the U.S. Department of Energy asked me to teach *Environmental Impact Assessments and Statements on Indian Lands,* which sent us off as a travelling family to Tribal lands in Montana, North Dakota, Wyoming, Arizona, and New Mexico, while I left the USC Cancer Center/National Cancer Institute pediatric oncology program on semi-cruise control. The point was to offer information about some of the rights Native Americans had on matters of law when it came to Big Oil's drilling or digging for energy resources on Native American lands, meaning, on Reservations, with little regard for the human and environmental health perturbations they were creating in the process.

Together with my family in an orange Volkswagen van, off we drove to Montana to the Flathead Nation, then to the Crow Reservation. Next, we went to North Dakota to engage with the Arikara, Hidatsa, and Mandan Tribes, the Affiliated Tribes of the Fort Berthold Reservation. All were successful meetings and presentations. They sought help in defending against Big Oil, Big Coal, the Bureau of Indian Affairs (BIA), and the Department of Energy (DOE).

Upon departure, one grateful Tribal member gave me a wild horse in appreciation for the support and information I had provided. He asked me to drive him to a local canyon. Pointing up to the rim, where stood so many wild horses, he said, "Pick one and he's yours." I pointed to a dark brown specimen with white feet, and he said, "He's yours forever."

"Thank you," said I, and off we went.

Heading to Wyoming, we attempted to approach the Wind River Reservation, which is shared by the Eastern Shoshone and the

Northern Arapaho. Upon approaching the Reservation entrance, we were intercepted from behind by a State Police car at warp speed, with red lights and sirens. We were informed that if we attempted to approach the gate, certain Tribal members stated clearly they would open fire. An immediate turn-around sent us off to Denver on the way to Albuquerque to engage with the Pueblo Nations. After all, these are sovereign nations. They have rights and they rule the roost, without argument. There are 19 Pueblo Tribes in New Mexico, each a sovereign nation.

At one particular location, upon approaching the auditorium where the presentation would be offered a few hours later, five young Tribal women in make-shift uniforms comprised of blue jeans and black T-shirts approached me in a circle, knives drawn.

"Get in the van," one ordered. "We will talk to you."

This was great, I had never before been kidnapped at knife point. They were wearing their very serious faces, but I couldn't help smiling. What could they want? We drove off in their on-its-last-legs, ancient, windowless, black van and parked a few blocks away.

"Who are you? What are you doing here? Who do you work for? What are you trying to do? Do you work for the Department of Energy? Do you work for the oil companies? How much were you paid to come here? What do you want?"

Ouch! Whatever happened to the alleged preparations by the advance team? That was my question.

"Nothing like that at all," I responded with a calm smile, and to five blank, nervous stares. "I work at the University of Southern California, Medical School, Cancer Center, in Los Angeles. Someone at DOE's Argonne National Laboratory in Illinois contacted me to see if I might be interested in providing a teaching program to Tribes in the Western United States regarding general matters of Environmental Impact Assessments and Statements. The point is that DOE wants to be sure the Tribes understand more about their legal rights as provided in some federal documents and laws. That's it!"

I doubted all that I said, but it worked, and they put away the knives. Kindness and understanding generally win over misguided hostility, or so I thought. They were serious. They were the local Tribal activists who trusted no one, for all the right reasons, most especially some white guy who represented DOE, the greatest monster of all after the BIA. Tribes for the most part all have deeply embedded concerns for their cultural traditions, as well as for any and all lands that provide for their sustenance, one and the same. I later encountered the exact same sentiment when working with First Nations members in Central British Columbia (BC), with the Canadian Government being even less sensitive to cherished spaces and traditions.

Sensitivity to cultural and spiritual traditions is a heritable human trait, no matter the misguided and errant behaviors of our political and commercial associates. By whatever means we got here, we are now well into a costly cleanup mode, including of our behavioral insensitivities. In the BC situation, the city of Vancouver trucked their sewage sludges east into First Nations' lands and dumped their toxic loads even into sacred waters where to this day, youths emerging into adulthood are ritually bathed in ceremony, just as they had been since before recorded times.

"Stop smiling," said one of the Pueblo activists. "We are serious, and this is a serious thing."

"Right on," I responded. "That's why I'm here."

"How much do you get paid to do this?" the leader asked.

"Nada," I responded. "I get reimbursed for travel expenses, and that's it. It was a chance to go on a trip and learn firsthand about the little-known Tribal issues with Big Oil, energy development, and the various agency conflicts and issues. You were given these resources by Treaty, and it must be in your hands to seek exploitation, or not, end of story. I had not been around here before and I wanted to learn about the Tribes, Tribal lands, and the energy development and pollution issues. So here we are. They may have selected the wrong guy for the job."

The tattered van sped back to the conference center, with me very much relieved inside, and I began the day-long program, which was well received.

Next, we headed off to Kayenta, Arizona, home to the Navajo Nation, to teach at a local community college. Again, we were successful. During the presentation, I mentioned the need to end coal mining and use, which was still fresh on my recently kidnapped mind, as a lasting environmental issue. The host then told me about the unexploited coal deposits on Navajo lands near where his elder family members still live. If I wanted to see the black strata in their original state, he would make arrangements for me to take a look on the way back to California. We did, and I remain in awe and shock. Those particular coal deposits on the Reservation exist untouched among the colorful, Grand Canyon-like strata and will remain so. We had connected on the matter of coal in general, and on coal deposits on sacred lands. I still can't get over the depressing fact that the U.S. Government allows mining in the nearby Grand Canyon.[10] Do they not hold anything as being sacred, even our National Parks?

Almost all Reservations have lodges or motels, and some are really nice. The time periods of our stays varied from a few days to a week. The upfront part was education, the back part was to: "Find out what's going on." Everywhere we went, what was going on by the Tribes was an effort to protect their lands, their heritage, and their cultural traditions in the face of some odd need to exploit pristine energy sources, at any cost.

"Not on my watch," or so said I.

As a whole, the environmental teaching programs were well-received, once they got the idea of who I was. I was teaching environmental sciences to Tribal members so they would know at least the beginnings of some of the federal and state laws about resource exploitation, in an effort to help them defend themselves against energy companies and the DOE, plus the Bureau of Indian Affairs. The DOE and BIA collaborated to achieve some common federal and industry objectives; however, and not so overtly, I was discovering what the energy companies were doing and what trouble they were inciting and having with Native American communities.

Sometimes you have to be in the middle of friends, enemies, and lovers, or all of the above, until you find out who's doing what to whom. I was finding out information on each and every side of the key issues and reporting everything, mostly to deaf ears, upstairs. The key issues included national energy security, energy industry commercial interests, Tribal sensitivities, and overall matters of public health, plus the far lesser concerns regarding environmental quality, protection, preservation, and restoration. The Tribes generally always lost, second only to the environment and us.

At the same time, I was managing this very large program at USC in Los Angeles to discover and evaluate new treatment regimens for children with cancer. We always made improvements in treatment regimens and in supportive care, targeted heavily toward management of infections, so as to save lives. The work always included improvements in the administration of yet more chemotherapy, surgery, radiation therapy, and immunotherapy, yet most of the children died no matter the well-thought-out strategy. The pediatric oncologists of all disciplines, the pediatric oncology nurses, the social workers, the clinical records analysts, the clinical data managers, the biostatisticians, and all of the support staff are and always will be the heroes of the story of pediatric cancer, yet it still remains as a long struggle ahead, until we unlock the causes, in the endless quest for prevention.

At some point, it occurred to me and many others that we needed to work more on the prevention of cancer, rather than on treatment. We do not get cancer by default. So, I flew to Bethesda, Maryland, to the home of the National Cancer Institute, and caused some serious trouble. I said, "We don't have any programs for the prevention of cancer, only treatment. What are we going to do? Continue killing small people forever? We need to find out what causes cancer and end it."

I was in big trouble. No one seemed to listen or care, not out of a lack of compassion, but because no one had any good ideas about what to do about it. The fundamental question had been posed for decades. No one in science and medicine really had any idea about what actually caused cancer, including in children, and how you could possibly intervene in the oncogenesis process so as to be in a preventive mode. After all, "people have to smoke, and we must apply liberal amounts of pesticides and fertilizers to fields if we are to eat. And preservatives, coloring agents, plastics, and metals are at the very core of our growth and commerce, as well as expansion and exploitation within our advanced societies. Without coal, gas, and oil, how could we turn on the lights?" Were these the actual quotes, or are we projecting how we got here?

So, I made it more of a political issue and was thereafter on the road to other cancer units and hospitals, giving my point of view. My suggestions were, "Let's find out who's interested. We will need what we will call a 'Cancer Control Program.' We need less statisticians and more epidemiologists, because if we're having 7,500 children per year with cancer entering these experimental treatment programs that means we're also having 15,000 parents in the picture. All of that means that the other 7,500 additional cases of childhood cancer per year and all of their parents who are not in our programs, are not even being calculated in just the U.S. national pediatric cancer system—so what gives? We need to interview the parents and find out just how the living environment of those parents and of their children may have contributed to their cancers."

"But you can't blame the parents," I was told emphatically.

"Ok. Let's blame the parents and perhaps even the grandparents, and also the aunts and uncles, and everyone else. Let's find out what's in their living environment. Are they living in Detroit next to a factory? Are they living in Newark next to the wastewater treatment plant? Are they farmers? Let's figure it out."

I had not long gotten back from the politically driven Antarctic research program, and gained further insight with the government influenced Native American Reservation debacle, so I said, "We were fighting garbage, toxic sludges, air and water pollution, contaminated foods, radiation, and ever-growing and ever-accumulating toxic wastes everywhere. People are getting sick, yet all is seemingly unreported. Don't tell me there's no cause-and-effect relationships, I've been there."

The outcome was this new, heavily funded Cancer Control Program, and as a result, we were able to collect sought-after epidemiological data for a number of years. The problem was that the program was just too big and the data became too non-evaluable, but the work did present a model for how it could be done better in the future. We must discover what causes cancer, and go after that, rather than endless and mindless searches for magic bullets to solve a seemingly unsolvable challenge.

The gut-wrenching part of what we had to do in the Cancer Control Program was to call up every parent: "Do you smoke? Did you smoke while you were pregnant? Did your parents smoke? Did your grandparents smoke? How about your sisters and brothers, and your aunts and uncles? Does your husband work in a paint factory or in a chemical plant? Is he an automaker? Is he a farmer using pesticides and fertilizers? Were any of you exposed to wood preservatives? Do you spray your house with DDT?" On and on and on. We were searching for clues, and how extensive were the commercial and political cover-ups, and just what is that leverage that industry has over matters of public health.

This was a battle with Big Pharma that was selling more drugs as more people got more cancers. But it all came down to the bare fact that the number one cause of cancer among smokers or among people exposed to smoking was smoking itself, no matter if you smoked or not. In the case of environmental epigenetics, the adverse effects of some chemical (and smoke) exposures may go on for as long as five generations post exposure. For example, if your grandparents or parents smoked, and if they experienced epigenetic modifications that affected the expression of certain genes, that epigenetic event or that series of epigenetic events may predispose a person in subsequent generations to malignant transformation, and then to the development of cancer. Those same, now heritable epigenetic modifications, may be experienced in children for as long as five subsequent generations. Such a scenario is well demonstrated, but also well suppressed until confirmed by succeeding generations of scientists.

That's the way it works when science and medicine are managed by politics and industry. Tobacco tax revenues continue to support city, county, state, and federal governments, and that means you must make more people smoke, more often, and don't ever say anything against it. The tobacco tax base is necessary to support any and all government services, or so they say, with a political and commerce bent. To the benefit of us all, the direct link between smoking and cancer is now well established after eighty years of the observed adverse relationships.

I'll tell you about one case, simply because it is indicative of the prevailing attitudes of the sixties, seventies, and eighties when, if you didn't smoke, there must be something wrong with you, despite the decades-long, known associations between smoking and cancer. During the process of contacting families to begin tracking toxic exposures, I ended up calling one woman, as a very bad but common experience, whom I should not have called. She was a mom. She had a baby. The baby had leukemia, and I had to go through the standard set of questions. You don't make them up. They're all listed because in an epidemiological study everyone has to be asked the same questions. The responses are captured for analysis.

"Do you smoke?"

"Yes."

"Does your husband smoke?"

"Yes."

"Did your parents smoke?"

"Yes."

"Did you smoke while you were pregnant?"

Dead silence. Then came the shouting: "Are you saying that I caused the cancer that is killing my own baby?"

I thought that was the worst possible response ever, until I heard it a few more times. It was a crushing question, but necessary for purposes of the study.

So, what do you do? Do you take away the question and figure out how to work around it? What is it you do at that point? We elected to continue, and we learned a lot, including about the National Cancer Institute and on to the greater National Institutes of Health approach to public health. Did the Centers for Disease Control and Prevention change Big Tobacco? Not! We were so dissed. No one would believe any of it or pay any attention to the study or to the emerging results, because it was almost un-American to oppose tobacco, without anyone even having to say so.

Was that my worst experience in these exploratory programs? No. There were lots of moms who were challenged with drugs and alcohol, or otherwise deprived of core human experiences and opportunities. It was the consequence of their disadvantaged life and income, their homelessness, and so on. You get the point once you have spoken with so many, but then when you turn the page and start talking to upscale, more affluent people, where the toxic challenges are more or less self-imposed, it changes your mind about the subject matter at hand. It still goes on. Even though that period was no more than a piece of time, there is no difference today, it still goes on. Today we are challenged with chronic agency and politically imposed exposures to our own toxic wastes that we will address later.

When I began the work at the USC Cancer Center, more than 50 percent of children with cancer would not survive five years. At the end of my ten years, more than 50 percent lived longer than five years, and not by a slight margin. We implemented immune therapies and infection control because people with cancer almost never die from cancer; they typically die from infections because their immunity is suppressed, primarily by cancer therapy, not by cancer itself. Once your immune system is compromised, you are susceptible to the multitude of infectious agents. Hospitals are the seat of infectious agents, most difficult to manage; therefore, infection control is paramount.

As an aside, and with regard to cancer triggers from transportation—trains, planes, trucks, and automobiles—we were able to make our transportation industry switch to lead-free fuels. Now we're headed for mostly electric, non-fossil fuel vehicles, although the high contamination costs for electrical power will fill in for the mystical toxic voids left behind.

With pesticides, many were banned in the seventies, although the rush to hide those toxics was intense. There are places near where I live in Washington State, for example, where landfills and huge forest tracks were filled with barrels of banned pesticides. Our local private and agency landfills were sites where flatbed trucks unloaded massive numbers of barrels of now banned pesticides because the disposal costs were too high. If you bury them and they don't leak for fifty to a hundred years, who's going to know? It's then someone else's problem.

I have witnessed and have witnesses who have seen them do that. An elderly gentleman in Orting, WA, for example, who was 93 when we last spoke, was in his local forest, way back when, watching this disposal into private forest-company lands. It was common. Such practices were common in Washington State, as they were all over the United States. What shall be the responses of the firms and agencies when the barrels begin to leak their extreme toxic loads into local waters? Ha! That was then, it's someone else's problem now.

When I went to the Orting forests five years ago, I was looking down certain canyons that drained to critical waters, and what I found were massive rolls of discarded artificial turf. When they take up artificial turf they're not allowed to landfill or dump it. Well, if you pay enough, you can instead drive your trucks in and dump it in the canyons and later fill it over. So the canyons are filled with artificial turf, and I went in there and took pictures, I was told where to look and that I would also see a large hillside of moved earth, and under it resided the pesticide barrels that were buried, not that long ago.

Though we may make noise about the ongoing mismanagement of our toxic wastes, those wastes we now know are scant, compared with the really big stuff—the big stuff people spend a lot of money to conceal, so as to avoid the costs of proper disposal, recycling or management, or of the fines that may result therefrom.

One of the reasons I also study landfill leachates, for example, which I do often, is because a lot of the legacy landfills now have parks, sports fields, elementary schools, high schools, housing developments, condos, and industrial sites built on top of them in an attempt to reap the benefits of development, while hiding what's beneath more securely. They all leak from the sides and from below, forever. The bad ideas of liners and cover soil containment over landfills are but myths of the toxic waste relocation and concealment industry, perpetuated by unsuspecting citizen victims, and revered by agencies and industry forevermore, to our collective detriment.

But moving on to explain how I became a sludge tracker, it was toward the end of my ten years at USC when I received another interesting phone call. I was asked if I would move to Seattle to run a start-up biotechnology company that had some fascinating technology that could have broad uses.

"Well, how much money do they have?" I asked.

"Oh, they raised a million dollars."

I said, "Yeah, but it takes twenty or fifty million to get anywhere."

Nevertheless, I came to Seattle and was at that firm for two years. After that adventure, I started my own company involving some beginning-stage DNA probe technology, specifically around the genomics and identification of exotic infectious disease micro-organisms, some of which I had studied way back in my UCR/NASA/JPL/Caltech/DOS/DOE/NCI days.

The field was the rapid diagnosis of infectious diseases using DNA probe technology. I did that for a couple of years until a friend received the Nobel Prize for inventing the underlying technology that superseded what I was doing. We were sitting on his front porch in La Jolla, having a beer. "I gotta tell ya something," he said. "My patent was just awarded."

That was the end of my company. What I could do in a day or two he could do in a near-nanosecond using rapid diagnostics based upon newer and more advanced DNA technology. He said he had the idea while driving on I-5 headed south of San Francisco. It just came to him.

These things are all interrelated, and right now when I work as a self-appointed sludge tracker on the pink-orange goo that oozes non-stop out of Cedar Hills Regional Landfill, and out of the FDA Food Safety Lab facility, and out of the NOAA site, and so on and in the parks, wildlands, forests, and farmlands around Washington State, shouldn't we know the chemistry or microbiology of that which remains as wholly unknown? It is wholly unknown, and what's worse, it's avoided. Knowledge of the danger of toxic waste is being purposefully ignored for reasons of then having to admit to the historical flaws, and to holding responsibility, and to then having to do something about it—to remove the whole toxic mess from our living environment.

Exposing people to toxic wastes holds unknown adverse consequences, including exposure to toxics of both chemical and microbial origin. Amongst the things we need to know about the human, microbial, toxic chemical interplay is that we don't yet even know what the right questions may be, and that's how we get to the key starting point of inquiry.

Humans have had no prior experience or interaction with any of the new chemical combinations presented in toxic wastes, and so the body has no inherent immunity to such toxics whatsoever. The body can't adapt. And what you end up with is epigenetic and genetic modifications of gene expression—two things that may cause normal cells to now transform into malignant cells. It's called malignant transformation; a healthy cell in one moment that in another instant becomes a cancer cell. The resulting cancer cells proliferate out of control and now you have a tumor, the cells from which may thereafter spread throughout the body as metastatic disease, and then you're dead.

We know so well now that cancer is caused by a mutation in a single cell that initiates cell proliferation, followed by the accumulation of additional mutations. Multiple factors influence the frequency by which the initial and subsequent mutations occur and the propensity of a cell that has acquired one or more mutations to then clonally expand. These factors include genetics, environmental insults, and a few intrinsic characteristics, such as age, obesity, insulin resistance, and inflammation. Inflammation is an essential factor in cancer initiation, development, and spread, and may also be incited by toxic insult.

We need to back up and figure out that point, and that chemical, and that microbe, and discover how their singular or collective influences result in malignant transformation. The study of cancer is not the study of oncology, it's the study of the biology of cancer, and that's what's different. Oncologists want to study tumors and chromosomes, but you must study the whole biology of cancer to find a solution. What initiated it? You can't prevent a disease until you figure out what started it.

There is a catena of deadly relationships that bears witness to the ill consequences of air, food, soil, and water pollution, among the toxic byproducts of industry that fuel disease and the current life-altering climate crisis, revealed as environmental degradation. The effects may be viewed as a public health crisis or as a pollution crisis, or as a petrochemical, agrochemical or transportation crisis, or as all of them combined, to form that singular reported climate crisis, which, according to science, could foreshadow the end of all life forms. We may talk about the toxic relationships, but it's just one thing—it is environmental degradation by the hand of us alone. The cause rests with those of us who tolerate and remain complacent about critical matters of overconsumption, degradation, and depletion of Earth's resources, and with those who worship the demigods of wealth, greed, and power.

Let me suggest how we arrived at such a dire situation. It's all about the ingrained reluctance of indifferent agencies to do anything progressive or productive about our vastly accumulating human wastes. "It's the way we do things," as King County likes to say, "It has to go somewhere."

Chapter 2

A Brief History of Sludge

The same people who were supposed to watchdog the wastes were helping move them around, confidentially.

Duff Wilson, *Fateful Harvest,* 2001.[1]

"Long ago and far away" is where everything started. Once civilization began, human settlements increased the density of people living in any one location. First, as single family units, then multiple family units, until all of a sudden it was realized that collectively we were creating more waste than we could tolerate in our living environment. Not long after, it was noted that disease was associated with our accumulating wastes. Living in our own trash was not good for us, or so it appeared. We were sharing our diseases through our shared wastes.

In Roman times, multiple seat potties were positioned over a stream as part of the public square. Thereby, everything was clean where you were, but those downstream suffered the consequences. This system worked for most, and became an established method for taking care of an individual's or of the whole village's human waste problem.

With the advent of animal domestication and greater-scale plant cultivation, there became good cause for people to live in ever more densely packed populations. People shared their foods and animals, and before long it became clear that the waste products from concentrated animal facilities and plant agriculture, as well as from the closer-packed people who tended and used those agricultural products, were more than could be tolerated. So, communities had to change or move. They moved themselves rather than manage their wastes, and this became impractical.

Smart people decided they needed to better manage their wastes, and that began by building human habitation sites along flowing waterways, so that all of the wastes could be carried away by the streams and rivers. Then it became someone else's problem.

In places where there were no waterways, people began to dig ditches in which to bury their wastes. They would dig deeper ditches, and soon these appeared as the latrines and outhouses that have lasted to today. Even now, there remain rural homes and farms that do not have plumbing. They simply move the outhouses each year or so, allowing the soils and waters to process and convey the various wastes. Though we may be talking about historical record, we're also talking about practices occurring even today in parts of the United States. There are families who wash their dishes, open the back door, and throw the dishpan of water out into the yard. It's the same with outhouses, they still exist across America, because many people are unable to afford the cost of putting in a modern septic system or connecting to the main sewage conveyance infrastructure.

As populations became more concentrated, way back when, it became apparent that infrastructure was needed to manage the issue. Now wastes became not only domestic toilet waste, but also wastes from kitchens and showers, as well as from medical offices and business centers, plus restaurants and industry, with yet greater systems needed for the conveyance of accumulated wastes in the form of collective raw sewage flows. This eventually led to our combined community sewage systems needing to be developed further. As towns became cities, wastes became an ever-increasing concern. In cities it became a serious problem.

As New York City grew from a mere colony on a rock, it became clear that people were suffering serious diseases as a result of raw human sewage flowing down the sides of the streets. The early cobblestone streets in Manhattan were constructed in a convex shape, such that the middle of each street was high, and the gutters at the sides low, so that all waste, human and otherwise, could flow into New York Harbor, or to other waterways for dismissal. In the thinking of the time, a waterway nearby to carry away all that waste was an essential resource.

In those days, it was expected and believed as a normal course at the heart of America, and as stated by the United States Senate in the eighteen hundreds, that the country's greatest natural resource was its rivers that carry away our wastes. That was written and stated by allegedly smart people. The key flaw in their thinking was that they thought only of tomorrow morning as the future, and definitely not from the perspective of short- and long-term consequences. Very bad for us. We still live today with their lack of foresight.

That was in the sixteen hundreds and on through the eighteen hundreds and upward, or so we are told. Such an approach continued into the early nineteen hundreds when my father, in 1919, got his first job for a dollar a day digging trenches that were to become the sewers of New York City. The resulting and subsequent sewer systems drained through the Battery Park region into New York Harbor. Now modern wastewater treatment plants can be found along the way, although still of limited outcomes. It was only later that those original open trenches, some partially covered, were replaced with cement pipes. Some of those earlier sewage conveyance systems still exist in New York today and are still functioning.[2]

Not long after, in the nineteen thirties and forties, garbage, rather than being burned along the streets, was collected in wagons and discarded in central locations, typically along the waterways of New York, including along New York Harbor, still to be burned. Only later did someone have the idea of putting garbage on barges to be towed out through New York Harbor each day and dumped into the Atlantic Ocean, most of which floated directly back into New York Harbor to again pollute the waterways and shorelines.[3] I saw it. I was there. Raw sewage and fresh garbage flowed in, then back out, then in again, until it disappeared, somehow, somewhere. I will spare you the grisly details, but that elusive somewhere was generally along the beaches of upper New Jersey, or in the other direction along the beaches of Long Island, and more. New York City, back then, was considered to be, in the vernacular of the time, a "Shit Hole."

The same situation occurred in the surrounding cities and counties. Staten Island became the target garbage dump for all users. Truckloads of human sewage, garbage, construction and manufacturing waste, everything imaginable, was put on wagons, then loaded onto barges, floated to Staten Island, and discarded. The trash piles kept moving further up from the beach. Until quite recently, Staten Island may still have been the largest garbage dump in the New York City area.

In the 1950s, my friends and I jumped off the docks into the Port of Elizabeth waterways, while attempting to avoid the lumps of raw sewage residuals, paper and otherwise, flowing directly into those waters that lacked any treatment at all. More dangerous were the thrown away pharmaceutical and personal care products sitting in massive heaps and piles waiting to be hauled off to the dumps, or somewhere. A great delight was tossing the limitless bottles of shampoos, baby oil, and other discarded personal products at passing trains that ran behind the manufacturing plants and factories, some of which remain to this day.

The trains of the New York Central, New Jersey Central, Pennsylvania, Baltimore & Ohio, Rahway Valley, and Lackawanna railroads were the near-endless pitching targets for the massive piles of jars, bottles, and packages of rejected new products. The possibilities for tossing cast-off, smashable items were boundless. Those railroads totally surrounded our homes and schools and they carted away everything that came off the ships, and returned with lots of Pennsylvania coal and hobos. With the advent of spray cans came the sport of exploding them singly or by the case in instant fires that emerged along the railway lines upon discovery. Anything and everything was thrown away, anywhere and everywhere, by the boxcar or by the case, without bias, and we created new uses. Hobos grabbed up whatever they thought they could use or sell door-to-door. Jobs were scant.

"Back when I was a kid," as they say, there were three main locations where the homeless and other disadvantaged people lived. They resided mostly in the garbage dumps where those toxic waste sites became their homes. One was the Newark, New Jersey, garbage dump along the meadowlands, which today is Newark Airport, among so many other venues, all the way to Newark Bay, and all of which rest upon the billions of tons of garbage it took to rearrange the life-sustaining wetlands, now renamed as meadowlands, into solid land for the development of commerce. Another was along the waterfronts of New York, and the other, the real Bel Air of all garbage dumps, was on the shore of Staten Island, as in, beachfront property. People lived there. They lived there for generations. They had their children there. All of their food, all of their clothing, every physical piece of their life came from garbage discharged from New York City and the towns and cities surrounding New Jersey.

The people of the dumps often spent their entire life there. They were sometimes dressed in khaki or olive-green clothes and boots captured from the parked or passing trains that left the ports carrying the tons of Army Surplus clothing and other materials that found their way back to the U.S. from the European venue of WWII. The trainloads of cannon shells (some unexploded), bullet casings, and other metals went to the smelters. The usable goods went to the many Army Surplus stores of the time. The web pistol belts, cartridge pouches, helmets, trenching shovels, and mess kits went to us kids, running off with as many hot items as could be carried.

Multiple family generations never left the garbage dumps. They made shacks out of cardboard and scrap wood. Discarded cars became luxury homes, doors or not, and in the nineteen forties and fifties, they never went to school. We saw them when we passed the dumps by bike, car, bus, or train. It was standard fare. There was nothing unusual about it. It was quite usual that populations of people lived in garbage dumps. No one bothered them, and no one batted an eye.

The lingo of the time was that if you didn't like someone, or if they insulted you, the thing to say was, "Send 'um down da dumps!" That meant, kill them and bury them in the dump, to be disappeared forever. No one even knew their names. The citizens of the dumps simply disappeared upon death, and the residents must have quietly buried them deep in the waste, never to be seen again. The mobsters had the same idea—if someone gave you trouble, just "Send 'um down da dumps," to be disappeared and decomposed among the rest of the garbage. Today, all take-offs and landings at Newark are over the many nameless deceased interred thereunder. If someone went away unexpectedly back then, we would simply remark, with a smirk, "He's prob'ly down da dumps."

In those days, another garbage dump heavily populated by the "disadvantaged" was located in Linden, New Jersey, about thirty miles east of Newark. The Linden dump was another former precious wetland, also considered at that time as being a useless wasteland. The selfish point was that if you filled the designated wasteland with garbage, you could then build on it, just as we did and still do here in Washington State and elsewhere, today.

As soon as the garbage started coming, so came industry and the dump's inhabitants. People moved in. We played there as kids, poking through everyone's discards, hoping for treasure, especially from the businesses and manufacturing plants. It was also a great place to catch pollywogs and frogs. As in all garbage dumps, frogs are generally toxin-tolerant and thrive on the masses of flies and other non-human vermin that flourish there, mostly dead and dying insects.

It was also a time when the Southeastern United States was being abandoned by people of color as a result of continuing segregation, persecution, and lack of employment. They moved north and to New Jersey because they could also easily move into and inhabit the luxury living quarters of Linden's primo garbage dump. When they tried to go to the Newark, New York, or Staten Island dumps, they could be chased away by others who had claimed those lands decades and generations beforehand for themselves and for their families.

From kindergarten through twelfth grade, we never saw or met any of those kids in school, they never attended because they just didn't exist, although immigrants from so many other lands were welcomed with open arms and smiles. That population tragedy is over, at least in America, but garbage dumps still sustain massive populations of marginalized people around the world.

As a teenager, I sometimes worked for a roofing company, lugging tar paper rolls, nails, and shingles up ladders onto the roofs of very old houses in the waterfront area of Elizabeth. My job sometimes included pulling buckets of hot tar up to the rooftops. The "Tar Man" was the guy who attended the tar kettle (aka "The Kettle Man"), so as to keep the tar both hot and fluid, in anticipation of being called upon to send up yet another bucket of the boiling, black liquid.

The most excellent Tar Man on my job was Bennie Bass, as he called himself, because his actual name, as he said, "Could not be pronounced by white people," then he laughed. He seldom spoke; he just smiled all day and sent up buckets of hot tar as needed. Other than his happy personality, his other defining characteristic was his very dark skin with white polka dots over every exposed area of his body not covered by baggy pants, suspenders, spotted T-shirt, and an old baseball hat. One day I asked Bennie about the white spots. He explained in marginally intelligible words that the surface of his skin was burned off regularly by drops of hot tar that fell from the buckets above as he hoisted them up to the roof tops.

I got to know this fine man, Bennie Bass, and eventually learned that he was from Charlotte Amalie, the capital of the U.S. Virgin Islands. He had never learned to read or write. His family origin was Africa, long ago, and his exceptional strong point was that he refused to live in a garbage dump on the mainland. He strove to keep it that way.

Bennie Bass worked hard as a highly underpaid laborer for this roofing company. He lived in the roofing company's storage shed and travelled back home to Charlotte Amalie during the bitter New Jersey snow period. We became friends over time, and I learned to understand the unusual lingo of U.S. citizens from St. Thomas. He and they detested garbage and waste, and he wondered why so many people lived in the dumps. Being a near expert in garbage dumps by then, I asked Bennie if St. Thomas Island had any dumps.

"No," he said. "We just toss it in the sea."

Oh, Boy! "So that's how you have a very clean island," I thought.

Skipping forward to today, to this moment, the Linden garbage dump is now the location of one of the most modern waste-to-energy plants in all of North America.[4] The garbage is gone, the dump is gone, the former inhabitants are gone, and the life-giving wetlands (aka the former "wastelands") are forever gone. On top of it now are industrial centers, and right in the middle is a modern, gleaming waste-to-energy plant. You almost want to say, "End of story." But not! While the people and the wetlands were critical, it was the management of the garbage that won the day.

Now take a turn and go across the Arthur Kill, the waterway that separates Staten Island, NY, from Port Elizabeth, NJ. The current Arthur Kill lift bridge was built by the Baltimore and Ohio Railroad in 1959 to replace the original Arthur Kill swing bridge that opened in 1890. The purpose was to transport municipal solid waste to that massive and historical garbage dump that dominated Staten Island, then and now, as a model for the far too late realization that our wetlands and other wildlands, wet or dry, must not be treated as repositories for our collective toxic wastes, but as precious wildlife sanctuaries for the life and health of us all.

The bridge supports a single train track used mainly to carry garbage out of New Jersey, and to transport freight to destinations in western Staten Island. In the 1950s, and from firsthand certainty, the Arthur Kill Bridge also happened to support a population of night-roaming teenagers who hopped the train late at night to get from New Jersey to Staten Island for pizza and beer, where the drinking age was 18, not 21. In actuality, the drinking age was based on your ability to pay the one dollar per pitcher price for any beer on tap. The bars stayed open until 4 a.m., and we became passengers, inside or atop the reliable, nighttime garbage train cars heading back to New Jersey by sunrise. If you've not smelled the inside of a garbage train, beware!

All garbage dumps (aka landfills), should have been ended and eliminated about 100 years ago. Yet, we still have people and agencies that cherish their landfills, underlying aquifers be damned. Hmmm? It must be the commercial rewards that keep them going, not any concern for life-supporting wetlands, or for any lands at all, or for environmental quality, or for public health, or for human life. As the saying goes: "Nothing cherished, nothing saved."

According to the New York Times article, *How the World's Largest Garbage Dump Evolved Into a Green Oasis*, by Robert Sullivan, the radical fix for a noxious landfill, such as Staten Island, was to bury the trash, plant some grass, and do nothing for 20 years.[5] Throughout the U.S., the lesson has not been learned at all. A similar pattern has emerged across the country. Lessons not learned—lessons never learned—lessons not wanting to be learned.

It must be the rats, stench, toxic fumes, toxic landfill leachates, forever impaired ground and surface waters, poisoned aquifers, destroyed wildlife, and the never-recoverable possibilities that are cherished. Yet the countless thousands of buried landfills of our far and recent past will continue to make us ill and kill us for centuries to come, while agencies and managers continue to report that their approach to solid waste (i.e., garbage) disposal is safe and beneficial. We used to burn it, now we bury it. We've progressed from air pollution to another form of air pollution, now with the added contributions of soil and water pollution. Of special note, we continue to have yet more toxic landfills in the planning and early stages of development right here in Washington State—lessons never learned.

We died of our wastes in the time of the Roman Empire and in colonial times, and we die of our toxic wastes today, no matter the continuing superficial attempts at waste management or infrastructure development. Just like then, our wastes have become an expected part of our landscapes and waterscapes, while we learned how best to conceal them. We have been cultivated over millennia to become part of our wastes, and them of us, 'til death do us part. We remain in the same dilemma to this day, if for no other reason than that our elected officials care not to provide safe and effective infrastructure, except for what may be provided by the scantiest of budgets.

Human and environmental health seem to have fallen by the wayside somewhere, as if rodents, insects, pathogens, toxic fumes, and contaminated surface and groundwaters are invisible, much less causing the continual release of Earth-threatening greenhouse gases (GHGs) and toxic volatile organic compounds (VOCs). Landfilling must end, as must the perpetuation of false ideas such as garbage for composting, and beneficial reuse of *Toxic Sewage Sludge*, plus the free and wholesale generation of the forever-toxic waters and gases from such operations—if not, welcome to our impending demise.

In Washington State, where I have conducted most of my investigations, there are many wastewater treatment plants, at least four hundred or more, as can best be determined. Not one of those wastewater treatment plants actually works. They do not treat wastewater. They separate the solids (*Toxic Sewage Sludge*) from the water, while the remaining toxic waters (toxic wastewater effluents) are returned to the environment, meaning in our case, to our rivers and to Puget Sound and the Pacific Ocean. The *Toxic Sewage Sludge* product then gets hauled to forests or over the mountains to farms to be land disposed where our foods are grown.

That should not be how this story ends. There's a difference between technology that was needed, technology that has been invented, technology that was developed, and technology that is deployed. Technology in use now could actually convert solid waste, such as *Toxic Sewage Sludge*, into renewable clean energy, with nothing more needed than the modified engineering inside. There's nothing in there that couldn't have been there in the eighteen or nineteen hundreds. In other countries, and in some enlightened areas of the United States, systems are being built more thoughtfully, with a mind to what comes out of the stack.

Infrastructure is not just about cell phones, internet, roadways, and bridges, it is also about public health. Our collective failure is not to have recognized the adverse impacts of our ever-generating wastes as one of the primary causes of our universally compromised health. Basic human health depends upon the ready availability of clean air, foods, soils, and waters not contaminated or polluted by partially reclaimed toxic wastewater effluents, recycled *Toxic Sewage Sludge* promoted as harmless composts, soils amended directly with *Toxic Sewage Sludge* from our minimally functional wastewater treatment plants, or our farms irrigated or fertilized with those same reclaimed toxic wastewaters and sludges. It's easier to prevent illnesses than to have to treat so many diseases incited by the failure of our agencies to admit to their misguided intentions in how they manage our wastes.

How did this happen? Who are these people who dump our domestic fecal, industrial, and medical wastes, and highly toxic stormwater runoff, into our living environment? These are not safe practices, and they ignore the basic instincts of survival. These are the lessons not wanting to be learned.

As one example of the consequences of that horrible thinking: On the east coast we have Chesapeake Bay, and on the west coast we have Puget Sound, where the surface area footprints are similar. The difference is that Chesapeake Bay on average is about 50 feet deep, whereas Puget Sound is up to 900 feet deep. The Atlantic Ocean has tides that roll in and out twice a day, so most of the Bay water is exchanged. The Pacific Ocean, on the other hand, also has tides that go in and out twice a day, but very little water is exchanged in Puget Sound. It sloshes around in the bottom of the deep pockets of the Sound, going back and forth, but fresh Pacific Ocean water never fully replaces the water that's in there.

Since the Seattle Puget Sound region was first settled by white people in the sixteen through eighteen hundreds, it is likely that the toxic materials they dumped in there are still sloshing around, as are all wastes that we continue to discharge into these precious waters even today. No, Virginia, it does not just go away!

If tomorrow you were to sample the accumulated sediments in Puget Sound, you will bring up dioxins and furans, among a plethora of other highly carcinogenic and otherwise toxic chemicals—and there are hundreds, if not tens of thousands or more of such toxic chemicals in there, not just one or a few. If you do the chemical analysis, these are the same dioxins and furans as were introduced decades ago. They are the combustion products of the forest industry because the lumber mills would burn sawdust, branches, bark, and chips in great incineration devices to get rid of wood waste. That predated the invention of particle board and all kinds of ancillary products that are made today with leftover trimmings from the lumber mills. Back then they were burned. Now there isn't a fleck of sawdust that goes to waste. It's all used. Even if it only goes to compost, it is used.

Combustion products are also still there in the water. If you turn the page, you go through the same story with the solid manure and urine wastes from cow farms, which were adjacent, near and far, to Puget Sound. Twelve miles from Seattle you were in Bothell and dairy country. All of those concentrated animal wastes went into the streams and rivers and eventually Lake Washington. No human being could go into Lake Washington at that time because it was so polluted. You couldn't go near the lake because no one could stand the smell. Lake Washington was a giant cesspool that drained into Lake Union and then into Puget Sound. Once all of that toxic waste reached Puget Sound, it could do nothing more than slosh back and forth for the following couple of hundred years, and it's still there today, as decayed residual wastes.

The same story holds true for the Snoqualmie River, the Nisqually River, and for all other waterways that drain to Puget Sound, which by default is the great recipient of our collected stormwater runoff and domestic, human, agricultural, and industrial wastes here in the great Pacific Northwest.

During the middle of the last century, or thereabouts, the land in this region became a target for development, as undeveloped lands seem to do. Someone I know well—he's 87 now—started buying up wetlands, then considered wastelands, around Bothell and the northern parts of Lake Washington in King County and in southern Snohomish County. At one time, the Bothell-Everett highway was the only road that connected Bothell to Everett, and wherever there was an intersection, there were wetlands that could be built upon, or so they assumed.

My friend bought the nuisance wetlands at reasonable cost. He started filling them with garbage, construction wastes, and waste soils, because why have wetlands at all, when you can simply fill them with garbage and other solid wastes, and then build strip malls, technology centers, the University of Washington, Bothell Campus, and all of that now-developed corridor that goes from Bothell partway up to Everett through Mill Creek? He bought it all. His joy now, he told me recently, while we were conversing on his forest easement lands, was to give back to the people by preserving the adjacent forest lands in as natural a state as could be expected for public use. We can work at healing, both of ourselves and of the lands we touched, but we must make the effort. My old friend was a hero of his time, back then, simply because he took on the task of filling and developing those precious wetlands.

I go hiking in his semi-restored wetlands and forests quite often, perhaps three days a week. It's all forested hills and trails that he had set aside decades ago for the benefit of the public. The preserved forests surround the technology parks, yet the springs and streams that emerge therefrom, formerly as clear waters, and of which I will never reveal to my friend, are now red, orange, and brown, for the berms and landfills that were constructed to defeat the wetlands were formed of solid and hazardous waste. Those once pristine waters now leach and flow most colorfully into North Creek, a "protected" salmon stream, which flows in turn into the Sammamish River, then into Lake Washington and Puget Sound.

All of these actions, past and present, are violations of the Endangered Species Act and the Clean Water Act, and because it all percolates into the ground, it's also a violation of the Clean Drinking Water Act. Who knew? Who permitted it? Who regulates it? No one! Still not to this day.

The state of Washington still cherishes environmental degradation as a mainstay of its culture and of its cultural heritage. Land-disposed *Toxic Sewage Sludge* and land filling of garbage are still considered as the way things are done, despite the extreme environmental degradation and the continuing human health consequences. Toxic garbage was spread all over the soils, which are now schools, parks, housing developments, technology parks, and shopping centers. Every kind of waste imaginable was put there. Those are blatant violations of the Resource Conservation and Recovery Act, known as RCRA. Anything you can imagine, whether domestic garbage, industrial waste, *Toxic Sewage Sludge*, cow manure, and the like, all of that was just let go to fill in these precious wetlands.

The salmon that once thrived in the area are now gone. That is the fate of nature and of the whole Earth, simply for the sake of the banal interests of transient real estate development, with the full knowledge and agreement of our elected officials and agencies.

My friend was the guy who bought most of it, in this case. He was a forward-thinking fellow, and in 1985, when the wetlands were filled in sufficiently so that you could put enough soil over top to construct some buildings, the Washington State Technology Corridor was established. New technology companies arrived, including the first ever Microsoft building, which is still there today. They were built on these former dairy lands, former wetlands, former Indigenous lands, and they're all filled with toxic trash, and they all still leach and drain red, orange, and brown stinking goo, and they are all in violation of at least the four federal laws mentioned above.

To assure that my statements remain valid, while completing this tome I captured a series of fresh digital images of the red goo and other tainted waters and scum as they flowed from the ground at several locations within the state's cherished technology corridor, including from my friend's set-aside forests. It's all still there, and it's all still flowing via "protected" streams into Lake Washington and Puget Sound.

Quite oddly, and almost to the point of hilarity, one of our U.S. Senators from Washington State recently announced that our salmon populations, both hatchery and natural alike, are the victims of climate crisis, and that global warming is the cause of the decline in salmon populations, including those fish allocated to the Tribes. "The waters are warming, and the salmon cannot survive," she announced. Nuts! Would she like a tour?

Yes, the waters are warming, and if you take a close look you can easily see that those same waters, once clear and cooling, are now colorfully tainted and absorbing ever more energy from the Sun that then heats them even more efficiently. The primary contributing factor to the colorfulness of the waters of our streams and rivers is toxic wastewater effluents from our many wastewater treatment plants and from the leachates and runoff from the land-disposed *Toxic Sewage Sludges* from those same sources. A primary feeder of climate change is in fact the improper treatment and management of our wastes—it's a matter of failed infrastructure.

Our once clear waters that have been sullied by toxic wastewater effluents absorb more sunlight, for sure, so the waters are warmer. And way too many dams in the rivers obstruct fish passage and restrict natural water flows, so they also cause greater absorption of sunlight and heat. The natural plant life removed from the banks of our rivers and streams had once provided necessary cooling shade, and so the exposed waters there are warmer, as well. Yet, despite these obvious climate changing factors, there seems to be a politically driven need for elected officials to jump on the nearest train of political vogue and correctness, damn the facts of science.

Yes, for sure, the Earth and its precious waters are in fact warming, but this Senator's advisors must have failed to inform her that our State Ecology Department's scientists, along with so many others, report publicly that the primary cause of the salmon decline in Washington State is the colorful and toxic wastewater effluents discharged from the state's 100 or more wastewater treatment plants that enter Puget Sound and other waters. These toxic wastewater treatment plant effluents are not "treated" waters, but in fact provide the trouble that is promoting salmon decline. Our state's agencies simply send their minimally processed raw sewage wastes onward to our surface and groundwaters, including most especially into our precious salmon breeding grounds and their migratory pathways.

At one time, a railroad line meandered along the Sammamish River to Lake Washington. It supported the lumber trains. The original Weyerhaeuser lumber operations began in the eighteen hundreds, and when lumbering diminished, these became redundant railroad beds. The rails were taken up and trenches were built, and a giant sewer system was constructed. For the first time, nearly all the nearby homes and businesses were tied into this sewer system, and the whole project was called Metro.

Metro became the sewer system that drained everything from the northern areas of Seattle into a pipeline that is below what is now the Burke-Gilman and Sammamish River trails. The Burke-Gilman Trail runs through Bothell, Kenmore, North Seattle, and the University of Washington, then down into Discovery Park. In the days before Metro, meaning in the 1960s, you couldn't get near Lake Washington, the smell was so overriding, because it was a twenty-three-mile-long cesspool of human, medical, dairy, and industrial waste, all on its way to Puget Sound and the Pacific Ocean. The system still doesn't work very well, always leaking, never with the objective of actually "treating" raw sewage, sewage sludge or wastewater effluents.

Lake Washington doesn't stink anymore because great effort went into limiting additional new inflows. Back then, all sewage drained to Lake Washington. The lake was considered a natural resource for the limitless drainage of all the raw sewage from the surrounding towns. No one had a sewer, and functional treatment plants did not exist, just as they do not exist in the vicinity to this day.

I was in the U.S. Navy on a submarine at that time and we used to conduct anti-submarine warfare operations with Navy seaplanes that flew out of the Sand Point Naval Air Station (1920 – 1970), that is now the Sand Point, National Oceanographic and Atmospheric Administration (NOAA) facility, as well as Magnuson Park and more. Sand Point used to be the Navy's seaplane base in the region.[6] Seaplanes would go up ramps and into the hangars, which today are indoor soccer fields. We used to tie up at the dock.

Our sub would cruise up and down the middle of Lake Washington underwater, and the anti-submarine warfare seaplanes would fly over and drop mock explosives on us. They claimed they could see us through the water, but not. It's really about a hundred feet deep, mean depth, and we ran shallow up the middle, nearly invisible amongst the filth. They dropped their mock explosive devices, referred to as PDCs (Practice Depth Charges), but they didn't damage anything without a direct hit. Instead of a large, functional, lethal, ordinary depth charge, they would throw out this hand grenade-sized bomb, which would hit the water with a bang, but we were already far away, headed for Kirkland. Boom! You could hear it and you could score a hit whether it was here, there or over there.

The sub was guided somewhat mechanically or intuitively by a DRAI (Dead Reckoning Analyzer Indicator), which meant "by fingers-crossed instincts in the close quarters of an opaque lake." It's a mechanical device that senses motion and tells you on a chart where you may or may not be located. In any case, that's how we plied the murky, smelly waters of Lake Washington, trying to avoid being whacked by a PDC. We would race up and down, but the trick was that the lake was twenty-three miles of human sewage waste, and you couldn't see a submarine or anything else underwater. It was that bad.

I know firsthand how Seattle's very bad history of failing to develop infrastructure looked up to the point when this beautiful lake had practically no real estate value on its shoreline because no one could live there—it was nearly uninhabitable through the fifties and early sixties. Worse yet, the shores of Lake Washington were also designated excellent locations for the city's and surrounding burbs' landfills, meaning adjacent wetlands—including at the University of Washington's shoreline sports fields and within today's NOAA facility—that were considered perfect for municipal solid waste disposal. And there are more, many more. I track them as they all still leach and flow into Lake Washington. You can't hide from your polluting and contaminating history, even decades later.

There's another point worth recalling, after having visited this area again recently. There's a botanical heritage center just to the northwest of the University of Washington's Huskie Stadium, and there are golf courses adjacent to the stadium with giant parking lots and ball fields that used to be Seattle's city garbage dump. This entire area was once life-sustaining wetlands inhabited by the Indigenous People, and then Seattle said, "Let's fill it with garbage and make something useful." Today you get to look at Huskie Stadium and the golf courses and the asphalt-covered parking lots, and you don't realize that these were historical salmon breeding grounds inhabited by Indigenous People for probably the last couple of thousand years.[7] Now part of it is hiking trails, and people go there birding and walking their dogs through the odd remaining wetlands and barely recovering woods, not forests. It's all covered-over garbage, and it all still leaks and leaches into Lake Washington, just as it will for centuries to come.

Lake Washington will never recover. That practice was not progress back then, it was planned and intended devastation, and it continues now, and likely will for decades or more after us if our approaches toward the management of our wastes are not changed.

Some of the other legacy garbage piles in the area resulted from when the I-5 was built over the Ship Canal just north of Seattle. Bulldozers, side-by-side, moved north from Seattle to clear every structure, every gas station, all of the streets, all of the schools, everything that existed along the way. They simply scooped the debris into trucks and dumped it in the wetlands adjacent to the University of Washington. They dumped it into the wetlands where I live in Kenmore, and they dumped it in Kirkland. They dumped it in Bothell where the dairies and wetlands were located, and elsewhere, because a few slick real estate dudes and their agency partners wanted to convert all of those "useless" former Native American wetlands and salmon breeding grounds into high-value real estate, because that was the direction Seattle had to move. That's the going storyline, but in reality, no one was smart enough to care about any living thing, human or other, similar to development practices still ongoing around the world and right here today.

Even where my grandkids went to school, Thornton Creek Elementary School, Jane Adams Middle School, and Nathan Hale High School, for example, are all adjacent to the origin of Thornton Creek, a treasured salmon breeding stream above Lake Washington.

There too we find legacy landfills. There's a large, old Seattle garbage dump still present, residing beneath the local schools and adjacent to the salmon waters that still squirt red, orange, and brown landfill leachates by surface and groundwater into that same Lake Washington at Matthews Beach Park. They just couldn't wait to fill in those wetlands and streams, which were inconvenient and in the way. Today it's a great high school atop a mountain of toxic waste.

Sometimes we have to wonder why schools and other public buildings are located by plan and intent atop legacy landfills, even when we know well that all legacy landfills are graveyards of highly toxic waste, continually oozing out into today's neighborhoods, schools, parks, and playfields. And isn't it quite oxymoronic that the toxics in these said landfills beneath today's schools are also known to be the inciting agents of learning disabilities, whether by genetic insult, epigenetic influence or by direct toxicity alone?

It's always a challenge to discuss these issues with city, county, and state agencies, as well with some agency-aligned environmental groups, because they all respond with the same gibberish that it's safe, while they remain unable to answer my basic question: "Have you conducted comprehensive qualitative chemical analyses of the colorful landfill leachates that enter these streams and Lake Washington?" which is to ask, "Do you know what's in it, what are its adverse effects on human and environmental health, and what is its fate in the environment?" Always blank stares—never a meaningful answer—end of conversation.

Our salmon are now in big trouble. A recent retiree who headed a very important Washington State agency told me that he and his wife had purchased a new motorhome and were leaving Seattle for good. He said, "I've lived here, and I worked at these agencies all these years until we retired, because I loved going salmon fishing right here. I sold the boat and gave away the fishing poles. We're leaving town. There aren't any salmon left to catch anymore."

I used to do the same thing. Sometimes I would get up at 3:30 a.m., go salmon fishing and be at the lab by 8:00 a.m. It was easy. They were everywhere. All over. You didn't have to leave the Sound. You could just go fishing right off Seattle, right off the Ship Canal, right off Edmonds, wherever you wanted to go there were several species of salmon, some wild, some hatchery-born. It's rare now, although there are still some migrating schools that come and go. And it's the same with the orcas. The orcas live on the salmon, but the salmon are declining so the orcas don't school anymore. They don't migrate as they did, because the water is disgustingly warmer, more toxic, and filthier, and they can't live in it, breed in it, or tolerate that environment.[8]

Sometime around 1986 I had a laboratory in this new technology center at Canyon Park, Bothell. Washington State's Governor, Booth Gardner, came out one day, along with a local TV news station. We were joined by the landowner and commercial developer. The Governor, developer, and I posed for a TV video celebration of the ribbon cutting for Washington's new technology corridor. The Governor got a bit peeved with me; as usual, he wanted to talk to the cameras about what he had done, and I wanted to talk about advances in technology. The developer stayed politely silent but wanted to mention how much fun he was having—he was the author of much of what happened in the development of the technology park, and his inspired works remain evident to this day. So we cut the ribbon to open the corridor, and we were all so happy. Back then, we believed it was just what was needed, and even today it is still expanding. Many biotech and other firms moved in. The early Microsoft manufacturing plant was there. It was all a very big deal in those days and I was a sorry part of that development.

Sadly, remnants of the original Native American places, the former pristine streams and wetlands, together with the subsequent landfills, dairies, and logging enterprises, lie buried below, some still producing colorful leachates as evidence of the former and present disengaged civilizations. The toxic sins of our past remain, and the resulting noxious pollutants will continue to infiltrate the landscape for hundreds of years into the future.

During that period in the late 1980s, representatives of a Japanese pharmaceutical company came to visit us in Bothell, to see our new lab and what we were doing. They became partners and investors in our biotech company. We were making DNA probes for the rapid diagnosis of sexually transmitted diseases, among other things. We had copied the name for the firm from some guy who built a technology manufacturing plant down the street. He called his gig "Microsoft," so we called ours "MicroProbe." We were out front laughing when the sign went up, jesting that we knew Bill was going to come up the street and say, "You can't do that!" but he never did.

Just out the front door and fifty feet across from the MicroProbe parking lot flowed North Creek, then relatively clear, now colorful. Our Japanese visitors would almost faint when we took them to see the migrating salmon run that covered the stream from edge to edge. Migrating schools of salmon were common back then, and they were a serious delight to our guests. "No," we said, "You cannot just grab one." Those plentiful days, those wild salmon runs, and those clean waters are over. We still go walking there several times a week, but the salmon runs were ended by the emergence of the brown and tainted waters. Recent photographs of the now dark brown, former salmon stream are frightening, indeed, for they signal the likely fate of us all.

The mounds with lush plantings and grasses that surround these former salmon streams are constructed of legacy toxic waste and garbage fill from multiple sources. The banks of the salmon streams were heaped into berms surrounding the former wetlands because you had to build berms to control the water flow so that when it rained and the water rose, it would still be in the streams and not in the parking lots and office spaces. Contaminated effluents still leach out of these berms. Absent any form of regulation or environmental or human health concern, this is what you end up with.

I have taken hundreds of images of these colorful and lethal toxics, mostly during the rainy season. It's my usual, "Let's see what's improved with all these berms leaking bright red and brown toxic goo from the garbage and other wastes from that many years ago." It's still there; it's still deadly; it remains qualitatively and quantitatively unevaluated; the toxicity and toxicology remain unknown; and it will be there for centuries to come, at least.

Who were and are the citizen victims? Who were and are the environmental victims? What are the consequences to public health? Who is responsible? Most of all, is there anyone who cares? Globally, other than from smoking-related cancers, we are witnessing an explosion of all cancers that I argue is primarily due to our poorly managed waste.

One of Washington State's and King County's mottos seems to be: "It just goes away," which they have recited to me many times, and which is to say, by implication, that *Toxic Sewage Sludge* and garbage just go away; toss it on the ground or in a landfill, or into a forest or on a farm, or into a wetland or stream, and it just goes away. Never!

"Microbes decompose and sanitize it, and it just goes away," or so they say.

Perhaps in several thousand years.

No, it doesn't "just go away," and I record videos and images of all this stinking mess that continues to leach and leak out of these berms and into our salmon streams.

If you ask the state's Department of Ecology about the salmon and the water, they tell you how wonderful it is. If you ask the scientists who work there and who actually publish papers about their findings, they say the salmon are dead and the orcas are dying because of the same *Toxic Sewage Sludge*, toxic wastewater effluents, and toxic landfill leachates.[9] The bottom line is, even within the same agencies they can't get their stories straight.

Some of the people I know and with whom I work today, and some of whom are my neighbors, know the systems well, for they are the scientists and managers of those same federal agencies who are forced to eat their own data. Seldom is their research ever published. Maybe once a year. But they have chronicled the entire evolution of toxicity. And one of the side stories they don't tell us about, but that they see, is that after decade upon decade of exposure to the toxic wastes, the toxicity in the waters does not diminish, it continues to increase. It doesn't just go away. Not only that, but because of the chemistry, or what happens underground or in an aqueous environment, new chemicals are being generated that man doesn't even know about. It's worse, not better.

"Why is the NOAA facility built beside this big, grassy mound?" I was asked when giving a tour to a group of local officials one day. The site is half a mile by half a mile and all the NOAA research buildings are built around it, with an employee daycare center on top.

"It's garbage," I responded.

"What do you mean?"

"It's one of the city of Seattle's legacy garbage dumps. Let me show you." And we walked around, and went up to the fence, and I said, "See that stream?"

"Yeah."

"I want you to notice that it's bright red. That is landfill leachate from these toxic, legacy landfill waste dumps. The Navy dumped its waste here. The city dumped garbage here later because it was considered wasteland. They referred to wetlands as wastelands.

Historically, it was an inhabited location for the area's Tribes because the salmon were here. Everything that you would need to live was here and this lake was full of fish. The salmon came in their migration and laid their eggs, right here, on that sandbar, later called Sandpoint," I informed them.

The stories our agencies tell and the history do not come together at all. I walked Sandpoint during those earlier times in the early 1960s. I was there. It was located at the end of the bus line on our way back from Seattle. The bus line connected Sandpoint Naval Air Station to Downtown Seattle, and being in the Navy, it was free. We'd go downtown and then ride the bus back. It was the end of the line. There were no other buses. Today it's a virtual hellhole of buried toxic waste, just as has occurred all around that beautiful lake and Seattle's many neighboring waterways.

Flowing waters as streams and rivers, plus ponds, lakes, wetlands, bays, harbors, and oceans were considered beneficial natural resources because they received all of our sewage, garbage, and other flowing toxic wastes. Seattle's pitiful history is but one bad example of what has occurred with human waste disposal throughout the entire country and into the waters beyond. As a result, we must appreciate that we are citizen victims of the toxic effects of our toxic waste, with full appreciation that ever-greater exposure to toxics yields the potential for greater morbidity and mortality—a self-inflicted journey into darkness.

Washington State's Toxic Waste Relocation Program remains in effect, and it continues to destroy both us and our living environment. Land-disposed *Toxic Sewage Sludge*, water-disposed wastewater effluents, and the land or water disposal of other toxic wastes, provoke a broad spectrum of adverse impacts on human and environmental health, noting well that nearly all chronic diseases are incited by long-term exposure to low levels of environmental contaminants and pollutants.

Toxic wastes contribute to the global crises of climate change, pollution of our air, food, soil, and water, famine, drought, compromised health, epidemics, deforestation, and biodiversity loss. No matter the source, kind, or concentration of toxics in our wastes, whether in sludges from wastewater treatment plants, wastewater effluents, stormwater runoff, landfill leachates, industrial or medical wastes, refinery sludges, or the runoff from agriculture as fertilizers or pesticides, or as wastes from concentrated animal feeding operations, the net result is environmental degradation. Key factors include a decline in readily available potable water supplies, coupled with population growth and mass migration, yet with only scant provision for renewable clean energy.[10]

It's about the reluctance of agencies to contain our accumulating wastes, and it's all about water. Earth is the only planet with surface water, and it resides at a steady state, meaning there will never be more or less water than we have now, and it's often in the wrong form or place. We need potable water for survival. Nearly half the human population is adversely affected by a shortage of fresh water, and such predicament incites famine, food insecurity, wildfire, human migration, and war, and therefore we must recognize that it's all about the ready availability of clean water for human and environmental health.

Norman Borlaug advised decades ago that we must figure out how to increase agricultural production on the same or on less land in order to reduce deforestation in an effort to save precious water. He stated that we must add forests, for it is forests that generate fresh water, not melting glaciers or additional aquifer-depleting wells, for all aquifers are in decline. Deforestation must end, he stated, for the fresh water provided by forests is the source stream of all life on Earth.[11] Declining forests and diminishing waters, amid an exploding population, impacted further by a changing climate, collectively set the scene for lost biodiversity, revealing that life on Earth is nearing a point of non-sustainability, and we did it.

So, where do we go from here? We can protect the air, food, soil, and water we have from willful contamination and depletion, or we can bear witness to the alarming decline of our life-sustaining resources, but for what purpose? The climate is changing, and the population is increasing, while we continue to contaminate and deplete our critical fresh water supplies, never seeking alternatives to the failed management of our growing waste streams.

The wetter areas of Earth are becoming wetter, while the drier regions drier, offering a deadly crisis situation for survival. The anticipated outcome equates with food and water insecurity, for we cannot desalinate the sea as a solution. While the forests we cut are the single greatest producers of fresh water on Earth, it is we who represent the core causal agent of the global water crisis.

Of note, while the amount of human-generated sewage sludge produced in the U.S. each year is about 5 million tons, cattle produce more than 500 million tons of solid waste, all of which pollute our lands and waters. Proper and beneficial management of all wastes is critical for our very survival. We can't simply sit on this knowledge and information, for as Rachel Carson told us so well, "Knowing what I do, there would be no future peace for me if I kept silent."[12]

CHAPTER 3

DON'T YOU KNOW CANCER IS OVER?

Man has lost the capacity to foresee and to forestall.
He will end by destroying the earth.

Albert Schweitzer, German-French theologian, 1875-1965.[1]

Cancer incidence and prevalence rates are climbing globally, and now we have escalating climate anomalies, ever-increasing CO_2 accumulation, and ever more polluting industrialization and consumerism to add to the challenge. All of these things go hand-in-hand with population growth. As well, the rate of new births, meaning, the rate of how many babies are born globally each year, is down, but because we're living longer, the standing population on Earth is remarkably larger. Also, there's a linear increase in the number of cancer patients alive this year over those who were alive last year; they are the standing population of people with cancer.

According to the World Health Organization, cancer is incited to some degree by "genetic factors and three categories of external agents, including: i) physical carcinogens, such as ultraviolet and ionizing radiation; ii) chemical carcinogens, such as asbestos, components of tobacco smoke, alcohol, aflatoxin (a food contaminant), and arsenic (a drinking water contaminant); and iii) biological carcinogens, such as infections by certain viruses, bacteria, or parasites."[2]

While *Toxic Sewage Sludge* is a mix of a vast array of toxic entities, you can't simply classify *Toxic Sewage Sludge* on a pie diagram as a cancer triggering agent. You must know what's in it—you must identify and know the classes of chemicals, if not the nearly inexhaustible and unknowable list of chemical variants generated and contained therein. The toxic chemicals in *Toxic Sewage Sludge* get into our foods and waters, and into everything else, but it's not a single and definable slice on a pie chart.

Tobacco (specifically those products tainted with chemical additives) is still the main cause of cancer, everywhere in the world, although declining in the United States. There's almost nothing else like it. People who chew tobacco, for example, may get testicular cancer. Most men generally do not die from testicular cancer because the lesion may develop just locally, not having had a chance to spread before detection and treatment. Chewing tobacco may also incite oral cancers, as well as cancer of the throat and nasal pharynx. Such head and neck cancers are generally accompanied by an unfavorable outcome. With smoking, carcinogens in burned tobacco may add synergistically to the oncogenic effects of other environmental carcinogens, resulting perhaps, in an even higher rate of cancer. If you live on a farm, for example, and are exposed to pesticides, fertilizers, and *Toxic Sewage Sludge*, individually or as multiple combined carcinogens, and if you are smoking as well, then your cancer risk may well be accelerated—substantially.

Worse yet, if your parents and grandparents smoked, if your aunts and uncles smoked, or if you lived as a child or beforehand in a smoking environment, and most especially if your mother smoked while she was pregnant, then your predisposition for any of several cancers is amplified. No, you didn't do it alone. It was likely the ongoing accumulation of multiple and various heritable or otherwise genetic and epigenetic predisposing risk factors in your life, plus any one or more of the many direct cancer risk factors in your living environment, including from persistent exposures to cigarette smoke, pesticides, fertilizers, industrial chemicals, pharmaceuticals, and personal care products, plus exposure to land-disposed *Toxic Sewage Sludge*, or to any one or more of the various environmental carcinogens that perhaps set the stage for your cancer fate. It may have been by inhalation of the resulting dusts or combustion products, or by ingestion of contaminated foods or waters, or perhaps by any one or more of the associated prenatal exposures whereby your predisposition to cancer was set.

Who knew? Actually, we all knew since sometime in the 1940s, so they, and we, all did it together. The message is that we must act now or we will kill even more of our subsequent generations of family members and others.

The largest global pandemic in history is just now ending. It caused more deaths than infectious agents, and we're just now coming out of it—cigarettes. When we count the number of deaths from cigarettes, that didn't go on for a year or two like the flu or SARs, it went on for about 150 years, and we are not even amazed or shocked. We expected it, and we're just now getting out of that predicament—without a vaccine. Who ruled the roost? Big Tobacco. Commerce ruled it. You need something addictive that will be adopted population wide and used in commerce. So only now we're coming out of it and this is not a trivial point. It's a model of a poor response to a healthcare emergency, just as we are witnessing today with *Toxic Sewage Sludge* disposal.

The epidemiology of cancer has been investigated for decades, covering the exposures, genetics, and lifestyles of near countless people, all as part of the quest to discover the causes of or the inciting agents of cancer, as well as of so many other diseases. "Where did you live? What was your lifestyle? What did your parents do? Did you live on a farm? Did they work in industry? Did they work with asbestos? Did they smoke? Who in your family had cancer?"

Now we turn to genetics to detect and evaluate the genetic and epigenetic markers that may presignal cancer risk or predispose a person to cancer, as well as to many other diseases. This is great work, relying upon great advances in technology. But can it help you? Yes, if for no reason other than to provide lifestyle choices to avoid specific carcinogens, or to achieve early diagnosis, well in time to effect a treatment for cure, rather than to just extend life for a while.

The curious issue is that if we now know the inciting agents of so many diseases, such as and including cancer, why would it not be a requirement of health agencies to have sway over those who would expose the rest of us to near certain diseases and untimely death? Toxic waste disposal is the dirty business of death, hardly regulated, all for the better interests of commerce in preference to public health.

Conjectural is a pejorative term that industry uses against the field of epidemiology. It's considered speculative and conjectural because you aren't able to control outcomes by evaluating data retrospectively. You have to do prospective, controlled clinical trials in order to have a definitive answer. Yes, of course!

Epidemiology is known as a conjectural science because it's always retrospective, it's never prospective. So you're always looking at data from the past. You couldn't know everything in advance, therefore you couldn't control the outcome. You didn't have a chance to set up different groups. You had to look at everyone after the fact and then let the ethereal algorithm sort it out. A lot of the data remain as being unsortable—we are not able to sort properly, so epidemiology is always viewed askance because it's a retrospective, conjectural science.

To an epidemiologist, tobacco was the easy part. Everything else resides in all of the other categories in our retrospective chemical toxicity pie diagram, and we know that the entire long list of individually known carcinogens, such as and including chemicals and metals, are found in *Toxic Sewage Sludge*, and in other toxic wastes, and that the list may be as long as your arm.

Hexavalent chromium [Cr(VI)] is one good example. Found in *Toxic Sewage Sludge*, Cr(VI) is one of the valence states of the common metal element chromium that is known to cause multiple forms of cancer. It also targets the respiratory tract, kidneys, liver, skin, and eyes.[3]

If you are exposed to Cr(VI), you may develop cancer, even by an inadvertent exposure from an environmental source, such as air, food, soil or water. If you are exposed at all, you are at risk. The adverse effects of exposure to Cr(VI) are similar to tobacco products, but fewer people are exposed to this highly toxic metal, and they generally do not inhale it, either by choice or by addiction, or by an inadvertent or unknowing exposure.

When you look at plastics (also present in *Toxic Sewage Sludge*), that are made primarily of carbon, plus some hydrogen, and little else, the dangers are generally undetectable and thereby unknowable. However, when you make plastic rigid, colorful, pliable, resistant to UV, or to chemical degradation, or if you make it safe for contact with your foods or liquids, or impermeable by the addition of chemical additives, for example, then the plasticizers and other additive chemicals become the things that kill us.

If you burn pure plastic in the presence of air, the products of combustion are primarily CO_2, water, and heat. And if you are able to capture that CO_2, there's little left but heat and some residual water vapor. That's what's so sweet about pure plastic, by its very nature and origin. Unfortunately, when the myriad of enhanced plastic products are incinerated, or otherwise burned, the combustion products cause big trouble, because they now also include the complex chemistries of the toxic products that are generated by combustion of those mixed plastic ingredients. This is the result from any form of thermolytic decomposition of these combined materials, and they are toxic.[4]

Now if you were to go through the same or a similar combustion regimen with *Toxic Sewage Sludge*, you would also have near-countless chemical compounds in the mix. What then is required is to figure out how to create the proper heat, pressure, and mixing environment, such that all of the various chemicals would be reduced to basic elements, such as carbon, nitrogen, oxygen, phosphorus, phosphate, sulfur, etc., and when you do that, you've solved the *Toxic Sewage Sludge* disposal problem. The resulting ash product has commercial value in the manufacture of concrete and other materials, as it renders concrete more structurally stable and resistant to degradation, such as by acid rain. Air pollutants in acid rain, such as sulfur dioxide (SO_2) and nitrogen dioxide (NO_2), can penetrate concrete, thereby reducing the alkalinity, and causing concrete carbonization and chemical corrosion.

One of the great challenges with *Toxic Sewage Sludge* is water. You have to reduce the moisture content, so that you get to a physical and chemical state where the sewage sludge matrix becomes an optimal British thermal unit (Btu) output feedstock. The physical and chemical characteristics of the *Toxic Sewage Sludge* feedstock material are very important, and the state of dryness also becomes important, so as to achieve a clean and efficient burn. *Toxic Sewage Sludge* is not necessarily consistent, it depends on the mix of the various contents, where it was made, what kind of facility, and so on.

If you ask King County and the Washington State Department of Ecology, "What's in it?" the blank stares and reply are along the lines of, "Oh, nine metals, but they're all in compliance." That's because they only test for those nine metals, plus a couple of microbes. I mean, why go to extra trouble when no one even realizes what you are doing?

In talking with the EPA after they had done analytical work to reveal about 80,000 chemicals in sewage sludge, they reported that it takes perhaps seven years to perform the needed toxicity testing for each chemical, so you can figure out how long that's going to take. If you look at revised EPA reports, it's no longer estimated to be 80,000 chemicals in typical sewage sludge, but is now about 137,000 known chemicals.[5] Yet, if you look at the global peer-reviewed scientific literature, you can find reports that count well over 400,000 chemicals in *Toxic Sewage Sludge*, collectively, and that's why it's toxic!

With well over 400,000 known chemicals in *Toxic Sewage Sludge*, it is obvious that you cannot have been exposed to just one carcinogen that may have incited just one adverse malignant transformation event. Therefore, when an environmental sample is tested for any single chemical, such as a flame retardant, pesticide, fertilizer component or combustion product, it is highly unlikely that there was one single chemical carcinogen that incited your disease. In reality, the inciting agent(s) of malignant transformation of a normal cell into a cancer cell is most likely not only unknown, but it is generally unknowable.

Combinations of chemicals foretell synergistic toxicity that in turn incite transformation of a cell from normal to malignant. And if you smoke, meaning if you live in a blanket of chemical carcinogens, that may push the process over to proliferative, malignant, and metastatic disease fast. As well, if you work in industry, if you use house cleaners, if you use coated pots and pans, if you wear synthetic clothes, and if you also smoke, why would you expect to get away with being cancer free? You almost can't, whether it's detected or not. (Note well that many cancers are detected first at necropsy, and those are never recorded as the cause of death. The actual death rate from smoking most likely far exceeds the reported numbers, to the extreme.)

If you were to investigate the vast array of possible permutations of the chemical entities found in any *Toxic Sewage Sludge*, not just the simple or possible additive combinations of the many chemicals, but the number of new, novel, and unknown chemicals that may result from the *in situ* chemical reactions, and all of their unknown toxicities, then the resulting blind guestimate could be well into the trillions of possible toxic chemistries. You can't identify them, simply because no such interest in investigating them exists, especially within the agencies or in commerce, and most especially with regard to conducting any revealing safety/toxicology investigations.

There is the physics, the chemistry, and the biology, and then there is the physical chemistry which gets you back to the biology. After that is the very complex and costly work of toxicology. As the U.S. Food and Drug Administration tells us clearly, the idea of "Safe" or "Safety," i.e., "Non-Toxic," can only be determined by costly, elaborate, and complex safety/toxicology studies, certified as being performed under the Good Laboratory Practices (GLP) regulations of the EPA, as required for new drug development, and all of which are way beyond the capabilities and capacity of agencies or non-GLP-certified laboratories. As a consequence, county and state agencies have no authorization or license to state that any solid or liquid waste or any other waste product is or may be declared as being "Safe."

Worldwide there is this big monster of chemical compounds, many or most of which are toxic, that can re-form in the presence of one another in the warm, moist, aerobic and anaerobic microbiological environment of any WWTP or POTW, and most especially in any form of *Toxic Sewage Sludge*, once revealed to each other. Therefore, the multitude of toxic outcomes will remain functionally innumerable and certainly uncharacterizable. Furthermore, the chemistries of the vast array of new chemical entities, and their functionalities, including their toxicities, shall continue forevermore as being unknown and unknowable.

What new chemistries do they make, those innovative microbes, and what are their toxicities? Wholly unknown! Never investigated! Avoided for fear of disclosure! Held close as a worrisome concern! It's just the way they do things.

We know enough math to know the difference between combinations and permutations. Permutations are simply ordered combinations, and as a result, there will always be far more permutations than combinations. If you look at the estimated 400,000 (or even at the 80,000 or 137,000), chemicals that are likely in *Toxic Sewage Sludge*, and if you calculate all of the possible or likely combinations, it's in the many billions. But if you tried to figure out the likely permutations of such chemical mixes and recombinations, and generally in a saturated moist environment populated by crafty microbes, then there are so many trillions of possible new toxic chemicals out there that there aren't enough spaces on the cosmic spreadsheet to accommodate the resulting data. That's how many. It's NOT safe! It cannot be safe, and it cannot be declared as being safe!

Where do they come from, those endless lists of novel and common chemicals? Is their origin the multitude of various combinatorial chemistries as they mix and join together, or may it be also by the work of the microbes that metabolize them? Ok, so clearly the testing can't be done. The point is that it's impossible to know—it's unknowable!

"How did you get to 80,000 chemicals?" one might ask those responsible for providing the number.

"Oh, we had these big projects and all these scientists, but we're not funded to do that anymore."

You see, they survive on federal budgets. An even worse question is: "How did you just get from 80,000 to 137,000 chemicals the day after you said it takes seven years to test one, or was that just for the toxicity characteristics?"

It's unknown. The question cannot be answered.

When you take the time to look at the cumulative world chemical toxicity literature, as may be performed by artificial intelligence, not by a person, you get to a point where you're in the range of perhaps 400,000 chemicals in any *Toxic Sewage Sludge* sample. That's a very difficult calculation, and it isn't done by counting noses. However, there is a machine learning form of artificial intelligence that develops pattern recognition capabilities for this type of work, including of chemistries, with the ability to continuously learn from and make toxicity predictions based on actual derived and reported chemistry data, and then predict toxicity relationships without being programmed to do so. These calculations and predictions of toxicity characteristics and of actual toxicities, then, would reveal the tangible dangers of exposure to *Toxic Sewage Sludge*, unfortunately, a science never performed.

If someone asks: "Are there dioxins and furans in your smokestack gas?" the response may be: "Oh yeah, there's TCDD and five more." No one ever asks: "Well how many dioxins and furans are there?"

"Oh, a couple of hundred," might be the correct answer.

"Well, let's just put them all on the list, but only eight of them are labeled toxic," could be the reply.

"No, that's not true. You've only tested eight of the hundred or more for toxicity," you would wisely point out.

"Well, it's going to take us about seven years to do each of the others," is the abrupt end to the conversation.

Please just bury your head in poop and get it over with. There's no rational discussion or explanation for what we're talking about here. Agencies know the chemicals exist, and we're all aware that nearly all of them are toxic, but they simply haven't been tested for their presence in sludge or for their toxicity, or for the type of toxicity. Some they have tested, and they are the few that are reported. The rest remain as "Mystery Toxics."

I had to analyze the structural determination of a compound isolated from a bacterium, and I was talking to the National Institutes of Health, and the National Cancer Institute. I was told I had to write a grant application for the next cycle, in the next year, to get into the schedule of the peer-review system, etc. So instead, I called a scientist at a lab in New Zealand, and said, "If I send you this little pinch of powder in an envelope will you test it?" He did, and the following week he called me and drew the structure. It didn't cost anything, much less years of delay, and it was accurate.

That's why you get totally bewildered. Is it important or not? Is someone going to get cancer from this and die, or not?

Dr. David Lewis, a former EPA scientist, and I, while presenting to First Nations members at a conference in Canada, both agreed in front of the audience that the number of toxic chemicals in *Toxic Sewage Sludge* is unknowable and in most cases remains as being unimaginable. So why would we dispose of something like that in our air, food, soil, and water? When someone says it's safe, please say back to them: "That's unknowable, and in many cases, it can't be known."

"Don't you know cancer is over?" a King County elected official asked me (he actually shouted that at me) during a requested presentation to the King County Council, with his sludge industry associates and their state and county partners by his side. Another elected official chimed in that he, ". . . knew ten people who had cancer, but they don't have it any more." Say, what?

"So what? What does any of that have to do with the reckless land disposal of toxic wastes, fully laced with carcinogens and other toxics?" I asked. They just did not care.

"Wow!" I thought, "I'll bet none of these guys has ever worked in the field of cancer biology, or of cancer clinical trials research or cancer epidemiology. We have a lot of work to do here." These individuals and their state and county colleagues seem to be functionally illiterate on matters of cancer, toxicity, toxicology, and disease etiology, yet they are ready and willing to offer opinions on matters of safety. Where do they get these ideas? Who coached them into this self-humiliating, political, and commerce-driven insanity?

The point of it all is that we must focus on prevention of cancer and other chronic diseases, not on treatment after the fact, for which there are few treatments that are functionally effective.

When I went to the World Health Organization (WHO) during an International Agency for Research on Cancer (IARC) program in Lyon, France, I collected their current data on what kinds and how many cancers there are over the Earth, and those stats defeat all that King County, and their councilmembers and university advisors seem to know or say. It doesn't matter if you live in Beverly Hills or in Uganda, you suffer the same cancers, simply because of global contamination and pollution, or as a result of your own bad behaviors, just at different rates of incidence (new cancer cases) and prevalence (living people who have or ever had cancer).

Those councilpersons are dangerous, and they asked me, "Don't you know cancer is over?" followed by, "I knew ten people who had cancer, but they don't have it any more." Totally naive and irrelevant.

It isn't that someone is behind this—it's that there's no one in opposition. I looked around the council chambers, and they sat grinning with their, "I ate the bird," thoughts. No, an increase from 18.1 million to 29.5 new cancer cases per year, and an increase from 9.5 million to 16.4 million cancer deaths per year, over an 18-year-period, is not a decrease in cancer, and you do not get to say that "cancer is over," most especially when you are likely the cause, and you seek places to hide your gruesome activities and toxic materials.[6] Your actions led to this, county councilmembers, and you did not do it by yourselves.

The land disposal of *Toxic Sewage Sludge*, the release of toxic wastewater effluents, and the release into surface and groundwaters of legacy and current landfill leachates is an occult but well-planned and intentional crime against our citizens—all for commercial gain. What's more, the ongoing microbial decomposition of all of these organic and not-so-organic materials adds to the extant climate emergency. The decomposition of organic matter, including the likes of *Toxic Sewage Sludge*, in conditions with oxygen (aerobic) generally yields carbon dioxide, water, energy, and heat, while in conditions without oxygen (anaerobic) it generally yields carbon dioxide, methane, hydrogen sulfide, and energy, so which greenhouse gas pollutant do you prefer, carbon dioxide or methane? In any case, introducing *Toxic Sewage Sludge* into the wilds or into agriculture generates global warming greenhouse gases—the practice is polluting.

We are in fact amidst a global explosion of cancer, a likely consequence of growing populations, in concert with exponentially increasing productivity of our manufacturing industries, along with the coincident and expected amount of domestic, industrial, medical, and septic-sourced sewage wastes, plus the solid wastes and other toxic wastes generated by us. Cancer is not "over," it's blossoming, and we're the cause of it—*Toxic Sewage Sludge* most certainly included. If King County and their collaborators wanted to provide a storyline to the county council, they should have started with first defining what it is, what's in it, what it does, and what is its fate in the environment.

At about ten-thirty in the evening prior to this memorable council meeting, my cell phone signaled a call, and I answered. It was a councilperson, asking quietly if I could testify at council chambers at ten the next morning. "Sure," I responded. "I'll be there." Somehow it smelled and sounded like a setup, but then I look for such opportunities to make the case. So, just who was the cat and who was the mouse in this case? I couldn't wait 'til morning.

Not long before this meeting, I had made one of my many expeditions to Central Washington, specifically to meet with a wheat farmer at his *Toxic Sewage Sludge* dumping operation. Coincidentally, just a few weeks prior to that surprise invitation by the council-member, I had decided to explore certain wheat regions where there would likely be thousands of tons of freshly dumped *Toxic Sewage Sludge* from King County and multiple other municipalities and counties in the state. I knew the sludge was trucked hundreds of miles from King County's five wastewater treatment plants, as well as sludges that had been sourced from thirty or more other *Toxic Sewage Sludge* generators. It was hauled long distances over the Cascade Mountains and across the Columbia River to our precious food-growing lands. Back then, as is still the case today, I was there to witness the misdeeds. Don't just ignore it, hear about it, or read about it, is the thought, go see it, smell it, image it, and walk in it as a primary, firsthand witness to the debacle, which I did and still do.

"Could we possibly change their deadly practices by public exposure?" I would naively ask myself. What did I know? I simply assumed they were ordinary civilized people, generally having concerns for their fellow citizens, but obviously that is not the case. Over time, I have learned that they are driven by selfish motivations and greed with the ability to displace themselves from responsibility by the implied separation in time alone.

This practice of land-disposing *Toxic Sewage Sludge* on agricultural lands is not by any stretch of the imagination to be associated with what has been cleverly named: "Resource Recovery;" "Repurposing of a Valued Resource;" "Beneficial Reuse;" "Application of a Soil Amendment;" "Fertilizer;" or "Nutrient Replacement of Depleted Soils." This is toxic waste disposal on precious and finite farmland, all under the guise of a contrived cover story to obfuscate the expected long-term adverse outcomes and to satiate public concerns, i.e., of the citizen victims, based on the flawed idea that this is good for us. Eating, breathing or drinking toxic waste is somehow good for us, they propose—it is not.

As an aside, while corn, rice, and wheat make up more than 50% of the world's caloric intake, wheat contributes about 15%, and in Washington State, we bury it in *Toxic Sewage Sludge* to pollute, foul, and contaminate forevermore the precious soils, as well as the underlying and adjacent groundwaters, surface waters, aquifers, and foods produced in the region. This is in fact a national security matter, so just who is it that provides authorization for such destructive behaviors in direct contradiction of national interests, and for personal gain?

I suspect that the core of the harmful idea and practice of dumping *Toxic Sewage Sludge* on our food production lands is based on the faulty notion that the actual time from exposure of a citizen to toxic wastes is expected to be long enough and far enough away from the first detection of an immunosuppression disorder, such as cancer, that no one will ever know the actual cause, or who did it. No matter whether a neurological disease, carcinogenesis, cancer cell proliferation, metastasis, diabetes, or deadly virus, in retrospect, no one will be able to relate the time of the application of the toxic inciting agents borne by the *Toxic Sewage Sludge*, along with exposure to endless applications of fertilizers and pesticides by counties, farmers, and contractors, to the lethal consequences to be later realized by the hapless citizen victims as disease. As stated blatantly, clearly, and well-practiced by our Washington State and county sludge operatives, including their university advisors, "You can't prove that!" Quaint indeed, albeit self-serving, insensitive, naive, and deadly.

In the case of an epigenetic influence by one or more indirect carcinogens as a non-mutation event, such as an indirect modification of gene expression, and similarly by one or more conventional carcinogens with a mutation event, both such cases of altered gene expression can be realized as lethal and heritable genetic characteristics from one to multiple human generations going forward. So, why is it that certain people make calculated decisions to injure other people, sometimes for generations to come, and to quote Clint Eastwood, "*... for a few dollars more?*"[7]

We all know so well that farm soils may be nutritionally depleted to various degrees after decades of agricultural use, or by historical glaciations, and that replenishment of organic content and micronutrients builds or rebuilds the basic structure and function of a healthy soil, but it must not be done with toxic waste. It is the microbes that thrive on organics in soil that foster soil rebuilding, soil structure, and soil function, and it is the toxics in sludges, fertilizers, and pesticides that incite dysbiosis (degradation or modification of microbial diversity and their functions) among the soil microflora and microfauna, resulting in depleted or dysfunctional soils.

Healthy soils are alive and biologically more functional when in microbial and nutritional balance, and it is the administration of toxics that incite soil dysbiosis and dysfunction. No soils, no foods. It's that simple.

We performed those good works way back at the University of California, Riverside, in the late 1960s and early 1970s, showing clearly that certain critical plant pathogens survived longer and infected host crop plants at an enhanced rate, when the soils had been amended with *Toxic Sewage Sludge*. Serious publications never did result from those studies, principally because the work was ended as the adverse agronomic characteristics of pathogen enhancement, not soil enhancement, were revealed. Some soil-borne plant pathogens actually fared better in sludge-amended soils because the elements of microbial antagonism focused more on the pathogen's antagonists than on the survival spores of the pathogens themselves.

Soils are considered a natural resource and a matter of national security. Therefore, anyone or any agency that disposes of toxic wastes on our precious and finite soils commits a crime that constitutes ecocide and impacts national security.

The toxics in toxic wastes, such as and including, *Toxic Sewage Sludge* from wastewater treatment plants and publicly owned treatment works, may accumulate in soils, thereby diminishing the favorable agronomic characteristics of the said abused soils. As well, such toxics may bioassimilate and/or bioaccumulate in the intended crop plants, now as plant-borne or food crop-borne toxics, to the detriment of any and all who may consume such food products.

The soil-borne, or the sludge-, fertilizer- or pesticide-administered toxics may also bioaccumulate in host plant parts, other than those that are consumed. However, that discovery work is the task of analytical chemists, but it is essential. In any case, toxics may bioaccumulate in the edible plant parts, even at very low levels, whereby the toxicologist's report may suggest that the toxics are at levels below the assumed, stated or revealed "Toxicity Level," which is typically unknown and generally unknowable. Regardless, the toxics are there, with ample opportunity to bioaccumulate and incite disease in us.

No, councilperson, cancer is not over—it only awaits us at dinner.

Stories about the beneficial attributes of *Toxic Sewage Sludge* are without functional public health value because the data and knowledge bases are slanted or scant, at best. So, for example, just how many milligrams, micrograms, nanograms, picograms, femtograms or molecules of a toxic chemical does it take to incite malignant transformation, or the early onset of a neurological disorder, and over what timespan of exposure, and at what cumulative dose, and by what synergistic combination will mutagenesis take place, or will an epigenetic event occur that may be realized clinically within this generation or even the next? In the absence of hard prospective data, precaution is best applied liberally.

So long as these critical topics lay stagnant as uninvestigated questions, we remain, as Alice cries in her *Adventures in Wonderland*, ever more "curiouser and curiouser!" with no answers to the riddle.[8] Why should anyone entrust their children or any children to the harmful fate of consumables grown in *Toxic Sewage Sludge*? Thus, we find ourselves ever more furiouser and furiouser as our same questions linger unresolved.

Go ask an agency representative, or a councilperson, or one of their aids or advisors for an answer. There are no meaningful or valid answers, just more insipid or insidious discussions, without reason or meaning, with regard to public health. These are not responsible people, in whom we can place our trust, or the health and safety of our children. Toxic is what toxic does—it's toxic.

"What's in it?" is always the cardinal question, meaning, "What's in raw sewage, *Toxic Sewage Sludge*, or toxic wastewater effluents?" And for sure, one sideways answer is that people develop cancer, and they take cancer chemotherapeutic agents, and those cancer chemotherapeutic agents are eliminated into the raw sewage system—a cycle that repeats itself over and over, *ad Infinitum*.

Nearly all cancer chemotherapeutic agents, for example, are known and powerful carcinogens, because that's how they work; they are intended to be more toxic to cancer cells than to healthy cells (i.e., by "differential toxicity"), including by the toxic mechanism of mutagenicity, hopefully targeted more so to the cancer cells. As well, all people with cancer continue to generate sewage. And further, all cancer chemotherapeutic agents are intended, designed, and formulated specifically to achieve a required level of *in vivo* or *in situ* stability, such that they will remain active in the body long after administration, and therefore are formulated to be stable in blood, in the GI tract, and in muscles, and thereby in the environment thereafter, by design or not. *Toxic Sewage Sludge*-borne and toxic wastewater effluent-borne cancer chemotherapeutic agents also cause cancer in people exposed thereto—it's axiomatic.

Cancer chemotherapeutic agents are discussed herein by example simply to suggest a common and typical model for what goes down any drain and into any *Toxic Sewage Sludge* and toxic wastewater effluent. The same story holds for many other classes of pharmaceutical and personal care products, as well as for industrial chemicals, pesticides, fertilizers, cleaning agents, road runoff, and the like. Toxic chemicals are not intended for administration to us, or onto our crops, or into our air, soils or waters, but they are, as a direct and indirect indisputable consequence of that irresponsible land disposal thing.

For probably 300 plus days each year, and over many decades, double-dump-flying-monster *Toxic Sewage Sludge* trucks trek over the Cascade Mountains from King County to dump their toxic loads either in the Yakima Valley or in any of about nine of the key counties that make up the agricultural central region of Washington State. It's a billions of dollars per year boondoggle to support farmer-centric commerce, at the peril of citizens, and any and all who would eat or drink the agricultural products cultured and generated in such toxic landscapes. It's a great business venture, for sure, that business of government-subsidized toxic waste disposal. Certainly, the risks far exceed the benefits, but in any case, please ask yourself, "Who benefits?"

Sitting in the front row of the council chambers at that aforementioned meeting, were about eighteen or so representatives from the Washington State Department of Ecology (permitters and inspectors), several people from the King County Wastewater Treatment Division (producers, haulers, and dumpers—one of whom worked previously at the Washington State Department of Ecology), a couple of *Toxic Sewage Sludge* advocates from the University of Washington School of Environmental and Forest Sciences (advisors, advocates, and enablers—one of whom collaborates with the permitters, inspectors, producers, haulers, and dumpers, and who is also a proactive advocate on matters of *Toxic Sewage Sludge* dumping), plus the very farmer from Central Washington with whom I had met recently at his digs (recipient, advocate, distributor, and beneficiary). A definite set-up. Just five of the nine councilmembers were in attendance. The County Council Executive and four others were absent—for sure, a far too politically charged and inherently "toxic" subject to suffer firsthand, for responsibility can be dangerous.

A pre-meeting session (without me) had been staged for the wily group so that the few selected, interested or concerned councilmembers could be told the woeful tales of my recurring objections to their decades-long practice of dumping *Toxic Sewage Sludge* in our forests, on our rangelands, and on our farms and foods. Let's wonder what was said. In any case, it must have been all about, "That good stuff that builds good soil, and that mitigates GHG emissions and climate change."

Having not been invited to the pre-meetings and discussions with the permitters, inspectors, producers, haulers, dumpers, spreaders, advocates, advisors, enablers, recipient, advocate, distributor, and beneficiary, I was at a definite disadvantage. I had been denied the likely rounds of incessant whining. They may have had long, detailed, in-depth discussions about the obvious advantages of *Toxic Sewage Sludge* dumping on our food crops, and of the financial benefits to the farmers therefrom, as well as about the errors of my ways. They may have had long discussions, and yet, I was offered just five minutes of testimony time, so I took 15, which must have sounded like one long, short, complete sentence, in hurried cadence.

This was knuckles on the dais at King County council chambers. When I finished my self-serving, self-extended, and targeted discussion about the toxics and toxicity surrounding *Toxic Sewage Sludge*, including with some focus on the raft of known carcinogens intertwined within any *Toxic Sewage Sludge*, that one councilperson leaned over the dais to shout at me, with twisted face, "Don't you know cancer is over?" Whew! He had just set the stage for me to unleash the core of my life's work. Not to happen, my five (15) minutes had expired. He and they definitely did not want to hear, or to have on the record, any information that could possibly contradict their contrived storyline about the human and environmental health benefits of land-disposed *Toxic Sewage Sludge*, much less about the adverse consequences therefrom.

Partway through, another King County councilperson got up and looked at me. You're sitting there and the TV camera is on because it's all recorded, and he leaned over the dais with a more relaxed mood and stated, "Well, I knew ten people who had cancer, but they don't have it anymore." I mean, a three-year-old wouldn't say such a stupid thing. We're in a global explosion of cancer, and these purveyors of a raft of known and unknown carcinogens upon the public never got the memo.

The most significant cause of all this cancer, in my opinion, is environmental contamination and pollution. It has nothing to do with, "Oh, my mother had breast cancer, so I have to have it too; it runs in the family." She, too, had obviously been exposed to toxic environmental carcinogens, so what's new?

Cancer is incited almost exclusively by factors from the environment, and to a much lesser extent by genetic predisposition or by some other factor of familial history of cancer, which is or was also likely initiated by outside factors in the first place, often even a generation or more before us. As well, cancer, or tumorigenesis or oncogenesis, is incited, or may be initiated, by multiple and complex mutagens, singly or in concert, or even by epigenetic factors, outside of our direct control—they are carcinogens not of our own personal making.

There's no predisposition, other than self-imposed predisposition, a generations-long, heritable trait, perhaps. For example, you and your mother were exposed to the same people who smoked, to the same household smoking environment. Or you lived near wherever toxic wastes were disposed, whether *Toxic Sewage Sludge*, toxic wastewater effluents or toxic landfill leachates, no matter whether current or legacy. You were exposed indiscriminately and without concern. You were all equally exposed to the same full set of carcinogens, and to the consequences thereof, and which consequences may carry forward over five or more generations, because that's epigenetics and genetics, and that's the way it works. It's environmental carcinogenesis, it's a political matter, and it's driven by commerce!

All of this business of council presentations and comments are recorded and on the public record, if you wish to dig in. This is just a basic discussion and theme of that event, odd at best.

On another occasion, a councilperson invited me to a coffee shop to be filmed by a local TV station, or so they said, with me providing testimony on the sludge topic. This scenario was simply, from my perspective, just another mechanism to coax me into telling them what I thought about *Toxic Sewage Sludge* dumping, so they could better prepare their defenses. Another set-up, for sure. And only for the benefit of the camera in this case. I really unloaded about toxicity, infectious diseases, cancer, and the like, to the point of their exhaustion, at which time they swiped two fingers across the throat, packed up the cameras and left, video and audio results never to be revealed—far too damning.

We can only imagine the hours of strategy meetings it must take for the agencies to then emerge with the next storyline for the public about the wonders of *Toxic Sewage Sludge* dumping, and the endless benefits to be derived therefrom for mitigation of climate crisis, reversing the depletion of glacially stripped and agriculturally depleted soils, greenhouse gas mitigation, providing nutrients to fish and wildlife, and so on—all so very degenerate and ill-fated.

Toxic waste is just that, it's toxic waste, no matter if it also may contain useful nutrients, it's still no more than simply toxic waste, and it must be managed or disposed of in light of its immediate and longer-term toxic consequences. Even if you are killing people and the environment long after you have retired, it's still your dark deeds and your personal burden of responsibility that will follow you.

One former employee of a county wastewater treatment division walked up to me at a meeting in the field and stated, "Please don't hurt me. I'm about to retire." "What on Earth was going on here?" I wondered. My objective has never been more than to encourage states and counties to develop some useful and effective infrastructure, so as to protect human and environmental health from their dangerous toxic waste materials, disposal methods, and practices.

I learned later from a few sources that the county had initiated a program to disrespect me that was also communicated to the state, including warnings that I was "dangerous." Maybe they concluded that I was out to get county employees who worked in the realm of land-disposed *Toxic Sewage Sludge.* That has never been the case! I understand they value their jobs and their long-earned retirements, and that they must follow the implored commands of their leaders, no matter the toxic consequences. The dangerous individuals are those who go to work daily with full intent to expose the public to massive amounts of toxic wastes, as their mission; in other words, the program leaders. It is they who are dangerous.

Another day, at the end of an hour-long presentation to the Washington State Department of Ecology, the director looked sincerely stressed and concerned, then responded by stating, "It's complicated!" at which point she upped and left the conference room, never to be heard from again. My grandkids do the same thing by saying, "Whatever!" implying that this discussion is over! Her lieutenants are likely all in the good graces of the sludgers, as confirmed by additional encounters.

It is "complicated," after all, there are endless megatons of *Toxic Sewage Sludge* to dump and long roads over which to do some serious, risky, and very costly long-hauling. We can only imagine the fuel, insurance, labor, repair, highway, contracting, liability, and lawsuit settlement costs of these expensive practices. So is the land disposal of *Toxic Sewage Sludge* performed simply as a bolster to the economy, or does it have some actual benefit to the taxpayers of Washington State? There is a likely explanation in there, somewhere, but as I say, the general response is, "That's the way we do things."

Land disposal of *Toxic Sewage Sludge* is a major business in our region, and likely without a serious consideration of the matters of infrastructure, sanitation or public health. Worse yet, at the end of my *Toxic Sewage Sludge* whining to a representative from the Washington State Attorney General's office, he stood and announced that their sole job was to "protect the state's agencies," not the public—end of meeting, he departed.

Note well, however, that if you peruse the website of the Office of the Attorney General of Washington State, you will discover the simple declarative proclamation that: "The Office of the Attorney General will provide excellent, independent, and ethical legal services to the State of Washington and protect the rights of its people." And further: "All staff in the Office of the Attorney General are guided by the following core values: We will deliver high quality legal services and remember that we serve the people of the State of Washington."[9]

On our side of the table we often muse about just who is it who pulls the strings on blatant toxic waste disposal, which is to ask, who supports commerce and cheap toxic waste disposal over public health and the public good, and who appointed them, and who trained them, and what may be their ill-guided motivation, and why do we tolerate them? Is this in fact state-sponsored violation or misinterpretation of codes, regulations, and laws designed and hard-fought initially to protect citizens and our living environment, and can they be prosecuted for such malfeasance, and held accountable? Or do they know something that we do not, such as, that human exposure to toxic wastes is good for us?

"Protect state agencies from what?" I wondered. "From having to endure limitations on their toxic waste dumping behaviors?" Agencies and Attorneys General are a mystery.

At the conclusion of a similar private presentation and meeting with the County Council Executive, his response was, "It's Friday afternoon; I'm going home to cut my lawn." No one wanted to hear me out. Even more interesting was the private response I received later from another councilmember, who told me emphatically, "I have my own advisors on this matter, and they work for the county." Hmmm? An obvious inside job! Now, Alice, things are really becoming suspiciouser and suspiciouser.

Yet another County Council Executive from a different county told me, before I could even present my issue at a scheduled council meeting, "We don't discuss controversial matters!" Their county council even has its own Agricultural Board, as a sub-board to the council that advocates and lobbies exclusively for the better business interests of the AG industry, (which includes the unfettered land disposal of *Toxic Sewage Sludge*), and never for matters of public health.

Isn't it notable that in this regard, not one county council in Washington State has a public health board, or any board that focuses exclusively on, and advocates for, public health? Or perhaps the prevailing county Departments of Public Health do not know that *Toxic Sewage Sludge* is contaminating our air, food, soil, and water, and that people are being harmfully exposed to these toxics statewide, *Toxicus ad Infinitum.*

At another event, when I broached the subject of air, food, soil, and water pollution as a result of radiation contamination, as it is carried in air, soil, and water, and that I knew where it entered the waters from a likely hot radiation source, the State Department of Health scientist revealed to me that it was his job to investigate and evaluate any such sites of radiation contamination, wholly at state expense. When he failed to follow-up with his investigation, as he declared he would, I inquired, and his response was that he had been advised soundly that if he were to assist me in any way with testing, no matter the reason, he would be "terminated" by his employer, i.e., "The State."

That hot site remains ignored, under evaluated, and excessively hot. "If you don't know, then it isn't there," seems to be the idea. Why sample, test, and evaluate if you really don't want to know the answer?

Having worked in cancer clinical trials research for a decade, including managing investigations on the epidemiology of cancer, a known black art, I had to mention to myself quietly that these people know not of what they speak, and cannot. Cancer remains as a great unknown to science and medicine, and as well, the lay public cannot be expected to comprehend anything about the relationships between dumping toxic wastes into our air, on our food crops, in our soils, and in our waters, and the initiation and development of cancer—few do. It is the very fact that they do not want to find out, or to know, that is both disturbing and dumbfounding.

Let's be clear on one thing at this point: local, county, and state agencies do not claim duty or responsibility for the health or welfare of their citizens, for their self-proclaimed task is to protect the agencies, as they so state. Therefore, we the public are on our own.

We know well the cause-and-effect relationships between mutagens, carcinogens, oncogenes, and cancer, and we know there are massive numbers of carcinogens in *Toxic Sewage Sludge* and in toxic wastewater effluents, as well as in landfill leachates from extant and legacy landfills, and in the fugitive waters from hazardous waste landfills. We know, as well, that said cancer-causing agents do in fact incite malignant transformation, and then cancer cell proliferation, often followed by metastasis, the actual killer of cancer patients. If we know all of this oh so well, why then do government agencies and their advisors and collaborators insist that "It's safe;" "We're in compliance;" "It just goes away;" "You can't prove that;" and, "That's the way we do things"? The logical conclusion is that they know well of what they do, but it is inconvenient to buck the system, even in the name of public health and safety, and of life itself.

Subsequent generations will have something to say about this, likely including: "We'll never let this happen again." Really? It has happened, and it is happening—*ad Infinitum*—again and again!

We know, as well, the "observables" of pathogenesis, meaning the clinical outcomes, and especially of carcinogenesis, while many of the actual biological mechanisms remain elusive. So, what shall we do? Shall we continue to dump known carcinogens in our soils and waters, on our food crops, and around our living environment until the cows come home? Or should we base our behavior on what is known? Shall we invoke "The Precautionary Principle," and exercise prudence and extreme caution until every last mutagen, carcinogen, and suspected and known inciting agent of cancer is defined, and including until their defined mechanisms of action of mutagenesis, oncogenesis or carcinogenesis are described? Perhaps one day we will, but what continues is a dangerous and deadly behavior for the sake of "a few dollars more" that may be saved by farmers from the cost of commercial fertilizers, rather than the cost of developing proper infrastructure for the thoughtful and safe processing and disposal of *Toxic Sewage Sludge*, as well as other toxic and hazardous wastes.

Note: The precautionary principle was provided as a directive or mandate to be abided by in the public interest, not ignored. The precautionary principle states "that if a product, an action, or a policy has a suspected risk of causing harm to the public or to the environment, protective action should be supported before there is complete scientific proof of a risk. In the absence of scientific consensus, the principle implies that there is a social responsibility to protect the public from potential harm."[10]

Clearly, we are in the midst of a conflict between public health and commerce, and we know well that commerce always wins. It's time for dramatic change.

Think, as well, that the business of cancer diagnosis and treatment is now a trillion-dollar business enterprise in America, and that, as we are told by certain agencies, if cancer were to be eliminated tomorrow, the United States economy would likely fail. We have become cancer-dependent (perhaps, "onco-dependent") as a nation, and the brilliance of toxic waste dumping and disposal into our living environment in fact supports the U.S. economy. Who can argue with that kind of sick logic?

The proper disposal of *Toxic Sewage Sludge* and other toxic wastes is accomplished easily, but requires the political will to do so. The planned and intended recalcitrance of and by our agencies to support the easy dumping of toxic wastes into our environment speaks testimony to our failure as a people to take action on our own behalf. The few elected officials and agencies who do so are ostracized by their peers and political obsessionists for having far too much concern for human and environmental health. It is the here and now personal gratification that wins the day, most especially in the case of toxic waste disposal.

All it takes is a simple search of the internet to become instantly overloaded with a plethora of publications about the incidence and prevalence of cancer. In light of the available demographic reports, billions of dollars are consumed each year investigating and cataloging cancer data in the endless search for causes and possible cures, including on outcomes.

Yet the conclusions are clear; exposure to toxics may assure incitement and/or death from cancer, no matter the economics, social status or demographics. The primary causes of the adverse experiences of cancer-incited morbidity and mortality are the failure of infrastructure and the associated lack of political will to make and keep us safe. It all starts with a willingness to learn the facts of the matter, not to dwell on the ballooned economic or commercial benefits of inappropriate toxic waste disposal. The science and popular literature on cancer, including local, national, and global reports, are there for all to see. The toxics did it; their purveyors did it; their authorizers did it; their permitters did it; the dumpers did it; their managers surely did it.

While *Toxic Sewage Sludge* is the known repository and "safe" harbor for hundreds of thousands of mutagens and carcinogens, and a raft of other toxics, the mere absence of laboratory validation does not negate the fact of the matter. Worse yet is the concept of synergistic toxicity, wherein two or more, or a multitude of toxics (mutagens or carcinogens) in concert or in any configuration, or in any possible combinations or permutations thereof, may easily incite malignant transformation and the onset of cancer. The unknowable unknowns come into play here, and by which the sludge-slinging agencies are playing a bad game of dice with our lives. If you don't know, and if you can't know, just say so, but don't say, "It's safe." By doing so you sound just like Big Tobacco: "There's no evidence that smoking causes cancer." Yeah, right! Go dig up some coughing, sputtering, breathless dead people and ask them.

We are just coming out of more than a hundred years of the "smoking pandemic," with far many more millions of deaths than could ever be imagined from the current "land-disposed *Toxic Sewage Sludge* pandemic," but we're getting there. How long does it take to learn? How many deaths will there be? The benefit this time is that we know the names, phone numbers, and email addresses of the culprits, and they may well be held accountable for generations to come.

Toxic Sewage Sludge is toxic, and it kills. Tracing back to the origins of the toxics is easy; just follow the *Toxic Sewage Sludge* trucks back to their WWTPs and POTWs of origin. Or read their publicly offered indoctrination works:

> Loop[®] is a fertilizer replacement produced by cleaning, recycling, and transforming 100% of the poop and food that goes down toilets and drains. We call it the 'Poop Loop.' Nature wastes nothing, and that's our goal, too! Loop[®] is used by commercial customers on farms and forests to grow bigger crops and trees. And some Loop[®] was processed further into compost for gardens and landscapes, called GroCo compost. Loop[®] generates healthier, more productive soil and is good for people, plants, and the planet—now and in the future. Unlike

> chemical fertilizers, Loop[®] is environmentally responsible and protect[s] fish and other wildlife because it does not cause nutrients to run-off into our waters. Loop[®] products help soil hold rainwater and runoff like a sponge, which keeps dirty water from flowing into Puget Sound, rivers, and streams. The Poop Loop is powerful, and it starts with you."[11]

Oh, boy!

No, I do not know that "cancer is over." What I do know is that we have a lot of work to do to create an understanding of the dangers of land-disposed *Toxic Sewage Sludge*. In the case of the applicable agencies, what becomes clear is that a closed mind is easy to fill with gibberish.

Chapter 4

Sludge Tracking 101

What do we live for if it is not to make life less difficult for each other?

George Eliot, *Middlemarch, A Study of Provincial Life,* 1872.[1]

My adventures as a sludge tracker in Washington State began around 1986. It all started when a professor of biochemistry at the University of Washington and his wife, who were our neighbors, suggested, "Let's go to the Snoqualmie Forest and collect mushrooms, and we'll have a great meal of all kinds of mushrooms." So off we went to a forested area east of Duvall, WA, that I had not visited previously.

"Don't eat these," I warned everyone almost immediately. "What you're looking at are mushrooms that have formed in response to exposure to *Toxic Sewage Sludge*, and it's loaded with toxics." You could see and smell the appalling stuff. They didn't know what it was. When I explained what sludge is, they then understood.

"Fungi grow mainly as a massive mycelium throughout the forest soil," I told them, "and only in response to certain triggering events, such as high moisture and chilling temperatures, or being buried in toxic wastes, do they produce mushrooms. Mushrooms are the reproductive bodies of some fungi. The actual mushroom, which is the above ground reproductive part of a fungus, may be just a minor fraction of the total weight of the whole fungus in soil, because the mycelium, which is an extensive subsurface network of fungal hyphae, exists throughout the soil, often for great distances.

"When toxic waste is disposed upon this mycelial network throughout the forest, the vast undersoil blanket of the mycelium uses the nutrients from the toxic waste as a food source, while also assimilating and accumulating those same toxics within the mushroom cells and tissues. The energy from the *Toxic Sewage Sludge* and other wastes supports development of the spore-forming reproductive structure of the fungus, which is the mushroom. You pick these mushrooms and that is the seat of concentration of toxins from the soils, perhaps for miles around."[2]

I gave that mini-lecture on the assimilation of toxics from *Toxic Sewage Sludge* while standing in the forest. Then I started investigating where it came from, and who else knew about it. I would go mountain biking in those regions, and I would find hunters, fishermen, hikers, and other mountain enthusiasts, and I would always stop to ask them, "Have you seen any sludge trucks? Have you seen or smelled any sewage?"

"Oh, we smelled some awful sewage over there. Who did that?" they would ask.

And so I started mapping out what they told me.

Then I started having meetings with King County. I'd call the wastewater treatment division and say, "I'd like to have a meeting. I'd like to discuss your forest disposal of sludges."

"Yes," they said, "come on down and we'll explain it to you."

They printed maps and spreadsheets, and they showed me with pride where everything went and how it supported the growth of the forest, and made the trees bigger, and that they grew faster, and how the lumber industry was thriving because of the land-disposed *Toxic Sewage Sludge*; all of which was totally made up.

That's just one example of the contrived stories originating and perpetuated from and by the University of Washington, King County's Wastewater Treatment Division, the Washington State Department of Ecology, and industry, all as an effort to justify accomplishing their lowest cost, most convenient methods for the land disposal of *Toxic Sewage Sludge*.

The University of Washington, and after twenty-five years of research in the Pack Forest near Mt. Rainier, demonstrated with hard data that no tree ever benefited from *Toxic Sewage Sludge* exposure.[3] The cells of *Toxic Sewage Sludge*-treated trees actually divided faster, and therefore the wood was somewhat punky, and didn't meet the minimal requirements in California, with regard to structural integrity. But it could be sold in Oregon and Washington where they have lesser standards for wood quality.

Someone at the agencies eventually decided I wasn't clear on how good they were at this, so the Department of Ecology invited me on a couple of tours. I would meet them at the DOE Bellevue office, and we would travel by pickup truck into the forest where sludge dumping was in process. The objective, with the double dump trucks and spreaders running back and forth in the forest spewing *Toxic Sewage Sludge*, was to unload the cargo, no matter what, and get back to base for more. They couldn't take it back, so it had to be dumped.

The DOE brought along maps and showed me their calculations on application rates per unit of land surface area (i.e., agronomic rates), because you have to monitor how much nitrogen in the sludge product is being spread on the forest soil in relation to how much nitrogen is being assimilated by the trees and other plants. While we were sitting on the ground with the trucks buzzing around and spritzing sludge everywhere, I explained to them how their calculations were somewhat off, and that they were without much meaning.

Everything in a forest is a function of topography and water flow, which is to say, the slope of the hillsides and the resulting flows from rain, snowmelt, and springs must guide the rate of application of any nitrogen source. If you were on a flat prairie in Central Washington or Kansas, and you calculated how many acres were to be covered, and how much nitrogen is in the sludge, and how much nitrogen the intended forest or crop plants would assimilate, then you could calculate how much nitrogen per acre may be applied. That would be the agronomic rate of nitrogen application. You must not apply more nitrogen than plants can assimilate, because you don't want excess nitrogen flowing into adjacent surface and groundwaters as a contaminant or pollutant.

Forests are multi-dimensional, and many of our forests in Washington State are formed on and adjacent to very steep and sharp ridges, with one logging road winding along the top of a ridge, for example. The way forests work is that every time it rains, which is quite often here, water runs down off the steep ridges and slopes, through culverts under the logging and access roads, into the many and various wetlands and streams that then enter one of the major rivers on their way to Puget Sound and the Pacific Ocean. These are rainwaters and spring waters in the Snoqualmie forests above the Snoqualmie River and other rivers that flow by their many paths to Puget Sound, without exception. Dump it up here and it flows down there, killing much of what we don't see between here and there, now and for many years to come.

The problem is that they would use a one-dimensional flat paper map, as though we were in Kansas, to calculate the planned and intended application rate for any load of *Toxic Sewage Sludge* in a specific forest tract at an acceptable agronomic rate. We sat in the forest, and they would explain to me how, after you drove a truck in there, they would look at a flat map and calculate the surface area of the forest. Even though there may be a stream or wetland directly below a steep and jagged slope.

When it rains, that load of *Toxic Sewage Sludge* washes into the stream or wetland below in nearly any rainfall. These are steep slopes that I can't hike up or down, and they would explain to me how the water somehow doesn't flow downhill. Yet all of those now polluted waters would then flow into the receiving wetlands, and all of those wetlands would then flow into streams, and then on to bigger streams that collectively then flowed to rivers and eventually through culverts under county roads and state highways to Puget Sound and on to the Pacific Ocean.

From my perspective, that is considered as being reckless toxic waste disposal into Navigable Waters of the United States and that is a violation of the Clean Water Act, at least.

"Oh no it isn't," they would say, "because the sludge doesn't flow in the waters downhill, and . . ." their next standard quote would follow, "we only put a fine dusting of dry sludge on the surface of the forest." Hmmm! If the *Toxic Sewage Sludge* that's sent to any forest or farm from nearly any WWTP or POTW is at least 72% water when delivered, how then is that "dry sludge," and how much rainwater then needs to be added to make it flow, most especially when it's already adjacent to so many convenient wetlands? In any case, the massive amounts of toxic chemicals and microbes therein would leach away and flow at the will of gravity and agency mismanagement, intended or otherwise.

At other nearby sludge disposal sites I would measure it with a yard stick, and I would say, "This is 21 inches deep, and that's 11 inches deep, and over there, that's 11 inches deep," and so on.

And they'd say, "Oh, those are isolated incidents. The sludge hit a tree and when it flowed down the tree it made itself deeper."

"But there are acres of it, 8, 10, 12, 14, 21 inches deep," I told them. "That's not a fine dusting. And by the way, everything here is dead. All of the insects, flies, gnats, and butterflies, everything's dead. And you're saying it's non-toxic or worse yet, that it's safe!"

What if they or some of their agency members actually studied the international crimes they commit by these heinous actions, commonly known as ecocide, at least by these beginning definitions as outlined by Stop Ecocide International: "…'ecocide' means unlawful or wanton acts committed with knowledge that there is a substantial likelihood of severe and widespread or long-term damage to the environment being caused by those acts. . . 'Wanton' means with reckless disregard for damage which would be clearly excessive in relation to the social and economic benefits anticipated."[4] They don't investigate, simply because they care not to know of what we or the rest of the world considers as being ". . . reckless disregard for damage which would be clearly excessive in relation to the social and economic benefits anticipated."

While out on these forest expeditions, I would photograph and make videos of flies and other insects in puddles of sludge runoff in the throes of death. And yet the agency members would only respond by repeating how safe it is. They always had a comeback: "Oh, this is an isolated incident," they would blurt out without concern, and then dump more.

"Let me take you to ten more such isolated incidents," I would say, "and then please tell me how many isolated incidents we must visit before the incidents stop being 'isolated.'"

On more recent encounters I have asked, "Do you know why people infected with SARS-CoV-2, the inciting viral pathogen of Coronavirus Disease 2019 (COVID-19) die?"

Everyone looked around until someone offered, "Well they die from infection with the virus. What do you mean? They die from the infection, it's the virus that kills them, right?"

"No, not really. They often die from a fungal infection of the lungs. The viral infection incites inflammation and cellular and tissue damage of the lungs and other areas of the respiratory tract. It may also cause cell or tissue damage to your brain or kidneys. What they die from is a consequence of COVID-19 infection, but the pathogens that disturb respiratory or lung function are often common fungal infectious agents, along the lines of ordinary molds, such as various *Aspergillus* spp."

"What!" They would exclaim.

"Yes, *Aspergillus*," I repeated. "There are a dozen or more species of *Aspergillus*, common bread molds, among others. If you've ever had a water leak in your house and it seeps into the walls or floors, and you may have had bread sitting around, the resulting black, grey or green stains on the walls or foods are fungal spores, and that's what people breathe in. They're in the air all the time. They're usually not pathogens, but when you have compromised tissue, such as may result from inflammation and immune-dysfunction within the respiratory tract, as may be incited by the SARS-CoV-2 virus, for example, and which can also compromise your immune system, then common or not-so-common molds may easily become human pathogens. That's what commonly incites death among COVID-19 victims. You can pump on their chest all you want in an attempt to bring them back, or put them on breathing machines, or respirators, or other kinds of devices, but it seldom matters, because the fungi have now taken over."

Most of the mold-like fungi that infect the respiratory tract of immune-suppressed patients, especially those with diminished lung function, are resistant to nearly any and all antifungal agents. Common bread molds are now common antimycotic agent-resistant fungal pathogens of humans. Survival is a challenge, for sure, and if you are or were a smoker, or if you lived among smokers, or if you are among the elderly, or if you are on chemotherapy, or if you may have a concomitant immunosuppressive disorder or other respiratory disorder, your health is already severely compromised.

It is clean air and clean water that we need for sustenance and survival primarily, along with clean food. However, when these vital ingredients of life are compromised by the introduction of, or contamination with, toxics, no matter anyone's implied or stated "compliance levels," your health is already compromised, and you are at substantial risk of disease and death. Compliance levels have little or no actual meaning with regard to life or death or with health matters, ever. Being "in compliance" with the stated criteria for *Toxic Sewage Sludge* dumping in forests or on farms, is wholly unrelated to matters of human health.

When microbial equilibrium is maintained in a healthy body, all systems remain functional. However, when we disrupt or destroy the microbial balance or natural equilibrium, such as by rampant use of antibiotics or chemotherapy agents, or by exposure to even low levels of toxins, toxicants or radiation in our living environment, chronic infections and chronic diseases find a gateway—end of story.

There are no demographics or special circumstances that confer safety or immunity against immune dysfunction incited by toxics, most especially in today's world of uncontrolled toxic waste production and disposal that are accelerating at rates greater than that of human population growth. If we do not intervene appropriately to preclude or manage immune suppression and infection, including as may be brought to us by these antimicrobial-agent-resistant and emerging new infectious agents, as well as by carcinogenic chemicals and biologics, generally in response to a crowded and ever more toxic society, we will bear witness to our own demise. The new infectious agents we see now, most especially those that are new to us as pathogens, have willing and accommodating hosts—yes, it's us!

Over time you come to realize that county councilmembers are similar to state legislators who just stare at you, because they don't know what you're talking about, and neither do their agency managers. Sometimes, our expectations can be too high regarding what people may or may not know about medicine, biology, chemistry, physics, infectious diseases, and cancer. The general public doesn't know a lot about the direct and indirect associations between chronic exposure to environmental toxics and the predisposition to, or the initiation of, disease. It's ok if you don't know, but when you're resisting finding out because you're worried about a farmer, or about fertilizer sales, or about someone who's driving a truck or disposing of toxic sludge as a solid or liquid in forests or on farms, and you don't want him to be denied a paycheck, that then is the troubling factor, because that is harmful foolishness based on harmful criteria.

My children and grandchildren could easily lecture informatively to the all too naive legislators, councilpersons, contractors, farmers, and agency staff on matters of toxic waste disposal. They have had to endure endless visits to *Toxic Sewage Sludge* dumping sites, sludge runoff into wetlands and streams, contaminated rivers and lakes, landfill leachate drainages into surface and groundwaters, dead wildlife, oil-coated waters in our forests, and more, much to their chagrin, disgust, and dismay, but they always go along. For them, there were way too many side tracks on the way to vacation breaks, when I would divert into an unsuspecting wastewater treatment plant or publicly owned treatment works, where they knew well, they were expected to get into character and ask some really dumb setup questions of the operator, at my urging, to wheedle out some unedited spur-of-the-moment comment in a game of "Gotcha!" There are more than 400 of these facilities in Washington State, and these kids have visited way too many of the belching monsters, in their opinion.

After we begged entry into one POTW for an unannounced impromptu tour, my ten-year-old grandson (at that time) got right into character and asked, "How often do you take BODs?" He just wanted to get to the point and get out of there.

That was it! Our host, the Plant Operator, knew right then she was being had, and we were asked to leave, after blurting out, "Who are you people?" She later attended one of my presentations, along with a former U.S. Representative from Washington State, simply to defend herself and her facility a bit more, but without any clear rationale, and together they failed in their red-faced rebuke.

So why would a ten-year-old know to ask about BODs at a POTW anyway? Because he had been there, done that, way too many times. BOD, or Biochemical Oxygen Demand, is the amount of dissolved oxygen needed by aerobic organisms, generally microbes, to break down organic material in water at a specified temperature during a specific time period. What ten-year-old doesn't know that? Kids and grandkids are great sleuths, with great minds and memories, most especially when they get set up by grandpa, even if they do blurt out the wrong questions at the wrong times, likely by devious plan and intent.

The challenges we face as citizen victims of toxic waste dumping in our living environment are heightened even more by officials who resist knowing about the adverse consequences of the dangerous practices they choose to support. Or do they know? Perhaps they do, which makes the blatant risk and threat of harm as presented to citizens even less forgivable.

In an effort to learn more, a small group made an appointment to meet with the Attorney General of Washington State (AG) to reveal the case of land-disposed *Toxic Sewage Sludge*, and discover if he and his office actually knew what was afoot, or not, and if he and they actually cared. We could only meet with a representative of the AG, we were told, and we could not meet in their offices. A nearby coffee shop was the best we could do, so that's where the meeting took place. Our objective was to discover if the AG and his staff understood that the toxic wastes disposed in our forests and on our farms also trickled, leached, flowed, and seeped into any and all surface and groundwaters on their way to the lakes and rivers beyond. And if so, did they know the relevant laws, and were they aware of the violations supported and performed by county and state agencies?

Upon our arrival, one good fellow from the AG's office was already there waiting. We introduced ourselves while he sat stoically and as disinterested as could be planned. He listened for a while, until I broached our key question: "Did the AG's office have any interest in the safety, health, and well-being of the state's citizens, and of its forests, prairies, and farmlands?"

Well, that was all he could tolerate. He abruptly stood up and stated: "Our job is to protect the agencies, not the citizens," and walked out of the coffee shop—end of meeting.

Mission accomplished. Set-up fulfilled. Truthful reaction experienced. My job was to confirm such behavior patterns as revealed by the state of Washington, and now displayed without question by the AG's representative. Clearly they showed greater interest in supporting toxic waste disposal into our environment than in any matters related to public health. Enough to make one wonder if they had ever heard of the Toxic Substances Control Act (TSCA) of 1976, which provides government the authority to require reporting, record-keeping, testing, and restrictions relating to chemical substances and/or mixtures. The official text of TSCA, as amended by the Frank R. Lautenberg Chemical Safety Act of the 21st Century, is available in the United States Code, from the U.S. Government Printing Office.[5] TSCA specifically addresses the production, importation, use, and disposal of specific chemicals including polychlorinated biphenyls (PCBs), as just one example, which is a common ingredient of nearly all *Toxic Sewage Sludges* from almost any WWTP or POTW, but then, I'm not an attorney.

Following this noteworthy experience, I was on my bike cruising through downtown Bothell, one sunny day, right where the Sammamish River runs through town along the bike trail. "Whoa, what is that?" I wondered, smelling the now familiar stench as I rode by a new luxury home built along the riverfront. Landscapers and contractors were busy spreading the stinking mix around the new home, so I stopped to inquire.

"Very nice. Where did you get that great mulch?" I asked. It was heavily laden with identifiable pieces of gauze, veggie stickers, and an assortment of plastic fibers and scraps. The stuff reeked of human poop.

"This is from that compost company over in Kenmore. They just delivered it," I was told.

Oh my, a mix of compost and sewage sludge was now a part of the landscape in someone's idyllic back yard. If only they knew. If only they understood.

It used to be, if you made a pie diagram of the disposal patterns of King County's *Toxic Sewage Sludge*, perhaps forty percent was dumped into the Snoqualmie forests, and about sixty percent was disposed on farms. Other amounts used in compost or sent off to prairies were never revealed. That's what they told you, but none of it was actually true, because they also dumped it, under the auspices of Washington's Mountains to Sound Greenway Trust, (and other such projects), along the State's I-90 corridor from Issaquah to Cle Elum.[6] *Toxic Sewage Sludge* disposed along the I-90 may be observed in both directions, despite the many lakes, rivers, and streams. The Middle and South Forks of the Snoqualmie River, for example, flow more or less along the I-90, headed west, then north as the main stem Snoqualmie River, through Monroe, and on to Puget Sound, having joined other contaminated rivers along the way.

Certain amounts of sludge are sent to forest and farm, as they confess, but what is sent to prairies is never reported, nor are the amounts that are found to be far too toxic for land disposal; those beauties are sent to landfill, an even worse alternative.

Landfills are immortal monstrosities that leach, leak, flow, and volatilize back to us, nearly forever. Grit is the name for what is screened at wastewater treatment plants when sludge is produced, and represents the smaller sand and gravel particles, *et cetera*, which are sorted out along with the physical inerts (manufactured inerts, plastics, metals, and other objects) that are destined for landfill or incineration. It's the same unfortunate story all over the United States—landfills and landscapes filled with dangerous byproducts from the treatment of sewage—*Toxicus ad Infinitum.*

It seems to be common practice that all sludge fractions and materials too toxic for forest, farm, prairie, compost, mulch or planter mix must go to landfill, whereby rainwater and natural processes thereafter cause them to seep and leach into that partial fraction of contaminated landfill leachate water that is returned to wastewater treatment plants for a second run through the system, or escape into the adjacent environment—I have tracked many of them. Non-accounted-for fugitive waters that constitute landfill leachates will continue their destructive flow into local salmon streams and rivers, then on to our precious lakes, Puget Sound, and the oceans. The philosophy appears to be, "catch what you can catch, then let the rest go," like chasing wild chickens out on the prairie.

The story told that forty or sixty percent of *Toxic Sewage Sludge* went to forest and farm simply wasn't true. I went to the King County offices where they pulled out their spreadsheets and showed me the maps, until they realized, as well, that it was all of little merit or value. They were just stories for the public. So, I started chronicling sludge dumping in wetlands and streams in the Snoqualmie Forest, and I invited several environmental law firms to join me. I brought environmental activist groups, and this really started generating interest, until someone turned up the heat.

None of the law firms or environmental groups really wanted to do anything about it; they wanted to "work with" the agencies to resolve the widespread and blatant toxic waste disposal issues, not fully comprehending that the state and county were in the business of the land disposal of *Toxic Sewage Sludge*, not in opposition to it, and they had no intention to be protective of any matters of human or environmental health. Their job was to defend the practice, for after all, "It's safe and it's in compliance."

The alleged environmental front groups were happy to know about the activity, and they would call the agencies and say, "Richard showed us this stuff, so is that true? We saw it."

"No; of course not! It's safe and in compliance," was the canned answer. All of their terms and answers became without meaning; just hollow words used as a salve to adversely affected or curious citizens.

Interestingly, agency officials, law firm principals, and large environmental group managers all tend to belong to the same 501c3 clubs as "Town and Gown" social get-togethers. The collective mission is to dispose of *Toxic Sewage Sludge* along our highways and waterways, while at the same time proclaiming the beneficial attributes. I have met with some of them personally, and they sit in both camps at one and the same time, as agency officials charged with *Toxic Sewage Sludge* disposal, and as elected officials challenged with assuring the better interests of citizens and the environment. It is that odd and institutionalized conflict of interest whereby they must wear both hats concurrently, while maintaining social status, and all the while wreaking toxic havoc upon the land. They pose as protectors of the people and of the environment, well within the law as common practice, but it doesn't work that way.

As a relevant example, a most favorable moment was an evening when one of the larger alleged environmental groups invited me to offer a presentation of my sludge work and findings to their local chapter members. Never trusting them at all, because of their well-known wily reputation, I handed the thumb drive with the PowerPoint file to one of the members and watched her insert it into her laptop. As I began speaking, I stood nearby and continued to watch. She immediately slid the file via email to a pal at the Department of Ecology. It appears some environmental groups keep in the good graces of the DOE by snitching on what other activists are up to, likely in exchange for grants or other favors. It's very clear, as is experienced time and again, that some of the greatest threats to the environment are in fact alleged environmental groups and their devout, albeit at times devious members.

As do little kids when caught red-handed, the agencies and their peculiar environmental group allies often create some interesting storylines to rationalize their behaviors. Matters of the environment are very serious, not a joke. Despite thousands of photographs showing otherwise they will tell me: "Oh, that was just an isolated incident, we don't usually apply sludge in forests that deeply." "Richard's always wrong," is the storyline. "You see," I've been told, "we live and work here, and we won't get grants or kudos from agencies if we don't 'play nice.'"

Often I can't believe what I'm hearing, and in any case, I'm not nice, this is serious business. "If you work with Richard," the environmentalists are told, "you're not going to get any more grants."

It's just like they told the woman who owns and manages a local environmental analytical testing lab, a chemist who did contract work for me. They called to inform her, "We're going to terminate your county contract for analytical chemistry work if you do any more work for Richard." This was somewhere between 2018-2019 when I was working on a project with local citizens along the Greenwater River. We were looking into possible lead in their well water and around their homes from illegal shooting. Some of the locals had to remove lead bullets from their roofs and gutters, and repair holes in the roofs and siding, or replace windows. They would call the police, and the response was that it was a federal issue, end of concern.

The area continues to be an assault weapons gun range for odd citizen and extremist groups, plus for the few remaining sport shooters who dare enter. The surrounding mountain homes and roadways are still coated with lead from bullets. My grandson, along on one inspection, quickly filled a large jar with bullet lead from the highway pavement, as we all wondered if there really was a law that restricted shooting on or adjacent to State Highways—obviously not!

In some cases, the shootists enjoyed shooting down whole trees in those federal forests, a clear federal violation, with some downed trees up to a foot in diameter. They would leave behind all of the junk carted from home for use as targets, including propane tanks, old TVs and computers, home appliances, bottles, jars, cans, lamps, mirrors, images of their favorite or not-so-favorite political stars, and kitchenware. Local citizens and environmental groups conducted cleanup sessions just to keep ahead of the mess, but the homeowners remained as mere targets of opportunity near those shooting pits from hell.

One legendary tale is told about the great day all the shootists had when one of their juiced-up members drove his junker car in the center of the shooting pit area while his colleagues shot the car to pieces, while in motion, and while driven in circles by the fearless fun person. After the fact, we stared at the shot-up former car in amazement, and in total disbelief that someone could have survived such whiz-bang fun. The dirt arena is still there today, home to megatons of polluting lead, plus all of the decimated appliances and other junk they had offered to the former pristine forest environment.

The U.S. Forest Service eventually came to the rescue with some novel strategies but whether it's sludge, lead or general shooting-pit trash, local agencies have no interest or commandment to resolve such issues.

The Feds came in to at least lessen the dangers from the literal tons of flying lead in those forested neighborhoods, hiking trails, and campgrounds, where people tend to migrate for quiet mountain solitude. As well, the agencies started responding to my pursuit to discover why they dump *Toxic Sewage Sludge* and other toxic wastes where they do, and why I was driven to reveal this practice, simply because it kills people. Toxics accumulate in air, food, soil, water, sediments, and dusts, and they destroy the environment and us. The point is to prevent the destruction of human and environmental health, not to go back and fix toxic insults after the fact, or stop something bad from happening, or to stop someone from doing something.

I have been able to devote my time to this quest because I work for myself, and the biggest mission of all is the prevention of cancer. That's what it's about. That's all it's about. Everything else is just so much clam chowder. In the course of this work, we have established that there is no one person in government or in any agency that has true regard for human or environmental health, most especially if it interferes with agency practices or any element of commerce, or with retirement. How did we get to this point? Most likely by a lack of basic education, or so I presume, plus bankrupt or non-existent moral values.

Eventually, more people learned what I was doing. Citizens who lived near landfills or hazardous waste disposal sites, or who lived near asphalt plants, would call and ask, for example, "There's this orange stuff running through my yard out of the landfill behind us. Can you take a look?"

People began attending my many town halls and presentations because they didn't know who to call or what to do. They had these terrible smells and oozing red, brown, and orange goo on their property. They would call and ask, "Can you come take a look and see what this is?"

I would go and report, "Yeah, this is *Toxic Sewage Sludge* from King County's Wastewater Treatment Plant," or, "This is landfill leachate and, yes, it is all from toxic waste."

When contracting with a number of analytical labs around the U.S., I have asked if they had ever been called upon to do analytical chemistry work on the red goo that oozes out of landfills and landfill leachate. The common answer is, "No, never." No one seems to care, other than directly affected homeowners, and they get brushed aside.

I can't recall a defined point in time when more of Seattle's *Toxic Sewage Sludge* was moved from Western to Central Washington. The timeline is variable because in some cases they just wouldn't dump it where they thought I would or could go. I could explore by mountain bike or by hiking only so far. I'd ride twenty-five miles in and twenty-five miles out, and then they'd go thirty miles in. There's no written document that says, "Richard did this, so we did that," but the more I worked, and the more citizen and environmental groups got involved, the more everyone tattled on me, not assisted. While they were trying to change the polluting behaviors of the county and the state, they were revealing my work by talking about what I was doing. That sensitized the players to the fact that I was on their illicit contaminating and polluting cases.

A brilliant moment was one balmy Sunday morning when I biked to a park beside a river in Redmond, WA, after responding to a citizen's call about a double dump truck they had seen there recently. As I biked in, by luck there was one of their trucks-from-hell, ready to dump its *Toxic Sewage Sludge* in this major public park. I rode up, stopped, and stared into the driver's window. When he saw me, he picked up his radio handset and announced, "That guy is here." Someone on the other end told him to take the load back, so he immediately started the engine, turned around and hauled the sludge back out of there.

Toxic Sewage Sludge trucks have always gone over the I-90 from the Seattle area through Snoqualmie Pass and the Snoqualmie forests to Central Washington on their toxic waste dumping expeditions. Upon arrival at the bottom of the pass east of the Cascades, they'd divide. Some would go south into the Yakima Valley agricultural regions, and some continued east or northeast into the great plains and prairies of the Palouse and its vast surroundings, north, south, and further east.

In the great pie diagram of toxic waste disposal, the roughly forty percent being dumped in the forests on the west side of the Cascades became less, meaning, less *Toxic Sewage Sludge* went into the beautiful Snoqualmie forests, with more now being disposed on the farms and prairies east of the mountains. The University of Washington had clearly demonstrated with their Pack Forest research program that no tree ever benefited from being doused with *Toxic Sewage Sludge*, and the public was becoming more aware of the blatant waste dumping practices, so the agencies had to lessen the load in the forests, to at least cultivate the grand illusion of beneficial reuse of a valued resource for city parks and farms—total BS!

Meanwhile, I provided yet more "Fetid Stink and Toxicity Tours" (as I called them), for more law firms, environmental groups, and anxious citizens. "See that stinking toxic waste that's running into that stream?" I expounded. "That's the planned work of King County, as permitted by the DOE, as exhorted and ignored by the state AG's office, and all as ignored by the EPA."

Oddly enough, not a single county, state or federal agency ever offered up an opinion, query or complaint, they just go along with the program, no matter the obvious adverse effects on fish, wildlife, and humans, and on the waters or wildlife of the state, federal or far oceanic domains. After all, pollution is what pollution does, it pollutes! Few cared about the details, it just stank, and what could anyone do to stop the behavior, anyway—treatment for health objectives is not within anyone's or any agency's purview.

However, evolutionary changes were emerging in agency toxic waste disposal patterns in response to greater public awareness. Unfortunately, sludge conspirators had yet more devious plans afoot. State agencies and the public didn't necessarily catch on at first, due to the inbred, indoctrinated belief system that declared that *Toxic Sewage Sludge* is a good thing; that it benefitted soils and waters; that it enhanced wildlife habitat; and that it provided essential nutrients lost when the last glaciation stripped the soils of the micronutrients that all trees need. Bull crap. No such thing! That was never the case. It is well known among glacial period-oriented geologists that nutrient-deprived, glacially stripped lands are first colonized by the very same tree species that we see growing, flourishing, and recolonizing here today; such geological modifications as glacial advances and retractions provide favored, new, non-competitive habitats for the generation of new forests and grasslands.

The agencies continued to invite me to the forests in their trucks and they'd say, "We're turning right here so we can show you . . ."

I would say, "No. We're going straight. Let me show you."

Even the agencies were at some point shocked. I would say, "See that hardly treated pile of *Toxic Sewage Sludge* over there, the one packed with veggie labels and other plastics? That's from the city right down the street. You're not alone in this. It's common behavior that if you're an agency, and if you have toxic waste you need to get rid of, you may dump it freely in these forests, wetlands, and streams without care, and that's what they do." Those sludges are now sent over the Cascades to Central Washington for land disposal, so is that an improvement, or is it simply hiding the toxic wastes elsewhere? Would it not seem as an expression of environmental sensitivity and caring for the Earth and its life forms to dispose of such toxic wastes by rendering them as being non-toxic, and perhaps even for actual beneficial reuse?

Some of that spurious toxic sludge forest dumping is ended now, but there was never a written document to order such changed behavior, and no person or agency was ever fined or put on the line for toxic waste dumping in sensitive areas. There was never a plan for forest dumping, they just did it, while the public barely looked on, trusting those in charge. We observed the subtle changes while I continued my investigations over the mountains.

At one point, screens at WWTPs were exchanged so that more of the physical inerts would be collected and sent to landfill, doing nothing of benefit other than enhancing the aesthetic appeal of their toxic sludge product—toxics can never be screened out. One county employee chided me by announcing they were obliged to spend roughly three million dollars to change the screens. I then asked, "What for? That has no impact on lessening the toxicity of your *Toxic Sewage Sludge*. The vast collection of extreme and persistent toxics are in that which passes through your sieves or screens." Frowns and scowls abounded!

Today they are land-spreading sludge on less forest land, at a lesser rate, or what they refer to as a more "in compliance agronomic rate." That, too, is all BS. No forest ever needed *Toxic Sewage Sludge* in the first place, and any forest-disposed sludge is nothing more than a lowest cost, most convenient disposal of toxic waste. Who ever heard of fertilizing a forest, anyway? The commerce surrounding land disposal of *Toxic Sewage Sludge* is just too good to pass up.

Somewhere around 6,000 *Toxic Sewage Sludge*-laden double dump trucks go over I-90 per year, barring a snowstorm, flooding or landslide event. If too dangerous to go over the mountains, the noxious load is temporarily stored, either north of Seattle, or trucked south to the Columbia River, then east, to avoid the mountain passes. Several emergency storage sites exist, as there must be other contingencies in the case of blocked highways or an unusual incident. One such storage site is located in Snohomish County where they dump and store, then reload and re-transport the sludge when the dangerous mountain passes reopen.

Toxic Sewage Sludge is generated every day of the year, no matter the weather or other emergency, yet there has been so little consideration given to highly beneficial, far more convenient, lower cost or safer alternatives to the hazardous and sickening highway sludge transport and disposal system. What will be the response should a great sludge double dump truck disaster occur? What will be the cost? There have been a few. I personally observed the cleanup actions of two such events, one in the Snoqualmie Forest (into a stream when the forest access road failed) and one on a curved, wet highway, north of Ellensburg, WA, also into a stream—yes, double dump trucks do flip over, and luckily, in these two cases, no one was injured.

When sludge trucks arrive in Cle Elum from the Seattle area via the I-90, they go north on US-97 to US-2, veering either east then north to the Chelan area when they meet the Columbia River, or east across the Columbia River into Douglas or Grant Counties. Or, if they continue east on the I-90, when they reach SR-82 at Ellensburg, they go south into the Yakima Valley. On the way south, there's a Beneficial Use Facility in Sunnyside, which is contracted by King County to receive then sell *Toxic Sewage Sludge* to local farmers for use in lieu of commercial fertilizers. The Sunnyside Beneficial Use Facility takes the *Toxic Sewage Sludge* from the county, and then they get paid by farmers for delivery to their farms and spreading of the toxic material on their crops, all an unnecessary transport cost in countless millions of dollars to King County's taxpayers per year. The fuel, maintenance, insurance, liability, and manpower costs are unimaginable.

One particularly sensitive storage site or disposal location I have investigated is in Horse Heaven Hills, Benton County, above the lower Yakima Valley. I have photographed tens of thousands of tons of *Toxic Sewage Sludge* being dumped in the area since 2019. To get there, you travel south in the Yakima Valley on SR-22 to Byron. Just past Byron is a graded gravel road called South Ward Gap Road that climbs about six miles upward to the west to a wheat growing area of Horse Heaven Hills. A popular wine is produced in this region. Once there, you can drive west all the way to Bickleton or Goldendale on rural farm roads across the tops of the plateaus, or south to Paterson on SR-221 to join Highway 14 beside the Columbia River.

There is no country in Washington State more beautiful than this, with brilliant views of Mt. Rainier and Mt. Adams to the north and west, plus expansive views of the hills of Central Washington to the east, and endless open views of wheat fields framed by other hills and mountains in all directions. It appears that King County plans to fill this extraordinary landscape with *Toxic Sewage Sludge*.

Without a doubt the disposal of *Toxic Sewage Sludge* in this unique location represents the cruelest and the most clandestine rangeland disposal contrivance for *Toxic Sewage Sludge* I have ever seen, and no, it's not safe, and it's not beneficial to anyone other than to King County. It's an economic and public health burden, obliged by all citizens of Washington State, for no beneficial cause or reason, other than for convenient disposal of toxic waste.

I happened to be on a certain satellite watching the action one day, when all of a sudden monstrous mountains of sludge began forming at the South Ward Gap Road site, near the juncture with Sellards Road, and I could see multiple and identifiable contract dump trucks discharging their toxic wares in formation. When I beamed down I could read the name of the trucking firm on the truck doors. So the next week I picked up some Tribal members and said, "Get in the car. We're going up there, and I need witnesses," and off we went.

The Loop® and contracted sludge hauling trucks leave the WWTPs in the Seattle area very early in a scheduled sequence to get ahead of morning traffic and be back over the I-90 by two or three in the afternoon. Some days, when I just needed a day out, I'd go up to Snoqualmie Pass and pull off at the ski resort and wait. I would sit there, with my video camera on a tripod, and here they would come. The drivers would look in alarm, and I would jump in my car and follow. We all knew where we were going, but I have found that personal, on-site verification is helpful.

Sometimes we would drive past Quincy, in tandem to Ephrata where they would pull into the truck stop. These are monstrously thirsty, sludge-hauling double dump trucks, most especially because their *Toxic Sewage Sludge* loads are about 72% water, a very heavy weight. Taxpayers footing the bill can only imagine the fuel costs for the hundreds of miles round trip from Seattle to the far reaches of northern Central Washington for so many sludge-haulers, every day. Ephrata is where they have to refuel before driving up past Soap Lake and Sun Lakes into Douglas County to the west of Grand Coulee Dam, miles and miles upon miles. Once in Douglas County, they'd be directed according to plan onto vast wheat fields where the dumping and spreading would begin.

Upon arrival, I'd be curious about just where all the sludge that day originated. They always helped me out because there'd be a giant sludge pile with a sign directing: "City of Bremerton," or "Mt. Vernon," or this one would say: "King County Sewer Plant." So I was there photographing it all, always from public property along the highway easements.

This horrifying story is just one example from one U.S. state that demonstrates *Toxic Sewage Sludge* dumping and disposal at its supposed "best." Lest we forget, the land disposal of *Toxic Sewage Sludge* is not beneficial reuse or residuals management, or the application of a soil amendment, or of a valued resource, or as a fertilizer substitute, etc., and what's more, land-disposed *Toxic Sewage Sludge* does not enhance wildlife habitat or embody resource conservation—period. *Toxic Sewage Sludge* transport, dumping, and disposal are no more than *Toxic Sewage Sludge* transport, dumping, and disposal, anytime, anywhere, and by any excuse or nonsensical reason.

"You can't be here. This is private property," I would hear the truck drivers shouting. And I would say something polite in response like, "Bull shit, this is a county road. Get out of my way." And they would get out of their sludge trucks and glower at me. Why would they be so defensive, one has to wonder, if what they do is as absolutely legal and beneficial as they claim? Must be something wrong here, I would surmise, perhaps dangerous or even criminal?

Sometimes I'd check old sites, then would tell my son, "Get in the car. We're going." And we'd go over the mountains and up to Horse Heaven Hills. Or we'd go over to Highway 24 on the other side of Yakima just to see if they were sneaking things back in there that I couldn't find otherwise.

Most recently, I happened to be spotted by farmers on a county public road viewing their just-sludged wheat fields. One drove his truck from behind to block me from exiting the single track, dirt road in reverse, while the other drove toward me in his truck, swerving at the last moment so as to threaten to crash into me and drive me off the road. My objective was to avoid being smashed and killed. This aggressive pair I recognized as business partners with the Washington State Department of Ecology and the King County Wastewater Treatment Division, who had been invited guests at the recent King County council meeting I had attended and mentioned earlier. Both were now guilty of aggravated vehicular assault in an attempt to terrify an ordinary citizen on a public road. Why? Because I was witnessing their commitment to spreading toxic waste on fields that produced the wheat our children consume. Therefore, as they say, how goes the critical infrastructure, so goes the country.

Thankfully, this incident did not spiral out of control, perhaps due to the presence of the young woman in my car who also happened to be filming and recording the entire hostile and aggressive episode. The threats of physical violence and the language choices of the toxic sludge-spreading farmers were so despicable, filthy, nasty, and aggressive that some people would have been terrified by the experience, but I was so gratified by being able for once to have recorded, by audio and video, their antagonistic behaviors, that I was pleased with the opportunity. Their toxic waste disposal practices, I will imagine, are solely for the personal financial gains to be enjoyed, to the detriment of the sacred prairie lands and of the people of Washington State.

King County has five wastewater treatment plants—three big ones: Brightwater, Woodinville; West Point Plant, Seattle; and South Plant, Renton, plus smaller plants on Vashon Island and in Carnation, WA. Sludges from Carnation and Vashon are dumped at South Plant, so three large WWTPs provide the collective sludge loads from five plants that go over the mountains to be unceremoniously dumped on our food crops. These three main facilities also serve many other WWTPs and POTWs in the surrounding region.

Brightwater cost something like two and a half billion dollars to construct. It is a membrane bioreactor plant—the costs are high, yet the outcome is but a minimal improvement. They were sold a bill of technology goods that provides no meaningful improvement, just ask the dead fish in Puget Sound. The facility costs about eight million dollars a year just to change the "trains" of bioreactor membranes that are installed there, but it does little to improve water quality. Everything sounds good, but the remaining sludge product isn't treated any better. Worse yet, the steady releases of toxic effluents and process waters into the surrounding streams and waterways in and around Bothell and neighboring towns flow directly into the Sammamish River, then on to Lake Washington and Puget Sound. I have filmed and imaged it all, non-stop, and it continues today as a persistent source of contamination and pollution. Who are these people? Most interestingly, the pump station leakages and pipeline seepages discharge beside daycare centers, along sports fields, and into urban neighborhoods, and those responsible seem to like it that way. What are these people about?

West Point Plant is a collection of old King County pipes, settling ponds, and digesters. South Plant has a methane capture plant where the gas can be shipped by pipeline and sold as "clean" methane gas, while claiming they're containing greenhouse gases. Their methane product is not clean, but is identified as such for purposes of public acceptance. This approach can be of benefit; however, Cedar Hills Regional Landfill (CHRL) produces toxic landfill leachates and effluents that also flow to South Plant. Therefore, any gains made in methane capture have only become worse by the fact that CHRL grows day by day, with not enough leachate transport capacity to staunch the ever-increasing flows of toxic landfill leachates into groundwaters and into the nearby Cedar River, a protected salmon waterway. It's a predicament, bar none, with no infrastructure improvements in sight, other than their always shifting 50-year plan.

CHRL can be improved in never-ending increments, but the only actual and functional solution to that massive source of pollution is to mine it down to bedrock and find a new place to stash the vast mountain of toxic waste as far from humans and from any waterways as may be possible. But is that really workable? Or just moving the problem elsewhere? Could there be a better way than endless toxic waste relocation? Unfortunately, it's likely now much too late for any form of remedial action.

Like many areas across the U.S., King County does not have any wastewater treatment plants that actually treat wastewater. They're old and they don't work. They simply process wastewater to separate the solids from the liquids, and then send the resulting toxic residual products back to the environment. *Toxic Sewage Sludge* is renamed "biosolids" and foisted upon the public as safe, yet it's that same sludge product that is alleged to be the resting receiving matrix for all of the accumulated toxics from the original raw sewage inflows. These are interesting verbal machinations here. "Treatment" is a word for the process, not a conclusion or an outcome.

Once upon a time, before CHRL was partially lined, that's where all the sludges went, as well—to the landfill. To this day historical sludge materials continue to percolate down into the aquifers, streams, and rivers below. Therefore, all natural process waters and rainwaters accumulated there today have to be collected and sent to South Plant where they are put through the wastewater treatment process again, including some fraction of the highly toxic leachate waters from the landfill.

CHRL, filled with decades of sewage sludge and all manner of solid wastes, plus reported countless barrels of toxic materials from industry awaiting decay, continues to produce its toxic volatiles and aerosols, along with countless millions of gallons of captured landfill leachates and many millions of gallons of fugitive leachates that escape to the environment. When hiking the rough terrain surrounding CHRL, you can see and smell the gases and the ooze, just as it has leaked out for many decades, and just as it will do for countless decades to come.

"Oh, we capture all of the methane," or so they say. Not! They can capture some, and it's dirty, not pure methane, and it can be burned as "natural" gas, but it stinks—just go ask the adjacent citizen neighbors.

Their claim is that ninety percent, or something like that, is captured. It may be just the opposite. Maybe only ten or twenty percent of the methane is captured. They don't know what leaks out. It's all detectable by air monitoring, satellite, and drone, but they don't do the studies. However, as with the DOE's latest proposal, they will say they sent out letters for public comment, and they may garner a response. That's what they've always done, but there is no plan at all for removal of landfills down to bedrock, plus restoration, for that would entail concern for citizen health, and cost, and deep concern for the nearby and downriver natural environments.

As with land-disposed *Toxic Sewage Sludge*, and water-disposed toxic wastewater effluents and landfill leachates, landfills should be removed and mined down to their original state and restored to functional, inhabitable lands. Long ago we decided to dispose of our wastes among our populations, or just out of town, but out of town is now in town, and thus we live in, among, and upon our own toxic wastes. This must end. The waste must be removed, and the sites restored back to nature. If not, they may contribute, in fact, to the end or otherwise unpleasant departure of us all.

What if the Department of Ecology and the King County Wastewater Treatment Division got together collaboratively to perform a comprehensive, qualitative analysis of just what's in the end product from their five wastewater treatment plants, plus from the other regional WWTPs and POTWs that contribute to the total *Toxic Sewage Sludge* dumped in Washington's forests, and on our farms and prairies? A comprehensive qualitative analysis simply asks that they determine "What's in it?" That doesn't suggest they discover the sources, or the amounts, or the toxicity, or what it does in the environment, or the adverse effects, or the outcomes on public health, it just asks: "What's in it?" As well, a comprehensive qualitative analysis does not ask how much of any one or more toxics or other chemicals may be in the sludge product, it just asks, "What's in it?" May it be that they just do not want to know, for it may be just too toxic to manage, singly or in configuration of any combination, and that then would likely require a call to action?

As things stand right now, and in response to persistent questioning, we simply get blank stares, then, "We're in compliance," or "We don't know," and best of all, "Why do you want to know?"

All of this mindless chatter then begs the question to the permitting agencies, "Why do you permit the disposal of *Toxic Sewage Sludge* in our state and private forests, and on our farms and prairies if you do not even know what it is or what's in it? And, how can you monitor the toxic consequences if you do not know the full spectrum of the tens of thousands or more toxics that are in there? Worse yet, why would you permit the land disposal of *Toxic Sewage Sludge* in our forests and on our farms and prairies at all, when much of it leaches and flows to our ground and surface waters, most especially when you have no idea what it is, or what's in it, or what it does, or what may be its fate in the environment?" The decision to proceed with these dangerous practices must be driven by political and commercial interests alone, because no responsible agency representative, law enforcement agency, scientist or healthcare professional would ever authorize or permit such pernicious activity.

This is the great *cri de coeur*—a passionate cry from the heart—or perhaps the single greatest *cri de coeur* we shall ever know, for we are forced to live in our own wastes, not by demons, but by all levels of government, from local municipalities, to counties, states, and on to the nation, and all as protected by, and in consult and collaboration with, our law enforcement agencies. As well, profit-driven citizens are motivated to join the agencies in their likely crimes, all of which have been witnessed. We suffer this fate for no reason other than politically and commercially driven interests. Infrastructure, the environment, and public health are sacrificed in the name of commerce, as though we have been conditioned as a society and as a nation to defer to the selfish interests of the one, rather than to the collective and better interests of the many.

Toxic Sewage Sludges, wastewater treatment plant effluents, and landfill leachates are toxic, as we profess, no matter the best treatment efforts of industry or government agencies. Yet they can be processed for subsequent beneficial uses, not as an aspiration, but in actual fact; however, we see no serious investment in doing so.

Chapter 5

Commerce vs. Public Health

The Constitution's commerce clause is one of the most effective weapons that waste management companies wield to force communities to accept hazardous waste or to prevent jurisdictions from developing or enforcing waste regulations that are more stringent than the federal ones.

Dorceta E. Taylor, *Toxic Communities,* 2014.[1]

There are about 300 cargo ships filled with plastic waste floating around the world's oceans at any time with nowhere to go. It seems there's money to be made in moving ships and their cargos around, but once you land them, it's over. They get paid per day to sit out there. Globalism has increased consumerism worldwide, which has resulted in all manner of waste-related commerce. So what do we do with all of the residual waste when commercial interests, in the form of hauling, storing, recycling, and disposing, thrive at the cost of public and environmental health? That's the core of the problem. What may be the beneficial, health-focused solution to this dilemma?

British Columbia is one example of the worldwide waste disposal problem, and it represents all of Canada, with regard to toxics and waste and public health. They do not want anything landfilled in Canada, and they do not want *Toxic Sewage Sludge* dumped in BC. Instead, their *Toxic Sewage Sludge* is transported to First Nations' sacred lands near Kamloops, far to the east and well out of Vancouver, yet still within BC. I went there and participated in a four-day program of presentations surrounding that murky "sludge-for-First Nations-issue," because it was being dumped on their sacred lands, an obvious crime against humanity, and a clear matter of environmental injustice.

As for British Columbia's garbage, they don't want to landfill it. They don't dump it in the ocean. They don't dump it in Vancouver Harbor. They put it on trains every day and ship it through Seattle to Chehalis, WA, where it is then put on other trains to either the giant

Arlington, Oregon, landfill, or to the other massive landfill in Klickitat County, Washington.[2] Landfilling of garbage is sold to the public and to the agencies as being the source of beneficial methane for home heating, and such, but no one needs methane, and the methane from landfills is hardly methane at all—it's a mix of toxic gases, much of which escapes to the sky.

Washington loves its toxic waste, yet Canada deplores its own, even while there are millions of dollars to be made. It's all about commerce. Washington State loves the cash that's generated simply by moving garbage and other wastes around and storing it or dumping it or selling it, because as long as you move and store toxic waste you get paid. If you prevent it from ever being produced, or if you convert it to energy, you're missing out on its greatest commercial value. As I like to say, almost in jest, but close to reality, "If you put all fifty states in a rank-ordered list from one to fifty, with regard to matters of public health and environmental sensitivity, Oregon and Washington appear to be fighting for fifty-first place." They are the absolute worst on any scale of environmental sensitivity or sensibility. Move it, then move it some more, and move it again; store it, then sell it, then dump it on or in the land. Never prevent its generation; never enable its safe reuse or elimination. And it's pretty much the same sorry tale throughout the U.S. and in many other places.

This is confirmed in the informative book, *Toxic Communities: Environmental Racism, Industrial Pollution, and Residential Mobility*, by Dorceta Taylor:

> In striking down ordinances in which jurisdictions try to block hazardous wastes generated outside their boundaries from being processed or buried within it, the courts have argued that the safe disposal of hazardous wastes is a *national interest* and they have signaled that any attempt to impede that interest—even if the ordinances are aimed at blocking wastes from states that have not developed adequate disposal facilities or even if the community receiving the waste is already inundated with hazardous wastes and has disposed of more than its fair share—will not be upheld.[3]

To summarize: The courts have argued that any attempt to impede or "block hazardous wastes generated outside of their boundaries from being processed or buried within" their boundaries will not be upheld. We are thereby forced by the courts to live near, on, in, and amongst toxic wastes, by anyone's definition.

On the other hand, California has so many restrictions they have nearly forced the dairy industry out of the state. Where did they go? Yakima Valley, Washington, because they don't seem to have any restrictions beyond the bare minimum. So, where there's a limit in California of perhaps 3,000 cows per dairy farm, as one approximate example, some of the dairies in the Yakima Valley have many times that number, and therefore their dusts, soils, and affected waters are heavily contaminated with nitrates and all else that drifts and flows from dairy waste, which is to say, from cow urine and manure.

The Yakima Valley is reported to have perhaps the highest rate of multiple sclerosis (MS) in the world, and they may also have an unusually high rate of anencephaly, a rare condition in newborn babies that prevents normal development of the brain and the bones in the skull. These severe medical conditions likely originate from cow manures and urine, and the associated nitrates and nitrites therein, often combined with toxic dusts from fertilizers, pesticides, and *Toxic Sewage Sludge*. No one wants to do anything about it because of the value of milk, and because of the cost of managing the wastes with care.

It's the same problem with *Toxic Sewage Sludge* because it's fecal waste that we don't have any interest in disposing of properly, and therefore the result is a contaminated environment, along with the many kinds of diseases associated therewith. California, through its onerous public and environmental health sensitivities and restrictions, is managing their state's contaminating industries very well by pushing the most polluting industries elsewhere, simply to relocate the problem—good for them, bad for the rest of us. Keep moving it around. Move it, hide it, don't fix it. Transporting and storing waste elsewhere is the issue, not necessarily the critical solution—a sorry game of commerce and politics overriding public health.

When we wonder about how all of this waste generation is impacting the environment we need only look to our oceans. Massive islands of accumulated human debris the size of Texas float around out there (along with the 300 cargo ships) and no one knows what to do about it. What impacts are these huge ocean garbage dumps, such as the Great Pacific Garbage Patch, having on ocean temperatures, ocean acidity, global warming, and the like?[4] If we are to consider how toxic leachates and effluents from agriculture, animal husbandry, industry, landfills, wastewater treatment plants, *Toxic Sewage Sludge*, etc., flowing into our waterways are affecting sea life, we need investigate no further than Puget Sound, Washington, as one unfortunate example.

The Washington State Department of Ecology states that it is interested in preserving the quality of Puget Sound for the sake of the orcas and their salmon-based food source. Sadly, there are a variety of salmon species in the Pacific Northwest, all now in critical decline, and all as a function of impaired water quality.

In bodies of water, some microorganisms grow in aerobic environments, while others thrive best in anaerobic environments. In either case, in the presence of over-abundant nutrients, they reproduce so rapidly that they deplete all of the available oxygen. The depletion of oxygen incites the decline of the salmon's food resources, primarily small fish, such as herring, as well as invertebrate food sources. These food resource creatures are killed by pollution, and then the salmon are killed in like fashion. It's that simple—kill the environment, kill the nutritional resources, kill the creatures, small and large up the chain. We fit in there somewhere, as well.

What does the title "Department of Ecology" even mean? Define the word "ecology" and it means the relationship of an organism to its environment. It is a field of science. It isn't something you have to do. Now if we had an actual Department of Environmental Quality, with "quality" having its own definition, we could possibly be in a healthier situation, salmon, orcas, and all. The mission statement would then be to preserve and uphold the quality of the environment and prevent its impairment and degradation, so as to support all natural environments of the state and their inhabiting species for the long-lasting, better interests of Earth. That presents a beginning example of what a Department of Environmental Quality would do as opposed to being a commerce-centric, toxic waste relocation agency, like the current Washington State Department of Ecology.

I mentioned the suppression of scientific data that I have witnessed in a meeting with some Lewis County, WA, councilmembers, because I am not just an environmental activist or advocate, randomly claiming these irregularities. The DOE has created a system of accounting to permit and monitor wastewater treatment plants and landfill leachates, and it's their own program that is killing the very environment they are obliged and chartered to protect. Those present had never heard this before. They didn't know that the Department of Ecology did such things as permit and otherwise support the land disposal of *Toxic Sewage Sludge*, and the release of toxic wastewater effluents from WWTPs and POTWs. This is the agency and the people who permit, authorize, monitor, and support the land and water disposal of these toxic wastes. The councilmembers were hearing me summarize the issue, and they were flabbergasted. I don't know what will come of it, but perhaps they have taken up the sword on behalf of their own citizens, a refreshing approach, indeed.

Yes, toxic wastewaters are discharged from wastewater treatment plants, and yes, *Toxic Sewage Sludges* are separated from those wastewaters in those same wastewater treatment plants, and absolutely yes, none of the separated wastewater effluents and sludges are actually treated in any effective way before being released back into our living environment—and that is my opinion.

The superficial approaches to treatment of the raw sewage inflows into wastewater treatment plants are designed solely to achieve some form of trite compliance, not to protect or preserve human or environmental health. The minimalist approaches to treatment of the toxic effluents and sludges are solely in the better interests of lowest cost and most convenient waste disposal, never in the better interests of us, and that is the story I have repeated over and over to the astonishment of some councilmembers all over Washington State.

Citizens must be included on review and oversight committees for all wastewater treatment processes, with a substantial budget provided for in-depth analysis of the most enhanced engineering systems proposed for waste processing. A focused mind toward public health should be the primary criterion for determination and selection of the safest and most effective waste management system to be deployed, and with equal weight required in favor of environmental quality and environmental health. Agencies cannot do this work alone, as they have so proven.

There's a great controversy, an ongoing battle inside the Washington State Department of Ecology, between management and scientists. For example, the scientists do studies which show that the greatest source of pollution in Puget Sound is wastewater treatment plants. There are more than 400 wastewater treatment plants permitted in Washington State, which implies that there are more than 400

WWTPs and POTWs that must be monitored for environmental quality and for potential adverse impacts on public and environmental health. It just may be that the State Department of Health should have an expanded role in waste monitoring, with a targeted eye on issues related to the release of toxics and toxic wastes, as well as infectious agents into our living environment. Such monitoring has not and is not done, simply because their implied charter is not greater than minimal analysis for purposes of ineffectual compliance. Someone or some agency must be conferred responsibility for oversite on matters of public and environmental health.

In 1997, I submitted a formal Public Records Access Request to the Department of Ecology. They subsequently violated the law and subjected themselves to a likely lawsuit by refusing to provide the information requested. So I persisted. I wanted to know. I wanted to see an excel spreadsheet that lists every wastewater treatment plant in Washington State. How many gallons of raw sewage does each receive per day? How many tons of sludge are created? Where do those tons of sewage sludge go? What is the disposition of the resulting toxic wastewater effluent? Nothing. They could not respond, because they had no such data.

Pursue and pursue, and what I discovered is what I suspected before I even asked the first question. The Washington State Department of Ecology did not know how many wastewater treatment plants they had under permit and review, or where they were located, or how much sewage went through them, or where the sludge and wastewater effluent products went. Four years later, the agency still couldn't figure out how many WWTPs and POTWs there are in Washington State, where they were, and what goes through them, what comes out, exactly what it is, how dangerous it may be, or where it goes. More recently, they enacted a program to figure out the number and kinds of WWTPs and POTWs, and what and how much goes in, and what and how much goes out, and to where. Simple arithmetic. So they have located most of them by now, but there are still some that remain in non-compliance, with a few of those now counted.

Someone I met had an idea of the actual number, because they were trying to do work related to fisheries and water quality, and they started researching the issue. They never got anywhere with the DOE. There were no records. The agency didn't know or respond. They didn't want to admit at the time that they didn't have the requested records, because all they could have provided were some partial records about the few they knew of. That would have opened up the can of wormy violations, so to speak.

But the real issue had nothing to do with that. That was just a ploy. The conflict is this: If you are the only agency in the state of Washington that is authorized to license and permit wastewater treatment plants, how can you license, permit, and monitor them, renewable each year, if you don't know where they are or what they do? So, they started a program and began soliciting the state's waste facilities with emails. "Do you have a wastewater treatment plant? Could you tell me how big it is and what it does?"

To this day I believe there are more than 400 WWTPs and POTWs in Washington State, with a few still not responding, because they don't want government interference in the first place, or inspections, and some definitely don't want permits at all. On the one hand there's cash to be made by receiving and disposing of toxic wastes, and on the other there are implied costs to be incurred by having to build new facilities. Both are matters of local interests of commerce superseding the greater interests of public health.

I have visited such facilities all over Washington, and there are similar obstructive attitudes across the state. It may be for the Governor to step in with a strong hand to staff, structure, and mandate that the state agencies become waste stewards in some form, so as to protect the state and its environment and people from continued and ongoing degradation. Note well that there are not any environmental organizations, or municipal, county, state or federal agencies willing to advocate for environmental quality or public health regarding *Toxic Sewage Sludge* disposal—it's too politically dangerous. Lack of courage wins the day: "It's safe;" "It has to go somewhere;" "That's the way we do things."

Washington State appears to be managed by farmers, operators, haulers, spreaders, waste firms, and their suppliers, contractors, and subcontractors, and not by the state of Washington, at all. The bad idea of, "That's the way we do things," seems in fact to be the code of conduct and of regular order, despite the needs of residing Washingtonians, or the best interests of those who consume our state's agricultural products. As a result, matters of public health and environmental quality are left to the wants and needs of commerce, not to the higher calling of citizens' well-being.

In more than a few cases, I was told by landowners, farmers, and their crews, all of whom were deeply embedded in the *Toxic Sewage Sludge* business, that I should go back where I came from because they didn't want people like me around. Of note, the agency people in on the discussions were of equal mind, which is to say, matters of environmental degradation and public health are not to be considered or discussed, ever, for such talks may interfere with commercial interests or their standard operating practices, i.e., "That's the way we do things."

Whoa! That's just like a script from a movie about the end of society, as evidenced by the self-interests of the few. I believe the greater interests of the many, as a function of the best interests of human and environmental health, must rule.

It appears that no elected or appointed official in Washington State, at any level, is responsible for the environmental degradation wrought by the land disposal of *Toxic Sewage Sludge* in forests and on prairies or farms. A quick search of a Washington State Directory of Departments, declaring the mission statement of each, reveals that not one of them professes or lives up to protecting the state's environment from the land or water disposal of toxic wastes, including from sludges, effluents, leachates or otherwise. They dodge the issue with canned statements, but not one has a stated emphasis to prohibit, prevent or remedy the land or water disposal of toxic wastes, including sludges, effluents or leachates:

- Commissioner of Public Lands: "…committed to ensuring our public lands are healthy and productive, both today and for future generations."[5]
- Department of Natural Resources: "Manage, sustain, and protect the health and productivity of Washington's lands and waters to meet the needs of present and future generations."[6]
- Department of Ecology: "Protect, preserve, and enhance the environment for current and future generations."[7]
- Department of Fish & Wildlife: "is dedicated to preserving, protecting, and perpetuating the state's fish, wildlife, and ecosystems while providing sustainable fish and wildlife recreational and commercial opportunities.[8]
- Department of Health: "works with others to protect and improve the health of all people in Washington State."[9]

- Department of Commerce: "… charged with enhancing and promoting sustainable community and economic vitality in Washington."[10]
- Department of Agriculture: "… preserve our Nation's natural resources through conservation, restored forests, improved watersheds, and healthy private working lands." [11]

From my personal experience of decades in the Snoqualmie forests, only the Department of Fish & Wildlife has ever expressed an interest or concern in preserving animal life and the environment, and that covers a lot of years and a lot of field agents. Specifically, the "commercial opportunities" of which the agencies spoke, and of which we discussed, did not consider the detriments of land-disposed *Toxic Sewage Sludge*, with a strong emphasis on, "did not."

I once met with the former Commissioner of Public Lands. He's a serious scientist and a very smart guy. He cleared out his office in Olympia saying, "I want Richard up here and I don't want any of you around to interfere with his story. I've got to hear this firsthand," or so he told me.

He and I sat for an hour, and I explained to him about King County and the caravans (he loved the word "caravans") of sludge trucks going over the mountains. He lives in Okanagan County and would fly back and forth to the State Capitol in his own plane. So I said, "Please fly over the I-90, but go early in the morning and just follow the caravans of sludge trucks over the Cascades from Seattle that then disperse their *Toxic Sewage Sludge* loads all over the farmlands of Central Washington."

He stated that he hadn't noticed or heard of any such activity on his past overflights. A while later he called me at home one night and said, "No shit! You're right. Those are caravans of sludge trucks. I followed them on the way home. Is that where they dump it?" You see, even the Commissioner of Public Lands wasn't told where this stuff goes, much less what's in it. So I was the great reveal even to him. The Commissioner retired shortly thereafter.

Communities throughout the United States are being confronted with similar issues. I know that because so many citizen victims have contacted me.

Sometimes, while out in the field on sludge-monitoring expeditions, I like to have a witness, so I have invited friends and colleagues to join me. We drive around and walk along the perimeter of the stinking, recently sludged fields. I take pictures and make videos. I've done that many times across the state. I may also pick up a Tribal member from one of the nearby Reservations. We drive up on the plateaus and prairies and I show them what has been done.

"God, I've lived here all my life and I've never been on this plateau. I didn't know that picture you showed of the hundred and seventy thousand tons or so of sludge was from here." This Tribal member has been with me on several occasions when I needed a witness. We would drive around, I'd collect samples, gather data and the like, and she would observe, and swear.

As well, when visiting some of the Snoqualmie Forest *Toxic Sewage Sludge* dump sites, I would be joined by local environmental advocates and citizens. We would drive along the forest access roads looking and smelling. Sometimes they wouldn't get out of the car because of the disgusting views and odors from sludge dumped 14 to 17 inches deep, including in wetlands and along streams and rivers. They'd look out the windows and see all the shit, and the flies, and they would smell how bad it was and roll up the window.

People are generally and especially disgusted by the nasty spectacle of *Toxic Sewage Sludge* dumped in wetlands and along streams. I'd attempt bad jokes, because there was just nowhere to get away from this and draw a fresh breath. They'd go along with scrunched-up faces, holding their noses, and enacting all of the shenanigans of disgust. I'd drive them around because they needed to see of what I spoke and had presented so often to the public.

A famous quote to me by a King County Wastewater Treatment Division staffer was simply, "We don't want you showing or telling people about these things!"

"Sorry, Lady, it is what it is," I responded politely.

One gentleman guest viewer to the forest who had studied, managed, and worried about the inappropriate disposal of solid waste, garbage, and sludge all of his professional life, and who had never seen these *Toxic Sewage Sludge* forest dump sites firsthand, joined me one day. Wearing nice business attire, he tramped about the sludged forest site with me for a while, viewing the agency-incited environmental disaster. From my perspective, if you had never personally experienced the sickening stench of forest- and wetland-dumped *Toxic Sewage*

Sludge, then you shouldn't talk about it, you just can't get the words right. Part way into the site he pleaded, "Can we go now?"

On another day, three women from a local environmental group were in my car and I drove them around the sludged forest. They wouldn't get out and investigate the site but they looked out the windows and saw the gunk and smelled how awful it was. "No! Don't open those windows! I'm not getting out. I've really got to pee but I'm not getting out of this car!" I'd laugh because there was just nowhere to go, literally, in this stuff.

By my personal experience, it is universally repeatable that if you can get someone to join you on a trek through a sludge-dumped forest, such that they can experience the actual fresh, visual, olfactory, and sensibility insults, and with all of the accompanying death and contamination, then you have done a good job with the educational aspect of the responsibility. We are obligated to inform the public of the human and environmental health crimes of which they are the citizen victims, by agency plan and intent. The list of tour guide experiences goes on, because if you don't get someone to see and smell it firsthand, then you're simply creating yet another victim of agency decision-making who cannot and would not ever believe that anyone or any agency would ever do such a thing to our forests, soils, wildlife, waters, and us.

The agencies in Washington State that are hip deep into the land and water disposal of *Toxic Sewage Sludge* in our forests and on our farms are well known and easy to identify by their attitudes and actions. However, the few employees of the agencies who are disgusted by the practices for the most part keep silent, but they do assist as best as they can—to remain unidentified.

Members of environmental groups seldom go with me to these forest or farm dump sites, because in many instances, they just don't believe it, or because they simply don't want to know. In many cases, as I have learned, such 501c3 organizations are beholden to agencies for their funding, so they want to make positive alliances out of obvious disasters, simply for the sake of continued funding. I have those groups and individuals marked on a list as Sunday morning environmental advocates, and not having genuine environmental or public health concerns, at all, and there are lots of them.

Why hasn't an organization, such as an environmental group or an environmental law firm found a functional solution to this dilemma, legal or otherwise? But then, why would they find a solution if it is this same process that provides their bread and butter, which would not be the case with a workable outcome. I've met with them so many times and said, "All we need to do is witness and demonstrate the fact of their toxic waste disposal behaviors, in wetlands, in forests, along streams, through culverts under roads, and on crops, and we will be able to make a strong case for prosecution. After all, federal laws are clearly being broken. Come with me and I'll show you what land-disposed toxic waste looks like. You can see it, you can smell it; you can even slog through it."

"Oh, we don't want to go there," they say, out of concern for the consequences. Fear of our government agencies seems to prevail.

Commerce, especially where tied to the movement and disposal of our ever-increasing toxic wastes, is responsible for generating so many chemical carcinogens that are of urgent concern. Why aren't environmentalists and health professionals looking more closely at what is occurring on such an alarming scale? "Perhaps this is just too big and far too entrenched an issue for people to wrap their minds around," I ponder.

Our biggest challenge today is commerce versus public health. Until now, commerce always wins. Yet, clearly, we have an important healthcare concern that has everything to do with the environment, politics, and commerce, as they relate to cancer and other serious diseases. That's why these engineering guys who were moaning to me one day about all the plastics in the ocean finally said, "Ok, let's stop it. What is this about, all of this insanity you're preaching?"

I said, "It's all about cancer. It's all about the fact that nearly all cancers are incited by environmental contaminants and pollutants. Cancer is a trillion-dollar a year business in the U.S. alone. Why do we make all of these people suffer when we can fix it? It's because there's no political will to do so."

"We get it," they said, "Let's build your waste to renewable clean energy facility with full carbon capture and utilization!"

"All we need is the first $150,000,000 to get it launched," I hinted.

Chapter 6

As Dogs Will Do

Every dog must have his day.

Johnathan Swift, *Gulliver's Travels,* 1726.[1]

If you go to Duvall, WA, and drive east on Cherry Valley Road, you get to Kelly Road and beyond. Make a sharp left off Kelly onto Swan Mill Road and it soon becomes NE Stossel Creek Way, then Stossel Creek Road NE, and now you're on your way due north to the King County-Snohomish County border. If you proceed through the Department of Natural Resources gate (DNR 1028) and keep to the left, you will cross Elwell Creek, then continue uphill, still keeping left, to a forested triangle, and you are finally there. I can be perfectly fuzzy about all of this; it's a jungle of single track logging roads through the forest, but that infamous forested triangle is still there, and the rains still flow through it, carrying all therein into Elwell Creek and the wetlands, streams, and rivers beyond, all the way to Puget Sound.

When north of the DNR gate, and after going forward about a mile, you cross a bridge over Elwell Creek. To the east of this bridge was the site of my first sludge investigations in the region, because I could not believe that King County would go into Snohomish County on DNR land and dump megatons of *Toxic Sewage Sludge* on a several-acre mounded hill that was once an Indigenous habitation site, fully surrounded by streams and a wetland that are salmon breeding grounds.

The waters of Elwell Creek divide east of this natural, mounded hill formation, surround the mound, then rejoin and continue west. These features explain why early Indigenous people lived here—high, forested ground above a spring-fed water source, a ten- or so acre paradise of contiguous wetlands, loaded with salmon, with clear running water on both sides snaking its way to join the Skykomish

River and later the Salish Sea (Puget Sound, the Strait of Juan de Fuca, and the Strait of Georgia are known collectively as the Salish Sea). I would have lived there too.

Elwell Creek is a vibrant, crystal clear forest stream emerging from the hills above that tumbles over a small, nearby waterfall. You hike forever to find one of these natural, peaceful places, and yet because of King County, with authorization by Snohomish County, and as surveyed, permitted, and approved by the Washington State Department of Ecology, along with the Washington State Department of Natural Resources, it is now and forevermore a pile of human shit and buried artifacts. While investigating the reeking, sludge-dumped mound, I noticed one day that when the sludgers had cleared the site for sludge, they had scraped up a variety of stone implements, including a perfectly rounded sphere of rock, about 5 plus inches in diameter, and a load of odd, red jasper chunks and pieces, not likely originating from there. These stone pieces were obviously in the beginning stages of being shaped into something of use by earlier inhabitants.

Now, if instead of following this stream, you go north, you're walking uphill into a giant, volcanic piece of geology that sticks up out of the earth at a very steep angle. On one side of this formation is an absolute sheer cliff that is hundreds of feet above the valley floor. Once on top, you look out over the valley and you can see the northward bound Elwell Creek running through the wetlands and forests below.

This massive rock formation is perched above the forest in a remote, wild, and beautiful place. It's a bewilderment to even imagine that King County or the DOE and DNR would select this most unlikely location to dump their *Toxic Sewage Sludge* loads. Yet they would open that yellow DNR gate for a few days to allow Loop® double dump trucks to drive up this steep incline, right to the edge of this super cliff and then dump thousands of tons of *Toxic Sewage Sludge* into one of the rainiest parts of Washington State. Somehow they seem to believe that when it rains it doesn't run down to the wetlands on one side or into Elwell Creek on the other.

Shit runs downhill, no question about it. They would load the trucks in Seattle, haul the sludge to the pristine forests, then dump the muck at will, spreading it over all life forms and along all life-giving waters, and leave—havoc played for the day, work accomplished.

How can this be? I'm not talking about a "fine dusting on the surface of the forest floor," as they say. I'm talking about sewage sludge six, eight, fourteen or more inches deep because I measured it. They dumped as much as they possibly could all over this elevated forest atop this marvelous cliff. It's a magnificent spot. I've taken people up there just to see it, and then they cringe in disgust.

I have visited this site many times and have far too often directly encountered the haulers, spreaders, and agency overseers in their gruesome work. They cannot deny their deadly practices, because I watch and film them in action, often while they are telling me of the benefits to humankind and nature by such exploits. Executioners are known to harbor the same attitudes and tales of justification, or is it rationalization, or is it the sheer joy of agency-sanctioned vandalism? In any case, the perpetrators either do their filthy job for the county or the state or become instantly unemployed, a not-so-unusual practice around here.

On one particular day, during a usual Cascade Mountain foothills investigation to observe the adverse consequences of *Toxic Sewage Sludge* dumping practices in this same location, I was out tracking the sludge trucks that I knew were going up there, and as I hiked past the locked gate into Marckworth State Forest on the county-maintained access road, I noticed one other vehicle along the roadside. It was a Subaru wagon, a not-uncommon choice in these mountains as a favored vehicle of hikers and bikers.

After about an hour's trek uphill, I reached the forested triangle near the top. There is seldom anyone there, but in front of me that day stood a young woman, perhaps in her thirties, wearing hiking clothes. She was just standing there in the trail, frozen in time, holding this big, long-legged, black Labrador retriever—a big one. She was crying, sobbing, and unable to speak.

Inquiring of the young woman to determine if something were amiss, and if I could be of assistance, all that came out was an even greater intensity of blubbering and clear discomfort, with a few audible sobs thrown in, "What is this? What can I do? Will he be alright? My dog ran . . . What did he . . . ? What is this? What can I do? Oh my God, it's all over me, too! What can I do?"

She was definitely not familiar with either the extreme human fetid stench or the slimy, black material that had been deeply dumped over the forest floor, or the inherent dangers that come with being covered with *Toxic Sewage Sludge*. Here she stood in Marckworth State Forest, with this noxious gunk all over her and her dog, without a clue.

I interrupted and blurted out that her best friend, aka Dog, had wandered into a typical roadside dumping ground for King County's *Toxic Sewage Sludge* that generally covers many acres of forested floor, forever, including any associated or adjacent wetlands or streams. A lot to get out in one, "I'm trying-to-be-helpful," blurb that was intended to be comforting. Not so! That didn't help one bit with the wailing and the stinking, filth-filled upset. The masses of flies didn't help either. Dog was complacent, as dogs are known to be when held in the arms of a shrieking master. Dog looked up at me and smiled.

Dogs do what dogs tend to do. If they smell any kind of fecal matter, they jump in and get into some earnest wading, wallowing, rolling, and sniffing. This Labrador had seriously long legs and had loped off into the deep sludge to investigate the sickening aromas and the flies and mosquitos, as well as all of the freshly killed dead things scattered about. When he returned from his excited romp, he just stood there up to his belly covered in sludge.

His master was in total shock. She had no idea what it was or what it would do. She had picked up Dog and went totally frantic on me when I told her what it was. Bad move on my part. She began by screaming, "WHAT!!! Sewage sludge? What's THAT? Who would do this?"

I offered, "This comes to us by courtesy of the King County Wastewater Treatment Division. It's sewage, ya know, toilet waste, and they dump it in the forest along our streams, wetlands, and waterways."

She was really getting upset now realizing she was covered in sludge, and so was Dog. Dog was no longer happy, and she was REALLY unhappy.

"What to do now?" I asked myself. She and Dog needed to be rescued, so I said, "Come with me to the runoff stream over here and we'll wash off as much as we can so you can get out of here. It's *Toxic Sewage Sludge*, typically waterproof, but we can get most of it off." Not quite the way I had planned the day.

She released her tight grasp on Dog, and I took the once clean and sweet-smelling black Lab over to the stream and let him trounce in the water with hopes that the greasy, oily goo might wash off, at least in part. Dog then did as dogs must do, he splashed in and waded about upstream and downstream for a while through the grasses and blackberries, with most of the stinking, slimy, sewage sludge washing or falling off in blobs. Not so good for her, however. By the process of rescuing and holding Dog, she was now slathered with smelly, slippery, human-originated, black goop from knees to chest, and it was not coming off—it had soaked in.

More crying out loud, once she got a good look. Dog jumped back in the water, still having a good day at the office. We had to do something, so we used grass and leaves to get the gobs of sludge off her arms and front, while Dog splashed in the stream some more.

While doing the wiping, washing, and cleaning, she was going through the bewildered gibberish of, "I've never heard of such a thing. Why would that be out here? Who put that here? Is that really sewage?" You know, this kind of thing. Not denial but incredulity. She was right on the money when she asked, "Who did it, and why, and what does it do?" and mostly, "What's in it, and will it hurt me?"

She was incredulous that it was all done by the agencies and with total disregard for anyone, everyone, anything, and everything in its destructive path. They just do it, despite the adverse impacts on people or the environment, or on ground and surface waters, or on animals, forests or farms. *Toxic Sewage Sludge* is toxic waste, and it harms and kills people, the environment, and the Earth. Forget ye not these infamous agency words: "You can't prove that." Unfortunately, they are, to a degree, correct, for so far we have not.

The questions remain: "Why not?" and, "Why do the agencies refuse to perform safety/toxicology studies on the very stuff they dump in our forests and on our farmlands, and in this case, on our citizen and her dog? And further, does not the state of Washington have an Attorney General to prosecute such crimes against people and the environment, permitted or not? And, as well, does not the state of Washington have a legislature to make laws to protect its citizens and the state itself, meaning in this case, from unmitigated toxic waste disposal on our lands?

In an effort to allay this innocent victim's fears and concerns, I decided to offer a mini-lecture to the totally upset, partially cleaned, and highly odoriferous young woman about what's in *Toxic Sewage Sludge*, who on Earth would ever do such a thing, why anyone would do so, and what was their ill-conceived and oddly proclaimed rationale for doing so. Didn't work! More wailing, now with the beginnings of swearing and anger. She was pissed!

We talked more about the hundreds of thousands of known and reported toxic chemicals and infectious agents in sludge, who is behind such despicable behavior, why they do this, and what are the known and otherwise deadly consequences of their malicious practices.

She couldn't believe what I was saying or what she had experienced. All she had really wanted to do was take Dog out for a walk in the forest, and then they ran into this. She wanted to get out of there and would probably never come back, even though she had purchased a Washington State pass to do so—she had state permission to get covered in sewage.

Off she went down the mountain, still with red eyes, very bad attitude, stinky clothes, and a fouled Dog. She was a very nice young woman who was totally bewildered and shocked by what had just transpired. I had assured her that this is common practice all over the state of Washington, much to the dismay of health-conscious people and those who revere forests and the great outdoors, generally those places that are not filled with highly toxic human waste. This is but one of so many sludge stories along a similar vein. Where do we go from here?

Agencies, legislators, and their members are undereducated and ill-equipped to manage our wastes, so they dump it on us as punishment for generating such wastes, and they bill us for their flawed practices, in any case. What if our infrastructure were developed in response to safety, sanitation, security, and matters of public health, and not for the best interests of politically driven commerce? What then? Human and environmental health could be restored, but that would be contradictory to, "That's the way we do things." Note well that only the toxic sludge industry professes that, "It's safe," and that singular mental aberration rules the day.

Note well also that the deadly *Toxic Sewage Sludge* into which Dog had wandered, as dogs will do, and which was subsequently transferred, in part, to the frontal regions of Dog's master, also as dogs will do, is the product of King County's wastewater treatment system, all of which is comprised of the lethal combination of human domestic sewage, human septage, medical institution waste, and industrial waste, plus the most toxic form of all, stormwater runoff, which is to say, all that drains and flows from any and all highways, roads, parking lots, stormwater retention ponds, and the like, and which holds within its chemistries nearly any and all toxic substances and materials known to man; and that drip, flow or are otherwise discharged or spit away by any person on this planet. It is the cumulative worst of the worst, and Dog gave itself and his master a nasty couple of doses. Nice day out in the forest.

CHAPTER 7

POLLUTION AS POLICY

Thousands have lived without love, not one without water.

W. H. Auden, *First Things First*, 1957.[1]

What would we do if the active polluting and pollution we see were in fact the result of policy that may well be a hidden political agenda adversely impacting public and environmental health? Why should citizens have to fight with government entities to protect their health and keep toxic wastes out of their air, food, soil, and water, unless pollution were driven by policy in support of commerce? I believe that pollution is policy, and that it is clearly articulated behind closed doors by agencies and elected officials, along with their commercial operatives, then foisted upon unsuspecting citizens as code, regulation or law, all under a falsely fabricated illusion that it is safe, that it is standard practice, and because "that's the way we do things." No, Alice, it's not, so please read along.

The second question begs us to determine if such policy may also represent the actual driving force behind agency science. These are related questions: Is pollution in fact public policy, and has science been politicized, and thereby emasculated by the process? Has science, in fact, been rendered void and ineffective? Yes, or so it seems, and therefore science has been diminished. It can't do its job because it's driven by the guidance and demands of embedded political and commercial interests. The hue and cry of, "I don't believe in anything based on science," has become the standard response of the "*Toxic Sewage Sludge* is safe," and, "It is beneficial for the environment," crowd. Cancer and death by cancer are natural and normal, they proclaim.

This all has to do with political theory, now conflated with political hypocrisy, and even with geopolitics, and with the social constructs and social engineering of acceptance. Unfortunately, the idea of representative government seems to have lost its way, for now we are getting into the political aspects of the issue at the municipal, county, state, and federal levels.

What citizens often ask after conversations with agencies and their political collaborators is, "Why do they do this when they know that it hurts people, much less the Earth and our living environment?"

It all goes back to those two leading questions: "Is pollution policy?" and, "Why do we have to fight the agencies for our health and quality of life?"

For example, when I meet with county and state agency staff, it is easy to perceive that they've all been instructed to repeat the same static catch phrases and storylines: "It's safe;" "We're in compliance;" "You can't prove it;" etc. They ignore science, deferring instead to the same canned and well-recited agency sham claims, just as instructed.

Do analytical chemistry test methods exist for performing the appropriate analytical testing work? Yes, of course—it's costly and time consuming, but it exists, and the agencies have both in-house and outside contract analytical testing capabilities readily available. What's missing is the desire to initiate analytical testing programs that would have human and environmental health as the singular key priority, above the bare minimum standard of simply being "in compliance," an unethical surrogate for rigorous science.

My local neighborhood friends of an environmental bent are good examples of the public's lack of awareness on this issue. I invited them one day to join me on an inspection tour of the Snoqualmie Forest above Duvall. They've lived in King County all their lives, and hiked many of the Snoqualmie Forest public trails. That's what they do. Yet they had never seen what I took them to see in those same forests, just a little bit farther in. Like everyone I take out there, they had the same initial, bewildered comments: "Who would do this? I've been hiking here since I was a kid. Why would anyone do this to our beautiful forests?" They asked the same questions again after looking around. No one can believe what they are seeing. It's a best kept agency secret from society; a deadly best kept secret.

Agencies are well-armed and prepared defensively. They practice terms, words, and speeches for when someone calls and reports, "While out in the forest with my family this weekend, we saw all of this terrible stinking sewage sludge. It was in the wetlands and along the streams, and there were dead animals all around. Who did that? What is that for?"

There's a moment of silence before the county or state person starts giving their canned lecture about how safe it is, and how it's been "treated and tested," and worse yet, that it's "in compliance."

Citizens never ask: "Treated how, by what engineering technology, and based on what criteria for safety," and "in compliance with what," and "being in compliance as conferred by whom?"

"Oh, it's the EPA that sets the standards," would be the queasy answer, if demanded.

"So it's the EPA that sets the standards," you say! No, not really, it's the state and local agencies, and they do not have standards for safety, as would be determined by the conduct of a formal U.S. Food & Drug Administration series of safety/toxicology testing protocols and as would be required for the development and regulatory approval of a new pharmaceutical drug product.

Then you delve into some of the dirty details. Someone set standards for the concentration of toxic metals per kilogram, per gram, per microgram, or per picogram, allowed in a *Toxic Sewage Sludge* sample so that the waste product may be considered as being "safe or in compliance" for land or for landfill disposal. The U.S. EPA, for example, lists thirteen priority toxic metals (antimony, arsenic, beryllium, cadmium, chromium, copper, lead, mercury, nickel, selenium, silver, thallium, and zinc, *sans* cyanide), plus the concentration of a couple of microbes by which they determine that a *Toxic Sewage Sludge* product is "in compliance." King County lists only nine metals, *avec* cyanide, along with the same microbes, so what happened to the other four metals (antimony [a neurotoxin], beryllium [a respiratory toxin], selenium [incites chronic systemic toxicity involving the nervous system, lung, heart, liver, and kidney] and thallium [incites chronic systemic toxicity involving the nervous system, lung, heart, liver, and kidney]), you may ask? Were they not toxic, after all, or were they simply too difficult for the county to find the source of the toxics or be able to assure their removal from the waste stream? No one asks! A major public health disclosure challenge for the agencies, indeed.

Not included on the in-compliance priority metals list, for example, is strontium-90 (^{90}SR), the radioactive isotope of strontium that is produced solely by nuclear fission, with a half-life of about 28.8 years. Radioactive strontium isotopes still saturate the central regions of Washington State, and beyond. They originated from "Fat Boy," the bomb that was built at Hanford to blow up Nagasaki, Japan.

Nor do they include uranium. Uranium metal is not on the list of priority metals, yet they had many accidents in Washington State during the nuclear bomb development process. As a result, dangerous forms of both strontium and uranium will remain around Hanford and throughout the greater Yakima Valley and most Central Washington regions for millennia, at least, including the natural or stable form of strontium that is found in native soils, aka, stable strontium, which does not degrade rapidly. It's in everything, including the air we breathe. No one can do anything about it, so life goes on in the region, as does untimely death.

Note: For clarity, uranium is a naturally occurring radioactive element that has no stable non-radioactive form or isotope, as does strontium, so any detected uranium will be either the natural radioactive form (generally U-238, with trace amounts of U-235 and U-234) or the refined radioactive form. U-235 can be concentrated in a process called "enrichment," making it "fissile" and suitable for use in nuclear reactors or weapons. Note as well that while uranium-238 has a functional half-life of 4.47 billion years, in case you may be counting, U-235 has a half-life of only about 700 million years, so what's to worry?

Under these EPA criteria, concentrations of any one or more of the priority metals that may be present in *Toxic Sewage Sludge* and in other waste products dumped in a forest are regulated, they say.

And citizens respond, "Oh, I get it. Then it must be safe."

"Yes, it's safe because it's in compliance," is the usual agency response—no validation or discussion required. No, in fact, being in compliance does not necessarily make anything safe. The ideas of being "Safe" or "In compliance" are but mythical social constructs, generated by agencies for self-serving purposes, wholly unrelated to the underlying principles of safety or of human or environmental health.

Citizens are easy victims because they are not familiar with the terminology, and they are trusting, at least for a while. The dangers, or the facts behind the sketchy responses, are never revealed. It's all a ruse to overcome valid and sensible public concern. Public health is never allowed to be a topic of discussion, and environmental health appears to be no more than a subject on a list from when they were in school, way back when.

Citizens should actually ask: "What are the chemicals in it? What added toxicity is there when those toxic chemicals are mixed? What harm can they do? What are the criteria for safety? Who did the testing, by what methods, and when? Did anyone do safety/toxicology testing on the chemicals that were found? What is the meaning of: 'In compliance?' and in compliance with what?"

Citizens aren't tuned in to ask these kinds of questions, so by default, they are victims by information deficit. We know well that it is impossible for *Toxic Sewage Sludge* to be considered as being safe by any fabricated or other contrived commerce-driven criteria, much less by the reality of the disturbing facts.

In Washington State, whether it's King County, the Department of Ecology, the Department of Natural Resources, the Department of Agriculture or the Department of Health, they all most likely attend information sessions whereby staff are told by their managers how best to respond to questions from citizens as part of a citizen indoctrination and public relations program. It's where they must learn the canned policy storyline. That's why everyone with whom you speak says only the same few recited words. Their mission is to protect the practices of land-filled municipal solid waste, unmonitored and unregulated open release of legacy landfill leachates, land-disposed *Toxic Sewage Sludge,* and water-disposed toxic wastewater effluents.

To give you an example, Cedar Hills Regional Landfill in semi-rural southern King County is the primary garbage pit of the county, holding decades-long accumulations of any and all forms of discarded toxic and other wastes, all of which leach and flow in part by many routes down and to the west into the nearby Cedar River.[2] An abandoned legacy landfill several miles to the south is located above yet another aquifer that also discharges into the adjacent Cedar River, an ancient and protected salmon stream that flows to the southern end of Lake Washington.

Making matters worse, this is an area that sports one of the highest rainfalls in the region. Is there a planning error here? No, not a planning error, but an intentional plan for harnessing the intense local rainfall, plus the convenience of abandoned coal mining operations beneath the site, to make it all go away by interment and dilution, or so they wish. The largest portion of the leachates from

CHRL are captured and returned to South Plant, Renton, for a re-run through the wastewater processing system, with the leached toxics precipitating into the resulting *Toxic Sewage Sludge*, while the remaining fraction of leachate waters percolate or flow into the aquifers below or into the Cedar River.

It all sits atop legacy coal mines with abandoned mining tunnels that receive garbage leachates to be funneled away directly into the Cedar River. Similarly, the abandoned legacy landfill nearby to the south is located within a spent gravel quarry atop another aquifer, so the solid wastes disposed there also continue to leach uninhibited into its subterranean aquifer, and on to the same river, forever.

Citizens' concerns expressed at local agency-sponsored meetings center mainly on emitted toxic and odoriferous fumes from the collective wastes at CHRL, plus the flocks of scavenger birds that drop waste debris across the neighborhoods. They worry about the landfill leachates that flow out of this municipal solid waste site into ground and surface waters, including into the waters of the Cedar River.

From the other or south side of the giant, fetid garbage mound known as CHRL, the waters leach and flow into Issaquah Creek, then northward to Lake Sammamish, the Sammamish River, and into Lake Washington at its northern end. Together the polluted waters from the Cedar River that enter Lake Washington at its southern end, plus the polluted waters of Issaquah Creek and of the Sammamish River that enter Lake Washington from its northern end, flow then collectively into Lake Union and on to Puget Sound. There is no respite afforded by the county or the state, for pollution here is cured by its conveyance elsewhere, but to where, and to what effect?

Adding to the toxic consequences is the scenario whereby the stated rivers and lakes along the way also harbor numerous other legacy landfills that add to the cumulative toxicity of the said waters, to no one's anticipated remediation. I have been locating, tracking, making videos, and imaging the massive number of legacy landfills and their colorful fetid leachates that leak into the protected salmon waters that flow therefrom for many years, all in an environment of denial and ignorance. In years to come we will state emphatically that people from past times did this, but in fact, we are those people from past times, and we are failing to act or respond as a citizenry or as local governments. And so it is repeated: "That's the way we do things," and not yet: "That's the way they did things," whoever they may have been.

The purpose here is not to blame, but to offer awareness to citizens for the benefit of their own self-protection, in light of historical and present agency complacency on the matter. Toxic wastes do not "just go away;" they persist and continue their toxicity characteristics for untold generations.

At a King County councilman's town hall meeting that I attended, a concerned citizen mentioned the clandestine land disposal of *Toxic Sewage Sludge* atop the local land reclamation area, which was flowing down to the same sensitive waters of the Cedar River. Suddenly, one of King County's landfill engineers, who had never worked with *Toxic Sewage Sludge*, exclaimed aloud to the crowd of over two hundred at the elementary school that evening, "It's safe!" She was referring to the *Toxic Sewage Sludge*, not to the garbage, the landfill, or to the landfill leachates, or to the compost or reclaimed soils. She was not a wastewater engineer, hydrologist, *Toxic Sewage Sludge* engineer, microbiologist, toxicologist, ichthyologist, environmental scientist, chemist, or a scientist at all, but a garbage engineer, a specialist in landfills and in dumping unsorted mixed solid waste. She was out of control, blurting out a statement she had been taught to repeat, like a trained parrot, such as: "It's safe," wholly in an intellectual and experiential vacuum.

Wrong! It's not safe now, never was, never will be, and cannot be. Toxic waste is toxic, never safe—end of story.

The people neighboring the landfill simply sat there aghast, nodding their heads in doubtful acceptance of this misguided and dangerous portrayal of the safety of *Toxic Sewage Sludge* that down the road would directly affect them.

When we reported this individual's exclamation to the EPA, the response was, "She cannot state that. Sewage sludge cannot be declared safe, because 'safe' can only be determined by formal, well-controlled, safety/toxicology studies, and they have not been done."

What these unfortunate citizen victims and many in the general public are not aware of, is that one of the most toxic metals commonly found in many waste sites is hexavalent chromium [Cr(VI)]. Of the various forms of chromium, Cr(VI) represents one of the most powerful carcinogens known to man. Cr(VI) incites genetic mutations in cellular DNA that transform cell function and gene expression, and may predispose a person to cancer, perhaps in this generation, or even far beyond. Cr(VI) is a toxic metal that incites toxicity, and is always in transit by the winds, waters, soils, and foods, or by the unpredictable outcome of inheritance. It's always on the move. Cr(VI) must be controlled to extremely low levels to avoid exposure, disease or death.

King County monitors local industries that contribute to wastewater flows to determine if they may be in the metals, metal plating or chromium infusion business. Such industries are responsible for treating their own wastes before they are discharged to the sewer system so that the various chromium species do not end up in landfills, toxic wastewater effluents, and in *Toxic Sewage Sludges* that go to forest, compost, or farmland, or into effluents that make their way to the Salish Sea. King County cannot accomplish that task alone, and it is far too expensive for many manufacturing companies to achieve compliance by themselves, even though they must!

The principal, underlying problems began when, not all that long ago, the EPA turned everything over to the states. In Washington State, as elsewhere, it then became the local agencies' responsibility to monitor these toxics. As the biggest *Toxic Sewage Sludge* producer in Washington State, King County created a laboratory called KCEL (King County Environmental Laboratory). If you were to delve into KCEL's history and obtain the lists of their testing capabilities and results, year by year over a lengthy period, you would see how that list has changed. The story is fuzzy, but Cr(VI) was on King County's list once upon a time, then disappeared. Chromium metal is still tested, but that's a separate test from the tests for Cr(VI) or even for Cr(III).

Cr(III) happens to be an essential nutrient for humans, which is why vitamin tablets contain Cr(III). It's essential because it catalyzes certain critical enzyme reactions. Cr(VI), on the other hand, is a deadly carcinogen—it causes cancer, it causes mutations—and therefore a major effort was undertaken for years to evaluate Cr(VI) in our food and waters.

Washington State and King County removed hexavalent chromium from their priority metals list (leaving insoluble chromium metal) because it couldn't be controlled, and now there's one less toxic metal listed. Although we are talking about Cr(VI) inciting cancer, there are other metals still on the list that may also incite cancer, heart disease, arterial disease, immune suppression, neurological disorders, endocrine disruption, and more.

In this discussion, another important point to make about chromium is that it is a characteristic shapeshifter. If you test for one form of chromium, and you don't detect it, that doesn't mean it wasn't there. It's just that if you have Cr(VI) in a water, soil, sludge or sediment sample, as opposed to a dry sample at warm temperatures, it may change from Cr(VI) to Cr(III) or to Cr(IV). Shapeshifting metals even confuse the agencies, because it has to do with when and under what conditions you are testing for chromium.

If you were to visit a chromium plating facility to test the water in their sewage outflow and found Cr(VI), by the time it was in a sewer it might not be Cr(VI) any longer. It may have changed to a different form, and vice versa. Everyone was confused, because it was a function of how you sampled, how the samples were managed, where you tested, and what conditions the wastewater itself experienced before it was tested.

Part of the argument was that if chromium passed through a wastewater treatment plant and there were microbes and other chemicals therein, you were ok, because the Cr(VI) in there likely changed to chromium something else once it became a part of the *Toxic Sewage Sludge* or toxic wastewater effluent. There wasn't any point in testing what came from the factory. You had to test what came from the wastewater treatment plant and then further on down the road to disposal. It all became very confusing. So, I went to KCEL and asked, "Why is Cr(VI) no longer on the list?"

"We don't test for it anymore because it doesn't mean anything," was the response. They were right, but they're also wrong, dead wrong, by their very failure to test, and thereby seek to protect human health.

Recently, I tested for a series of toxic metals in dust samples from public places in Central Washington State. Cr(VI) was on the list, but turned up negative at all three sites tested, even though the sites are in the direct air flow from the Hanford Nuclear Reservation, a known source of high concentrations of Cr(VI). Had the contaminating and toxic Cr(VI) shapeshifted to Cr(III) or Cr(IV), or yet another form? It can't have disappeared, but it can have shapeshifted. What next dilemma shall we face? That's the nature of chemistry and of toxicity and of testing. Where did it go?

When tests are performed in a wastewater treatment plant in Washington State they don't test the sludges and effluents that go into Puget Sound or to the farms or mountains. They pick the disposal target location for testing, and they always select a target where they know it will not be, so what's the point? I mean, these were face to face arguments and conversations back then. You'll never find it in the literature.

King County claims that sewage sludge is important for farms and forests because it's so high in carbon, as organic carbon, and when you dispose of it in soil it takes CO_2 out of the air. Therefore, they say, it's beneficial for mitigating greenhouse gases and climate change. "The fibrous or polymer forms of carbon from *Toxic Sewage Sludge* build soil structure," they would proclaim.[3]

"So, what?" I would ask. "The CO_2 and methane generated from the thousands of double dump truck transport vehicles that go to the forests and prairies of West and Central Washington, and to other distant land-disposal sites, far outweigh the minuscule amounts of CO_2 that may be sequestered in a sludged forest's soil. What's more, as the enhanced microbial activity in a sludged environment increases, so does the release to the atmosphere of CO_2, methane, nitrous oxides, sulfur dioxides, and other greenhouse or toxic gases, which remain as fugitive gases, totally unmonitored."

They realized that people were starting to see that if you have a lot of organic carbon, the question then becomes: "What kind of organic carbon? Was it dioxins and furans? Polybrominated diphenyl ether flame retardants? Per- and polyfluoroalkyl substances? Polychlorinated biphenyls?" These are toxic organic compounds, and they really didn't want anyone looking.

In soils or sludges, the number of possible carbon compounds is virtually near-infinite because of the ability of carbon to form long-chained, branched or aromatic molecules. While typical chromatographic methods, such as gas chromatography (GC) or high-performance liquid chromatography (HPLC), are able to make quantitative determinations of specific compounds, the user must first know which compounds to look for. Total organic carbon (TOC) is a non-specific test. TOC will not determine the specific chemicals present, because most environmental samples are complex mixtures of near-countless thousands of organic compounds. But TOC will inform the user of the sum of all organic carbon within the compounds and samples, just not which ones.

An environmental attorney with me on one of my scouting trips was trying to have a conversation to figure out, "Why does the county's list of analytes [the substance to be identified or measured] change?" It changes when people start arguments. I was starting to look for things, so they changed the list.

"Oh, we don't do that anymore. We don't measure Cr(VI) anymore," proclaimed the county.

"Why?"

"Well because it doesn't mean anything, because it changes."

They always have these diversionary schemes or excuses, as if whatever they may do should be of no concern to the public. That's why I get into these conversations about total metallic chromium or the more soluble Cr(III) or Cr(IV) forms.

Environmentalists state, "I want to test for chromium."

"Which form?" I ask them, because they really don't know much about it or how it morphs in the environment or over time.

In the case of lead, for instance, when you find an old bullet outside that's all lead, it has a white coating on it. That's lead oxide and if you lick it, it's as sweet as sugar. During WWII, when sugar was scarce, they actually manufactured lead oxide to use as a sweetener. It's not necessarily always toxic. You can eat a bullet and you'll pass it, and it won't be a toxic form of lead at all. But the form of lead that gets into water supplies from lead pipes, for example, that is toxic lead. It's mildly soluble in water. Lead can enter drinking water when plumbing materials that contain lead corrode, especially where the water has high acidity or low mineral content that corrodes pipes and fixtures. The most common sources of lead in drinking water are lead pipes, faucets, and fixtures. In homes with lead pipes that connect to the water main, also known as lead service lines, these pipes are typically the most significant source of lead found in water.

When you have all these different kinds of analytes combined in a sample of water, soil, sludge, garbage, sediment, household or industrial waste, and more, they may actually change. They're shapeshifters, just like in the movies. They change in response to the environment. They merge into various ionic forms and may become toxic and cause cancer, or they may poison you, while others may be of value. People who do sampling and analysis for agencies know that, so they only test when they know they're going to get the results they want. I only test when I am prepared to get the worst possible results, the extremes of toxics and of toxicity, not by plan or intent, but because it's actually there.

We have a serious public health crisis on our hands, and it's a fool's deadly mission to attempt to intervene or otherwise interfere with the politics or commerce associated therewith, as public health has not been a major issue of concern to anyone.

CHAPTER 8

TOXIC INSULT

As a writer, you should not judge. You should understand.

Ernest Hemingway, *Esquire*, (1935).[1]

Quite recently I became really sick, among numerous other times, whilst collecting certain dust samples in Central Washington in areas that are open to public exposure, especially to children. The respiratory insult of my personal experience, in this recent case, resulted from the toxic dusts I was breathing in through my mask. Later that same day, when sifting the same materials for shipping to a lab for analysis, I experienced similar respiratory trauma, again even while wearing a respirator type mask. That's because there is a form of dust throughout the Central Washington region that is so fine that it hangs in the air, fully suspended, somewhat like a black fog, awaiting your next inhalation.

These toxic dusts are within the realm of microparticulate to nanoparticulate in size or perhaps even smaller, or so I speculate. No one knows, and no one wants to find out. I will muse that the ultrafine particulates are the sources, carriers, and in fact inciting agents of many respiratory and systemic diseases, based on their insidious, ubiquitous, and penetrable nature. Nanoparticulates are of and by themselves toxic, and even more so when they function as carriers or delivery vehicles for adsorbed toxics delivered *in situ* in the process, most especially in Central Washington's agricultural lands where they bond with a vast menu of known toxics.

After collecting the samples, I awoke late that same night and found blood on my pillow. It was coming out of my sinuses and I had sores on the inside of my nose. It was obviously from handling the toxic dust samples collected in Central Washington that day; from sites exposed without end to decades upon decades of applied fertilizers, pesticides, industrial chemicals, and *Toxic Sewage Sludge*, and likely also from Hanford, not far away to the south.

Hanford, considered the most contaminated nuclear site in the United States, is a decommissioned nuclear production facility located on the Colombia River in Benton County, WA. Established in 1943 as part of the Manhattan Project, the site was the first plutonium production reactor in the world.[2] Hanford's operations released significant amounts of radioactive materials into the air and into the Columbia River. In 1989, the state of Washington, EPA, and DOE entered into an agreement to set milestones for cleanup of this highly toxic area after decades of plutonium manufacturing, some of which was used in the first atomic bomb. The plutonium manufacturing project left 53 million gallons of high-level radioactive waste and 25 million cubic feet of solid radioactive waste to be dealt with by future generations. Billions of additional gallons and cubic feet of toxic wastes are known to be "disappeared" forevermore. Eventually, some of these pollutants became wind-disbursed toxic dusts. If you go there, don't breathe.

The people who live and work among these dusts must suffer greatly, and not as some spurious relationship, they live in it. It's in their air, homes, schools, workplaces, foods, soils, waters, and bones—it's everywhere, forever. We have a public health crisis throughout Central Washington State and no person, no citizen group, and for certain, no municipal, county, state or federal agency, has any useful plan to do much about it. We are in some measure of good luck because there are some serious citizen-originated environmental groups that endeavor to keep tabs on the work efforts of the agencies and their civilian contractors, but full containment of any form of hazardous waste is at least many decades and billions of dollars from here to there. As a general rule, it can't be cleaned up, so it must be contained.

The soul-corrupting proposal now is to build a new series of modular or package nuclear power plants in the same Hanford location, plus a few other sites. The attitude seems to be: "It's already an historical, extant, and deadly radiation contamination hot spot, so why not use the same site for more catastrophic, radioactive waste opportunities, rather than contaminate new areas?"

Is that some form of a muffled confession, a statement of no interest or ability to clean up the historical and already contaminated mess, or is it that no one really cares, especially those who do not live there? In any case, it is a clear demonstration that we remain unable to learn from our past mistakes, no matter the likely consequences. We didn't learn, because it's all about the cash, never about the people or the environment. We and the Earth are not expendable, no matter the real or imagined political or financial challenges or benefits therefrom.

I have chased this down the environmental testing pathway debacle to the point where I actually drove over to King County Environmental Labs more than once, and with an attorney in tow, to sit and ask questions. It's as if they're taking a deck of cards and shuffling them. Every answer is mixed up, covered up for purposes of intended obfuscation—it's just the way they do things. The public or any form of public concern is simply a nuisance.

As an example, while working with a group of citizens and environmental activists in Central Washington who engaged me to sample soils, sediments, and waters in a defined area, I had a meeting with them to ask, "What's the point? The analytes you're looking for, such as arsenic, mercury, lead, and so on, are there, just as they have been for decades. And they are also likely to be detected in the sediments in the bottoms of those same streams and canals because they will be preserved there even longer. And they are likely to be found in the adjacent soils, as well, not washed away. All of the toxic analytes of your interest are always here, just as they have been for so many decades, so what will you expect for a productive outcome?"

No one will do anything about any of this because they cannot—these contaminants and pollutants have been there since the beginning of modern agriculture, now just worse, and with an ongoing accumulation over time, with no person or agency having any interest at all, because that's just the way things are.

King County takes its lesson from the DOE, which takes its orders from the EPA. They are not autonomous bodies. The thirteen priority metals listed by the EPA are all known toxic metals and they incite various manifestations of toxicity, while the compliance list for King County now includes just nine heavy metals, plus cyanide, a carbon atom triple-bonded to a nitrogen atom that kills.

If King County were to take a random handful of the tens of thousands of tons of their ordinary *Toxic Sewage Sludge* and test for each of the priority metals alone, and if each one of the metals individually were to be within the EPA, DOE or King County safety zone, then what does that finding actually mean, other than that they're in the sample, and that they're in the mysterious "safety zone?" So what? What do those findings mean from a toxicological perspective? Does that mean it's safe? No, definitely not, because we have other factors that the EPA, the Washington State Department of Ecology, the State Department of Health, the King County Department of Health, and the King County Department of Natural Resources and Parks, with its wastewater treatment division, all fail to reveal. It's the concept of not just whether there may be identified toxics or toxicity in, or associated with, the *Toxic Sewage Sludge* material, but if there may be synergistic toxicities effectuated between and among the vast array of toxics in the subject sludge materials. There are various and multiple pathways to toxicity, not yet fully known and not to be ignored, so focusing the public on just one or a few toxics gets us nowhere. A comprehensive qualitative analysis of the chemistry of *Toxic Sewage Sludge*, and the toxicities associated therewith, would be a good place to begin, but that would likely be far too revealing.

For sure, an absence of positive results (i.e., toxicity or toxics at or below compliance levels) by superficial compliance standards, is not proof of an absence of safety. "An absence of data is not data of absence," or so the saying goes, especially when less than appropriate testing methods have been engaged in less than competent hands. As was so clearly stated to me by the EPA: "Safety can only be established and reported as a result of subjecting test sample material to an EPA panel of safety/toxicology tests conducted by an EPA-certified lab." Not done! Never done! Won't be done!

Note: Toxics testing not done is toxics not detected. Simply stated: "Good for commerce, bad for us!"

Let me give you another example regarding this highly toxic insult: If we have chromium plus lead, or chromium plus any one or more of the other nearly inconceivable number of combinations of the individual toxics or combined toxics, metals or otherwise, the answer cannot be predicted, or even evaluated after the fact. Taken within a greater context, now with the combinatorial expansion of the toxic possibilities, how should we view the same sewage sludge toxic mix when it's clear there may also be tens of thousands of organic compounds in the same sample, including many organic toxics?

And, by further extension, how should we view adding inorganic chemical compounds that reside in *Toxic Sewage Sludge* with organometallic compounds that would contain at least one chemical bond between a carbon atom of an organic molecule and any of one or more metals?

The various manifestations of toxicity from any one combination or series of combinations of toxic chemicals, metals or compounds is nearly immeasurable, much less the number of toxic compounds that may reside or be formed in *Toxic Sewage Sludge* by natural, normal, *de novo* chemical synthesis possibilities in any or all stages of the wastewater management process, ahead of any land or water disposal. The answers or outcomes proposed by agencies, with regard to the functionality of any wastewater treatment plant, are incalculable and even inconceivable and without merit, from the perspective of the plausible toxicity of the resulting materials, and more so in the absence of intelligent or comprehensive qualitative analysis. As a result, the reports and attitudes revealed by any wastewater treatment operation or by any regulatory agencies must be regarded as being without merit.

No amount of filtering, aerobic digestion, anaerobic digestion, aeration, settling, adding of sodium hypochlorite and lime, or UV light will ever diminish the toxicity characteristics of any sample of *Toxic Sewage Sludge,* whether it may be there by default, denied or otherwise. And that's because the numbers simply cannot be counted, and the toxicities cannot be measured.

Lest we forget the biological parts of the equation: Toxicity is a biological issue on assay, so focusing on any one or two of the three sciences of biology, physics, and chemistry alone, denies us the additional bewildering features of our living world that may result in the untimely demise of any one or all of us in the presence of any toxicity from any source. Comprehending toxicity is beyond the limits of toxicologists themselves, and defining the toxicity of toxics is, as well, a bewilderment to science. Therefore, we resort to doing the best we can, but we must do something, rather than standing by and waiting until someone dies of sludge-borne toxicity, origin identified and confirmed or not.

Physical chemistry is the branch of chemistry concerned with interactions and transformations of materials. Understanding the immeasurable possibilities for the creation of novel toxics is critical. Understanding the innumerable physical chemistry principles that influence the myriad of new chemicals, organic and inorganic, plus the immeasurable combinatorial possibilities of the candidate metals, organometallics or metalloids, which may be created or otherwise generated in any mixed waste environment, and in the presence of water and microbes, all in a favorable aqueous environment, is somewhere beyond an ordinary calculation.

For example, organometallic compounds contain at least one chemical bond between a carbon atom of an organic molecule and a metal, meaning water-soluble compounds can penetrate almost anywhere and end up thousands of kilometers from the point of initial spilling or disposal. What's more, organometallic compounds represent a special case for toxicity studies. Due to the presence of metal-carbon bonds, several transformations can occur under varying physiological conditions, such as upon contact with biological fluids. Thus, the toxic effect may be caused not only by the organometallic compound itself, but also by the corresponding degradation products, and there are many![3]

Dr. David Lewis, author of *Science for Sale*, is a stalwart, meticulous scientist who worked for the EPA.[4] We've done some programs together. He was released from the EPA after a thirty-five-year career because he was becoming a standard-bearer for the critics of the *Toxic Sewage Sludge* problem. That's how bad it is. Everything David does, writes, and presents is prefaced by how badly he was treated because of the out-of-step findings of his research programs. Science reveals the answers, not the scientist or the agency, and science, the scientists, and the citizenry pay the price.

The way your cells and the rest of your biology react to one or more toxics alone or in different combinations is not only not known, but, as Dr. Lewis and I agree, it is wholly unknowable. If the toxics, toxicity, and toxicology of the fetid sewage sludge environment remain unknown and wholly unknowable, how could any person or agency state that "It's safe," having never done the experiment? Let alone the fact that the revealing "experiment" for safety may not by itself be doable.

In any case, while I am not a toxicologist, biochemist, biophysicist or physical chemist, I do feel certain there is no sane or practical relationship between the tests that are performed by King County or Washington State and the toxicity of the specific analytes tested. The selection of test analytes and the specific tests to be performed seem to be engendered by the need to simply achieve the lowest cost and most convenient route to "compliance," not to meet the needs of health and safety, for safety in terms required by the field of *Toxic Sewage Sludge* analysis remains ethereal at best, and unapproachable, at the very least, but it can be done when actual science rules, not commerce.

The failure of agencies at any level to monitor closely the dispersal of chemicals from WWTPs and POTWs, either as wastewater effluents or as sludges, as well as from industrial and agricultural operations, provides a general and overwhelming risk and threat of harm to affected human populations and to environmental health. Beyond simply the toxicity characteristics of the individual chemical elements themselves, rests the expected adverse consequences of exposure to the vast myriad of organic and inorganic chemicals within these disposed wastes.

The concept and adverse consequences of synergistic toxicity by the *in situ* combining of chemicals within any wastewater system seem to be ignored, or the result of a widespread lack of awareness of the chemical reactions that can and do take place within any WWTP or POTW. Pollution thereby becomes policy; due either to ignorance or neglect, or by intentional design.

In terms of toxicology and toxicity synergism, or synergistic toxicity, we are referring to the adverse effect caused when exposure to two or more chemicals in combination at one time results in health effects that are greater than the sum of the toxic effects of the individual chemicals alone. And further, the concept of *in situ* neosynthesis of novel chemicals, including of new forms and chemistries of toxic chemicals, within any WWTP or POTW, as well as within the resulting toxic sludges and effluents therefrom, present a class of wholly unevaluated toxicities for which assays do not even exist.

A lot to take in, but the key point to understand is that agencies or other entities have no license or other good cause to state that any *Toxic Sewage Sludge,* wastewater effluent or landfill leachate, extant or legacy, may be safe or otherwise suitable for land or water disposal, or for any likely or unlikely human exposure, without comprehensive safety/toxicology testing to assure a determination of actual safety, an overwhelming and necessary task never performed.

To emphasize the point, and in case you may be wondering, "How can this be?" the *EPA Biennial Report* – No. 8, dated February 2021, confirms what we are saying:

> EPA found 18 new articles that provide relevant data on chemical pollutants that may occur in biosolids in the United States. The articles identified 116 new chemicals in biosolids: 50 polychlorinated biphenyls (PCBs) (39 congeners, 10 homologs and total PCBs); four pesticides; 19 flame retardants; eight perfluoroalkyl substances (PFAS); three antibiotics; one metal; two inorganics; and 29 other organics. The articles also identified new data for 48 chemicals that were previously identified in biosolids. Concentration data for biosolids were found for 61 of 116 new chemicals and for 34 chemicals identified in previous biennial reviews.[5]

For perspective, we know there are about 118 chemical elements listed in the Periodic Table of Elements, and that about 94 of those elements occur naturally, while 24 are synthetic elements produced in nuclear reactions. We know also that about 93 of the 118 elements are metals, with 18 being non-metals, and seven referred to as metalloids. Carbon, hydrogen, oxygen, and nitrogen, the primary elements of all life forms, are nonmetals, and make up about 96% of the mass of a human body. Sensitivity to excess metals is the issue.

As well, we know that 35 of the 93 metals are of concern because of environmental, residential or occupational exposure, i.e., they are found in the environment and diet. Also, 23 of the 93 metals are heavy metals, including, antimony, arsenic, bismuth, cadmium, cerium, chromium, cobalt, copper, gallium, gold, iron, lead, manganese, mercury, nickel, platinum, silver, tellurium, thallium, tin, uranium, vanadium, and zinc. And nine of the most toxic heavy metals are often listed in their order of relative toxicity as: Hg > Cd > Pb > As > Ag > Ni > Co > Cu > Zn.

Toxic metals, including the heavy metals, may be found in wastes as individual metals or as metal compounds, any of which may negatively affect a person's health. In small amounts, many metals are necessary to support life, but in larger amounts, they become toxic. They may accumulate in biological systems and become a significant health hazard, including as inciters of disease and/or death.

Many toxic heavy metals are bioassimilated and bioaccumulated in plants, including in crop plants, but are seldom monitored. They are common in air, dust, soil, sediment, and water. The cumulative effects over time may include malignant transformation (cancer), neuro-degeneration, immune suppression, and so much more.

Testing for toxics in *Toxic Sewage Sludge*, wastewater effluents, landfill leachates, industrial effluents, stormwater runoff, CAFO wastes, septage, and the like, is very complex, but so is public health, and public health must hold reign. It's not just about cancer; it's about endocrine and immune disrupters, neurological dysfunction, and the list goes on.

As an example, if I were to go to any laboratory that does analytical chemistry testing on environmental samples, and say, "I'm going to bring in a sample of *Toxic Sewage Sludge* and I want a test for dioxins and furans."

"Do you want to test for the core 17 and nine compounds in the class, or do you want to test for the congeners as well?" they would reply. Which means: "Do you want to test for the seven hundred or so chemicals in the total class of which we know?"

Recently I was on the phone with an environmental attorney who's been in the field for about 18 years, and we were doing Chemistry 101 by cell phone. "I've got this client," he told me, "it's an important case, and they're saying all this about toxins. Can you explain it to me?" he pleaded.

"Well, it would take about a year," I responded.

"It's PFAS," he added. "That's what they want to test for."

"Why?" I asked. "There may be a thousand forms in any one sample, of perhaps 8,000 or more forms overall, so you would be best advised to test for one important group of PFAS chemicals."

"Well, I've got to have it for this case and it's important," and this and that.

"But not only are there perhaps a thousand of them," I continued, "anytime you take from a class of chemicals and you put it in the presence of sunlight, air or moisture, it changes, and now you've got new compounds, or no compounds at all. So, if you test for PFAS-1 you're likely to get zero because it's now PFAS-2, or something other, or many others." (Note: PFOS, PFOA, and other PFAS—fluoropolymers or per- and polyfluoroalkyl chemicals—persist in the environment and are commonly described as persistent organic pollutants (POPs) or "forever chemicals.")

So then he says, "What do you mean? How can that be?"

I answered, "Well, think of DDT. Do you know what DDT is?"

He says, "Yeah."

"Well, if you take DDT—as in the old days, they don't use it anymore—and imagine you spray the flies in your barn with it, and someone says, 'That's terrible! I'm going to test your barn for DDT.' And they test the barn, and then they say, 'There's none there! It's gone. It's safe.' They're right, the DDT didn't linger in the environment—it killed the flies, but it didn't linger—because once you spray DDT into the air and it reacts with sunlight and water vapor, it becomes DDE, and more. It no longer exists as the parent chemical. That's not an unreal or unusual event. That's common with many chemicals, including the PFAS, PFOS, dioxins & furans, PBDEs, and PCBs. Their mission is to change their chemistry and to bond with any other compatible chemical, at the least expenditure of energy. That's what they do. And the story from the agencies is, 'Oh, you spread DDT. Now it's not there. It's safe.' DDE is also toxic and it is not safe," I offered, "so think about where this project may be going. It's a rathole."

All chemicals and groups of chemicals are different, as are the tests there for, so the best answer for his purpose, I suggested, was to contact a few EPA-certified testing labs and inquire about the most appropriate panel of PFAS forms that are being tested at present, and use that panel. Experienced labs do this work frequently, and can anticipate what's out there, what's important to the agencies at present, and what may be the outcome as a function of the location and matrix (matrix meaning: the environment or material in which something exists). In my case, I want to know what the agencies, industry, and environmental groups are testing for, because they are the critical stakeholders, and those who need to develop the most defensive postures as a result of testing outcomes.

All that DDT/DDE testing was done in the sixties and seventies about lingering DDT from World War II: "Oh, it's all gone." No, it's not. It's just in a different form, and just as deadly or even more deadly now. It's everywhere.

What we were talking about with DDT is simply an example for PFOS or PFAS. There isn't any sludge without PFOS or PFAS.[6] Why is that? Why are these in all wastewater effluents and why are they in all *Toxic Sewage Sludge*? Well, there is a billions of dollar industry with PFOS or PFAS in every fabric you wear, unless you wear pure hemp, bamboo, wool, cotton or nothing.

Clothes today are made of plastic fibers, and many of them are coated with flame retardants, waterproofing, and wrinkle resistance for safety, durability, and look. All of these PFAS and PFOS shed every time you do your laundry, and they go down the drain. Some of the highest sources of PFAS and PFOS in North America are from laundry cleaning procedures and operations, and so where does it go? Into the same sewage. It goes into the same *Toxic Sewage Sludge*; into the same toxic wastewater effluents. It doesn't just go away. It lasts for hundreds of years as it flows into the soils, wetlands, rivers, and streams, and on to the oceans and into the fish and the whales. It's nothing new—since waste was dumped above waterways, it's always been that way.

In 1993 the Environmental Protection Agency introduced the 503 rule to regulate the treatment of waste.[7] The 503 rule is simply permission. According to Dr. Lewis: "It does not regulate pharmaceuticals, pesticides, growth hormones, flame retardants, and other potentially harmful pollutants, which EPA strictly regulates in air and water. Many of these chemicals are carcinogenic, mutagenic, neurotoxic, or otherwise harmful to public health and the environment in trace amounts." And further, "This rule allows municipalities to collect industrial pollutants at wastewater treatment plants and spread them on farms, forests, school playgrounds, and other public and private lands without monitoring any pollutants other than nine metals and the two nutrients, nitrogen and phosphorous."[8]

I've presented on this EPA 503 rule with PowerPoint displays that break it all down. It simply gives permission to King County or anyone else in the country to dump sludge. When you read it, keep in mind a little phrase that begins with a comma, "*..., except for sewage sludge from wastewater treatment plants.*" The 503 rule supports industry, not people or health, because if you're going after pollution in wastewater effluents or sewage sludge, and their toxicity, just remember: *"..., except for wastes discharged from a wastewater treatment plant or publicly owned treatment works."* Dr. Lewis clarifies the purpose of this policy in more detail:

> Although it was not apparent at the time, EPA and the WEF (Water Environment Federation) planned to achieve public acceptance of the 'science and substance' of EPA's

> 503 sludge rule by almost any means necessary. Their overarching goal was to overcome any 'barriers' by the Year 2000. In retrospect, it's clear now that 'over-coming barriers' meant just four things: (1) covering up pre-1993 studies that documented adverse effects; (2) discrediting post-1993 'biosolids horror stories'; (3) defunding scientists who oppose the 503 rule; and finally, (4) creating a body of supportive 'science,' which EPA scientists refer to as, 'Sludge Magic.'[9]

You could spend the time, and I know people who have done this, "Oh, well, I'm going to take on the 503 rule and we're going to have a meeting about that next week and I'm going to study it and go in there." And everyone comes in and all of a sudden they realize, comma, *"except for what is discharged from a wastewater treatment plant or publicly owned treatment works."* The rule is about toxic waste, but with the exception of wastewater treatment plants. It's about what runs off agricultural lands, etc., but is absolutely without meaning. That's what they did—they emasculated it by intent. When 503 came into effect now all this raw sewage had to be run through a wastewater treatment plant, whether they worked effectively or not.

In this business of sewage disposal, whether it's toxic raw sewage inflows, *Toxic Sewage Sludge* or toxic wastewater effluents, it's not a joke. When people glide over the toxicity and other safety and public health issues, let me tell you who's concerned, the Director of National Intelligence. All intelligence agencies are up to their hips in this very thing we're talking about because this involves national security. The deniers are the agencies: local, county, state, and federal, except for ODNI, the Office of the Director of National Intelligence.

Overall, we need more transparency and accountability across the entire spectrum in the management of all of the various forms of our domestic, industrial, agricultural, and medical wastes. And further, we must build sustainable waste-to-renewable-clean-energy (W2RCE) facilities to consume both new waste and legacy waste from extant and legacy landfills, as a combined effort to restore past errors and to move forward with the proven idea that renewable clean energy can be generated from waste streams destined only for landfill or for other disposal into the environment—a serious matter of public and environmental health policy, and of national security.

Our critical infrastructure hangs upon the delicate thread of access to reliable, continuous, and renewable sources of potable water and clean energy. We need reliable energy that assures clean water inflows and non-polluting wastewater outflows. Therefore, reduction or prevention of waste generation must preclude any conventional means of waste disposal into our living environment, before the fact, and become the essential feature of any waste management program. This grave responsibility must be in the hands of private entities. The public sector cannot perform this work as a functional business, for while governments may be adequate as regulators, and as the "toxics police," WWTPs, POTWs, and landfills should be managed as businesses, by businesses, and separate, one from the other, and under the watchful regulation of those who are able to appreciate the national security issues surrounding our critical infrastructure.

The challenge is that WWTPs, POTWs, and landfills are owned in large part by municipalities, while counties are operated and managed as lowest cost, most convenient, tax revenue-based operations. It is clear they do not have experience operating or managing for-profit businesses, which is why private waste firms are generally more successful—they are more beholden to shareholders and to profitability than to taxpayers.

When we state that we have conflict between commerce and public health, and that commerce always wins, that is an observation. And when we ask, "What must we do when pollution has become policy?" that is a valid question. It also offers the answer, in that: "Science has been politicized, thereby declaring the fate of our wastes, and the fate of us all, with a multiplicity of adverse consequences."

So what must we do? We must first recognize the problem in order to determine innovative, workable, and non-polluting waste-disposal policies, coupled with new technology solutions that are, remarkably, already available. "We must never let that happen again," would be the most highly applicable, after-the-fact term to consider here, and which is the standard and most often stated cover-your-ass phrase offered by agencies and elected officials. We obviously have work to do!

Part II

Toxicus ad Infinitum

Chapter 9

The Microbiome of Sludge

It would be naive to believe there is plastic everywhere but just not in us.

Rolf Halden, *The Guardian*, 2020.[1]

When I worked in Antarctica, way back when, we flew by helicopter into remote areas of the Dry Valleys where it was thought no human had ever been. While isolated in the full containment of an aseptic hazmat suit, the plan was to venture forth and take soil, rock, and ice core samples to discover what kinds of microbes might inhabit these hostile, saline, and desiccating environments at about minus 50F. It never rains or snows in this barren landscape because it is generally too cold, and if it does, the desiccated snow crystals blow away to other areas or out to sea. The high wind speeds preclude any raindrops or snowflakes from ever touching the ground. Anything that enters these Mars-like valleys stays, unless carted off or blown away.

Our mission was: "We're going to Mars someday soon and we have to figure out how to do sampling, preservation, and shipping." But, what if you didn't find any microbes? Is that an absence of data, or does that represent data of absence? That is a very old question in science, most especially in field science, but the idea remains valid and holds truth. The underlying tenets of science will never provide an opportunity for you to say: "You're right, there were no microorganisms," just because you didn't find any.

Science says: "No, no, the samples were contaminated;" or "You didn't use the right technique;" or "You missed them;" or "You were in the wrong place, so try again—they're really likely there." Life does that, and field science knows that an absence of data is not the end result.

Back in the old days of the early 1970s, we used primitive lab culture techniques—utilizing several types of selective media with both aerobic and anaerobic culture at multiple temperatures and salinities—in an attempt to isolate microbes from new or challenging sites because those were the traditional standard microbiology methods of the time. The routine isolation of DNA followed by whole genomic sequencing was not yet in our tool kit, much less anything that encompassed molecular genetics, genomics or microbiomics.

Today we wouldn't look for any specific microorganism by itself. Instead, we would collect a handful of soil and then back in the lab we would perform DNA extraction and sequencing, followed by an analysis of the derived DNA sequences using artificial intelligence (AI), along with the aid of a custom algorithm, to scan global databases and determine what kinds of microbial and other DNA examples were in the sample. The revelatory objective nowadays is to determine how many of the DNA fragments align with any, some or none of the various and multiple taxonomic groups or clades in the multiple and various global databases. The point is to identify the origin of the DNA so as to identify the genus and species, or at least the larger taxonomic groups of microbes that may have evolved from a common ancestor. Even if we had found some spurious DNA back then, we would have doubted all that we did, simply because the said DNA could have blown onto the Antarctic Plateau from just about anywhere on Earth, as are the pollens and other dust-laden materials that have been detected.

Back in those days, we'd likely have asked, "Why are they here, and how did they get here, whatever 'they' may be?" Today we have even more bizarre questions to pose, such as, "Are they adsorbed to microplastics and nanoplastics? Have they been degraded by industrial metals, flame retardants, and phthalates? Have they been mutated by exposure to various industrial chemicals or by pharmaceutical and personal care products, or are they bonded to any of the various radionuclides that may persist for millions of years?"

But wait a minute; we now know that any and all samples of nearly anything and everything are likely to be loaded with DNA, mutated or otherwise, from so many historical and current sources—microbial, human or other—and perfectly preserved in that freezing and desiccating South Polar environment. So how do we unravel all of this from a single grab sample from that distant region, and what is the relevance of such information? Where would the well-preserved DNA found in a South Polar soil sample have come from?

Perhaps it came from dust in the atmosphere that blows from the Sahara Desert over the Pole to be dropped with at least the resilient DNA sequences still intact and bonded to the said dust particles. It could have been derived, as well, from any one or more of the pinniped mummies that are found throughout the interior Dry Valley regions of the Antarctic. Perhaps humans and seals carry the same or similar microbes, for example, and perhaps with the same response to exposure, and that is not a stretch.

How to answer these questions? You just work hard at it and report the results, but you must do the work. Work based on inference or intuition does not tell the proper tale, so you do the work, including both the field work and the lab work, and then make statements of the findings, but not in response to the needs of convenience. Nothing should be reported until the appropriate studies have been done, repeated, and then verified yet again by outside referees.

Curiously, as an aside, and in line with what we preach, the deep cores we made back then into 60,000-year-old Antarctic ice did reveal microbes, of sorts, but the preliminary findings were "frozen" by the State Department over objections by the academics, NASA/JPL, and the National Science Foundation as likely being a flawed result of contamination by a lowly new field microbiologist, aka, *moi*. We just did not have the necessary technology back then, or so I excuse.

So how interconnected are we? This is the era of microbiomics, and science today understands that life is not an isolated experience. Life is intertwined through a microbiome that extends from the larger environment, such as DNA on Saharan dust spores settling in Antarctica, all the way to what makes up our human bodies.

This human microbiome (i.e., the combined genetic material of all of the microorganisms in a body or environment, or the collective microbial genome) is defined as the genetic material of the microbes, including all of the viruses, bacteria, fungi, protozoa or other microbial sources that live within and upon the human body. The environmental microbiome refers to the collection of genomes and their individual genes within and among all of the microbes of the subject environment, while the human microbiome refers to the 10 - 60 trillion or so individual microbial cells or genomes on and within a single person.

By investigating all of the cells within a human microbial population or community, it is revealed that most cells in any single human are microbial. Which is to say, while there may be more than 100 trillion cells within any individual human being, most of the cells are microbial cells, and they far outnumber the human cells. For clarity, the number of individual genes that represent just the microbes within a person's microbiome may be perhaps 200 times greater than the number of genes that represent that same person's human genome. (Note well that these numbers and ideas are subject to constant revision as the technology and knowledge base evolve.)

Consider then the key question: "Do the intact microbes deposited in *Toxic Sewage Sludge* disrupt normal microbial populations that reside within us? Could exposure to the great genomic unknown of everyone else's microbiota and its collective microbiome, as may be found in any *Toxic Sewage Sludge,* incite disease by disrupting the normal microbiota and functions of the commensal and synergistic microbes that reside inside us?[2]

When considering these questions, how could it be that agencies report that exposure to *Toxic Sewage Sludge* is safe, and that spreading toxic sludge and toxic wastewater effluents into our natural living environment is some form of "beneficial reuse," while the science of the human microbiome and the associated science of sewage microbiomics are just now newly emerging as sciences, as yet unconfirmed and unsubstantiated?

As a current example, I have been exploring the microbiomics of urine from women who have a clinical diagnosis of cancer. On the first pass, we easily surpassed 17,000 or so DNA fragments of the various and many assorted taxa of microbes in just their urine. So what? Where do you go from there? Well, if all of the urine from all of us passes into and through the sewage systems of the world, and into the *Toxic Sewage Sludges* that are land disposed, what are the genetic messages of all of those uncountable microbial DNA codes, and are they the precursors of cancer, or the initiators of malignant transformation, or of cell proliferation, or of metastasis, or of other genetically predisposed chronic diseases?

In that no one knows, who then has license to authorize any or all of us to experience the risk, threat or harm presented thereby? In that no local, county, state or federal agency has ever performed microbiomics experiments to evaluate and define the eminent risks of disease by their planned exposure of us to these threats, the day of accountability must be defined. Our infrastructure is insufficient to mitigate the public's risk of disease by these shortcomings. The science is now available; and therefore, toxic waste disposal practices must follow closely and address these revelations of risk, threat, and harm. It's not done; it has never been done; it's unrevealed and unsettled science; and, by being oblivious to the potential adverse outcomes, it is decidedly irresponsible and life-threatening behavior.

"Ok, so you detected the microbes, now what? What is the storyline to be constructed by those findings," one might query. It's a start, and it's a work in progress, but we are unwinding some astounding relationships between urine-borne microbes and certain forms of breast cancer and other cancers. Where is this going? Unknown for now, but exploratory research based on a sound hypothesis is a valid beginning.

The point is, if the urine from a billion or more women is collected by wastewater treatment plants, and if their urine-borne microbes are carried along for the ride, and if the wastewater treatment plant has never been shown to do more than reduce the number of a couple of common urine-borne or fecal-borne microbes therein, then what form of science is it that makes the bold and wholly untested statement that "It's safe?"

Does the urine-borne or fecal-borne microflora from those one billion women harbor the very microbes that may have incited their cancers, or was it specific microbial gene sequences, or was it perhaps the vast array of chemical or biochemical substances excreted by those countless trillions of microbes that prescribed their disease? And what of the other hundred or more cancers that may, or may likely have, any one or more microbes as their inciting agents? This likelihood is wholly unknown to science or medicine, yet is frequently denied. We must consider, as well, that many of the sewage-borne microbes are also antibiotic-resistant, which means they carry the gene sequences that confer resistance to one or more antibiotics, and many also carry genes for virulence, meaning the genetic-based capacity to do harm in many ways.

The point to take home is, raw sewage, wastewater effluents, and *Toxic Sewage Sludges* have all been touched by the urine of us all, and all of those urines harbor many of the genes and gene sequences of the microbial and human populations sourced therefrom. Why, then, would anyone believe it is a good idea to dispose of these wastes on land or in waters, or in composts, mulches, or planter mixes, when none of the needed and revealing studies of infectivity, antibiotic-resistance, virulence or oncogenesis have ever been performed, and all in light of the current global explosion of new cancers, cancer deaths, and more virulent and resistant forms of infectious agents, and yet while the main inciting agents of cancer, such as smoking and other tobacco uses, are in global decline? Look to our wastes, for the ill fate of carcinogenesis likely resides therein and in plain sight.

Our beginning and unsophisticated analysis of an association between certain identified urine-borne microbes and specific cancers indicates that elimination of the said microbes tends to presage a diminution of some cancers, pending the stage of disease at diagnosis and the health status of the subject. More to be discovered here, but the associations are becoming clearer. At least in this case, a direct association between the microbiome of a subject's urine and their cancer disease outcome appears to be related directly. This is not a new or novel idea, but it awaited the proper technology for the informative analyses and conclusions to emerge.

Viral, bacterial, and other forms of microbial-incited cancers are now well researched and documented—as are radiation-incited and chemical-incited oncogenesis—with more yet to be revealed as a function of both detectable mutation events and the more subtle epigenetic influences that are being explored as inciting agents. At this still primitive stage in the science of carcinogenesis or oncogenesis, who then dares to even suggest that known carcinogens may in any way be safe?

It is well understood, more so now than ever before, and as revealed during recent National Academies of Science toxicology workshops, that variability in results in early science provides unreliable outcomes for good cause and reason. And with regard to valid animal models of cancer and other diseases, it is clear that mice are not necessarily mice, and that mice are not rats, and that rats are not humans, and that none of them are infant or adult humans, confirming that toxicity data are not usually translatable across species!

Using that storyline, let's take a look at microbes and their presence in *Toxic Sewage Sludge*. The EPA also has a priority list for microbes in sewage sludge that includes just two microbes, generic *E.coli* and *Salmonella* spp., suggesting that the presence of these two sometimes pathogenic microbes in *Toxic Sewage Sludge,* or in toxic wastewater effluents, at levels above some poorly defined mark may pose a risk, but what may be the risk, and when, and to whom, and by which strain?[3]

Beyond the few strains that do incite human disease, most *E.coli* and *Salmonella* serotypes are either innocuous or good for us. The EPA has little or no supporting data based on any analysis of the microbiota or microbiomics of wastewater treatment plant effluents or of *Toxic Sewage Sludge* that reveal much of anything of use or value about which strain is present and harmful or not. Superficial testing by antiquated methods holds no validity and is generally incorrect and misleading. We are all at extreme risk simply by agency ideas and practices alone, and more so as they are passed down to local labs or technicians for performance.

To further this discussion, we need to consider the nature of microbial cells versus human cells within any human body. The human body has perhaps 40 trillion human cells of nearly 200 types, with each type having a distinct set of functions. The greater human microbiome has an average of perhaps 60 trillion or more cells per person, with a total of about 100 trillion combined microbial and human cells per human body that is comprised of all of the human and microbial cells in one single carcass. That's a lot of cells, so who can understand the implied dynamic? Likely not yet anyone!

The microbiome of a human may contain perhaps 100,000 kinds of microorganisms (kinds of microorganisms this time, not individual microbial cells), and that would include the likes of viruses, bacteria, fungi, oomycetes, protozoa, and more, not counting parasites and prions. That represents a lot of non-human cells or non-cellular entities, or self-replicating proteins or microorganisms in a human body, nearly all with marginally known and mostly unknown functions.

Some scientists may say these microorganisms represent perhaps only 50 percent or more of the cells in our body, and that they rule the roost. Yet no one can deny the influence they have. They produce hormones and provide messages to cells on cellular function, and, for example, when to divide and reproduce or not. If we didn't have that roughly 50 percent microbiomic load we would be dead, because human cells by themselves would be dysfunctional.

So, are we really human after all? Or are we merely masses of microbes supporting and messaging human cells in some form of a variable and consolidated body? It seems that no human cell functions without proper instructions from microorganisms that make up the human microbiome. They're pulling the strings, yet we alter them with the toxics in toxic waste and other non-beneficial exposures.

When there is an exposure to toxic waste, microbial cells may function to protect our cells from such toxic exposure, including those that may result in toxics-incited mutations and the resulting altered cell function. However, and this is important to understand, the volume of toxic waste from *Toxic Sewage Sludge* and from all else that is discarded into our living environment is now so great that the usually protective commensal microbes can become overwhelmed and no longer able to protect us.[4] Hence, I propose, this is one of the main causes for the alarming increase we are witnessing today in global cancer afflictions.

By this increasing and overwhelming over-exposure to toxic wastes, we are now charged with altering the definitions of the associations and relationships between and among the microbes of the human microbiome and of our cells and tissues. The alien toxics now challenge our cells, tissues, and organs with greater tasks of accommodation and response to toxic exposure than at any time previously. The very definitions of the interrelationships between and among our cells and with the inhabiting microbiome members must be reevaluated. So let's then examine the general interrelationships between and among them:

Commensalism between cells, such as between a bacterial cell and the cells of a host organism, refers to a relationship whereby neither cell benefits from the other, nor does either provoke harm upon the other (often the category of default when the benefit or harm have not been recognized or fully defined).

Mutualism among cells refers to a relationship whereby the cells of both organisms benefit by the very fact of their association with the other (often the category of default when the harm has not been recognized or defined).

Parasitism is the case whereby a microbe benefits by harming the cells or tissues of a host organism, including as an infection.

Parasitic Beneficial Mutualism is the case whereby, while some microbes may incite cancer or other harm to the host, or to other microbes or to the same microbes, both may obtain and provide benefit by protecting host cells from cancer or by eliminating the cancer cells.

The point is, if you happen to take a kilo of *Toxic Sewage Sludge* to a toxicologist, you should ask not only how many microbial cells are in it, but what are the kinds of microbes, and what are their functions, and are they infectious agents, and are they hemolytic or toxigenic, for example, and what other dangers may they present?

The question then becomes: "Why on Earth would anyone select just *E.coli* and *Salmonella* spp. to test, if there may be 50 or a 100 million microbes representing 100,000 or more kinds of microbes occupying the human body and thereby our sewer systems? Is cholera (*Vibrio cholera*) in there? Is anthrax (*Bacillus anthracis*) in there? Is methicillin-resistant *Staphylococcus aureus* (MRSA) in there?[5] Are the gene cassettes for virulence or antibiotic resistance in there?" Where do they not exist? The relevant questions are nearly endless, yet seldom asked—correction, never asked.

So, agencies came up with a new term in response to this dilemma: "Oh, those are 'indicator microbes' or 'indicator pathogens,' and if they go down, then they all go down." Well, all of "those" have never gone down, and therefore all of them never will go down, most especially if you have no clue what they all are, what they do, where they came from, and how to detect and count them.

"If we heat it, digest it, or treat it with lime or sodium hypochlorite," they say, "all of the *E. coli* are killed, and if they go down, then everything else goes down." That bad idea is in total contradiction to known, confirmed, and settled science. No, not so, no they don't, they never did, and they never will, because there are many kinds of bacteria, fungi, and other microbes in *Toxic Sewage Sludge* and in toxic wastewater effluents, and some are spore formers that resist heating or drying, or more.

In field microbiology, when we collect a soil or *Toxic Sewage Sludge* sample and take it to a lab for testing, we can't just throw it out. We have to autoclave it. We have to put it in an autoclave at high pressure and temperature, not once, but at least twice, because when heated under pressure, the spores of certain pathogens simply germinate and grow. Allow me to provide a colorful and hypothetical example of why this is relevant.

If you and I were to travel to Kansas along the old trails where cattle were driven from Texas to a railhead in Kansas City two hundred or so years ago, and then shipped east on trains, we could still isolate the resistant spores of anthrax from those soils along the trail.

Cattle have anthrax, as do moose and other animals, and if you go there today, to those old trails, and collect a scoop of soil, you will not find anthrax, unless you heat the soil sample first, then culture specifically for anthrax. That's because anthrax spores go dormant and persist in hot, dry desert or prairie soils, awaiting favorable environmental conditions, such as a warm body, to become revived. Anthrax spores silently lay about waiting to be inhaled into an animal exposed to the long-ago contaminated dusts.

In the Southwestern United States you may develop the respiratory fungal disease coccidioidomycosis (cocci, or Valley Fever) in similar fashion. "Cocci" is a respiratory infection incited by either of two common, soil-borne fungal pathogens, *Coccidioides immitis* or *Coccidioides posadasii*. It's in the soil and it is perpetuated in the environment by rodents. You can plow the desert and make a new farm, but if you breathe in cocci spores, you've got coccidioidomycosis.

Cocci is also well entrenched in Washington State, mostly in Central and South Central Washington, generally in the more arid farmland regions of the Yakima Valley.[6] The usual victims are farm workers. Should we not wonder about what the influence of land-disposed *Toxic Sewage Sludge*, fertilizers, and pesticides may be on the perpetuation of cocci in our soils? As well, should we not inquire about what may be the contribution of susceptible rodents in our farming areas that serve as reservoirs of infection and as alternate hosts for the fungal pathogens? After all, it is spreading.

Becoming infected with *Coccidioides immitis* is easy, dangerous, and often fatal. Cocci is treated with antifungal agents, none of which work very well, except for amphotericin B, a highly toxic polyene macrolide antibiotic. A cocci patient is generally placed on life support while amphotericin B is being administered intravenously.

Of note, the therapeutic dose (TD) and the lethal dose (LD) of amphotericin B are nearly the same, and thereby the therapeutic index (TI, or the LD/TD) is nearly 1, which suggests strongly that you may be at equivalent or greater risk of dying from the treatment of choice as you are from the fungal infectious disease itself. As well, if you were to add the immunosuppressive and mutagenic effects of dust-borne pesticides, fertilizers, and land-disposed *Toxic Sewage Sludge* to the mix, you could die simply by breathing in an affected agricultural district, and no agency will claim responsibility—citizens be damned.

And what of the adverse effects of the ominous climate crisis, with the added aridity and warmth, and with the accompanying increased winds that disseminate all toxics and pathogens far and wide? Is our immunity then also challenged as a result of endless exposure to the combined adverse effects and compromised immunity that results from inundation by *Toxic Sewage Sludges*, pesticides, and fertilizers? Many human microbial pathogens, including viral, bacterial, fungal, or parasitic infectious agents, may persist in soils or in other matrices for decades in desiccated form, awaiting activation by the artificial nutrient exposures from *Toxic Sewage Sludges*, fertilizers, and pesticides or even climate change.

These are not unusual events; this is the way it is. When some agency person tells you: "Oh, those are 'indicator microbes,' and if you kill those, all the rest are gone," please be advised that cannot possibly be true—ever. The agencies have a lot to do, but unfortunately, no one is doing it.

We await any one single agency that will initiate a broad-scale testing program intended to detect and prevent disease, rather than simply understand the risk, then later commit resources to treatment of disease after the fact, much of which could have been prevented. The idea of prevention is real, and can become policy, despite any contentious responses from the agricultural industry and their agency supporters. Yes, we must eat, but we need not die because of it.

If we compromise a few trillion cells within us by toxic exposure, is that a minor impact? If we alter microbial function, are we thereby also altering human cellular function? This is not just an off-base rhetorical question; these are real numbers and actual risks to human health. What then are the individual or cumulative adverse effects of exposing our 100 trillion collective human and microbial cells to a vast cocktail of toxics? Unknown, but we are learning—the hard way.

Why did I bring this up? I surface the issue because if you were to collect a sample of *Toxic Sewage Sludge* from the Snoqualmie Forest that originated from the West Point Seattle Wastewater Treatment Plant, as just one example among so many, what would be the microbiology of that sample? Would that represent the *Toxic Sewage Sludge* microbiome, i.e., the DNA of all of the microbes of the sludge sample, and not only of the microbes, but all of the DNA of the human, animal, and plant DNA contributions contained therein, as well, now and perhaps even historically? What would it be? Would there be one, two, five or ten microbes—not numbers of cells, but numbers of kinds of bacteria, viruses, chytrids, protists, oomycetes, fungi, parasites, and the like? If 100 trillion is the number of collective human and microbial cells in one human body, what then would be the numbers and kinds of collective human and microbial cells in that single sample of sewage sludge, and what do they do, and are they essential, or are they just casual inhabitants?

Further, what is the number of toxic metals in that sludge sample, and what is the number of chemicals, known and unknown, and what are the toxicity characteristics of any one or some of or all of the collected toxins in the sample? And what then are their singular and collective functions? Unknowable, so once again, why do agencies state that it's safe?

We can't know the total number of microbial cells, or the kinds of microbes, in a *Toxic Sewage Sludge* sample, because no one knows the number, but there are far more than had been imagined until recently. We do know that the number is greater than a hundred thousand. Not numbers of microbes, but kinds of microbes. So, what do the agencies do? They only test for *E. coli*, and *Salmonella* spp. If they find low numbers, not an absence of the targeted microbes, just low numbers, again they say, "It's safe," totally ignorant of the potentially vast number of infectious agents therein. This is crazy!

"What *E. coli* did you test for?" I asked during a meeting with King County officials several years ago.

"What?" they responded.

"*E. coli*? Which one? There are over a thousand serotypes. Some of them kill you like the Jack-In-The-Box *E. coli*. Some are enterohemolytic *E. coli*, some are enterotoxigenic *E. coli*, and some express Shiga toxins. Testing for generic *E. coli* has no meaning. What *E. coli* did you test for?"

"Oh, well, we don't know," they said. "It's just general *E. coli*. If the numbers are below a certain level it's safe, because they are 'indicator microorganisms' or 'indicator pathogens.'"

These are terms of convenience without any scientific derivation or meaning. Once again the lowest cost and most convenient methods are used to support wholly contrived standards to achieve agency agendas, without regard for human or environmental health. If it's otherwise, then let's see the objective, repeatable, disciplined scientific data—end of story.

So, I said, "There's no such thing as 'indicator microbes,' and what about *Salmonella*? There are thousands of serotypes among the many species and subspecies of *Salmonella*, many of which are pathogenic, while others are innocuous. Which ones did you test for?"

"Oh, same thing. If those are low then we know it's safe," was the answer.

"But which ones are still there?" I persisted. "The ones that kill you or the ones that kill cows?" It's a simple DNA test to find out, and they won't do it. What happened to the other hundred thousand kinds of microbes in there? They have no idea.

As the virus SARS-CoV-2 and the resulting disease COVID-19 appeared at the beginning of 2020, people started testing for the infectious viral agent in raw sewage, sewage sludge, and in wastewater effluents. They had one hundred percent success—it persisted and was detected in all three matrices, all over the U.S. and elsewhere. If any segment of the population has it, inevitably the virus shows up in sewage, and it is persistent and infective. SARS-CoV-2 was there. This single point tells us that everything these agencies do on a routine basis has little or no meaning. No public health meaning at all. No science meaning. They do what they do as an exercise based on a false criterion of public safety policy. The presence of SARS-CoV-2 in *Toxic Sewage Sludge* has been reported everywhere, as have so many other infectious agents.[7] Who knew? We awaited the technology to inform us. Now we know! When politics informs science, there is death to pay.

History proves that the biggest epidemics or pandemics were incited by infectious agents derived from our wastes, which often resulted in cholera, influenza or the plague. Millions of people died, so why wouldn't the same scenario be repeatable? It is, and our poorly treated wastes may very well be the vectors for other, more serious epidemics, silently awaiting willing infectious agents and lax surveillance.

As a case in point, the 1918 influenza pandemic was perhaps the most severe pandemic in recent history. Incited by the H1N1 virus, it was disseminated by contact and in wastes and waters throughout the world during 1918-1919. It is estimated that about 500 million people, as many as a third of the global population, became infected. The number of deaths is estimated to have been at least 50 million globally, with about 675,000 deaths in the U.S.

Dr. Rolf U. Halden (Center for Environmental Security, The Biodesign Institute, Security and Defense Systems Initiative, Arizona State University, Tempe, AZ) and colleagues tell us that, "Processed municipal sewage sludges (MSS) are an abundant, unwanted by-product of wastewater treatment, increasingly applied to agriculture and forestry for inexpensive disposal and soil conditioning. Due to their high organic carbon and lipid contents, MSS not only is rich in carbon and nutrients but also represents a 'sink' for recalcitrant, hydrophobic, and potentially bioaccumulative compounds. Indeed, many organics sequestered and concentrated in MSS meet the U.S. Environmental Protection Agency's definition of being persistent, bioaccumulative, and toxic (PBT)."[8]

Just as the fire-ravaged Notre Dame Cathedral is now being restored as closely as possible to its pre-fire state, i.e., in restorer's terms, to its "last known state," could we ever imagine restoring our disrespected forests, farms, prairies, and rangelands, and the receiving waters adjacent thereto to their last known, pre-*Toxic Sewage Sludge*, state? Not a chance, for the massive tonnage of toxics harbored within the toxic wastes discarded upon those precious lands and waters, as the insensitive works of our agencies and their colleagues in commerce, have rendered those lands as being forevermore contaminated and polluted by such misinformed and dangerous behaviors.

CHAPTER 10

THAT'S THE WAY WE DO THINGS

Of all the concerns about changes in our food economy, none gets our attention so quickly, or brings home the paradox of modern food so sharply, as that of food-borne disease.

Paul Roberts, *The End of Food*, 2008.[1]

The definition of *Toxic Sewage Sludge* is nothing more or less than that: It's sewage sludge; it's toxic waste; and it's toxic to just about any and all living things by the very nature of its physics, chemistry, biology, and biological activity; i.e., by its toxicity. Couple that with the fact that wastewater treatment plants do not actually treat wastewater and we're in for a load of endless toxic exposures, including all of the solid and liquid forms of toxics within sewage sludges, plus the released volatile organic compounds. So, where's the good news?

The challenges we face, as citizens concerned about life-long exposure to toxics, are that our expressed self-interests in personal health and well-being are just that—self-interest in the health of ourselves and of our family, friends, others, the environment, and more. To the contrary, here, where I live, amidst King County's Wastewater Treatment Division and Solid Waste Division, the Washington State Department of Ecology, and the Washington State Department of Health, as well as among the domains of their allies and co-conspirators, they all appear to care more about maintaining a clear and easy road to retirement before the world catches on to their sickening and lethal schemes than they do about the actual matters of ecology and health, which is to say, about the matter of public health. What a horrible thing to have to say aloud and in public. Their singular objective is the unfettered land disposal of *Toxic Sewage Sludge*, hopefully without any citizens getting in the way. Scientific research, newspaper articles, and my conversations with individuals nation-wide, reveal that this is not just a local affair, but rather a national problem. Personal stories told herein are simply intended to illustrate what is occurring throughout the United States and elsewhere.

Lest we forget, the United States Environmental Protection Agency started this mess and continues to bless all of this, for reasons yet to be revealed or discovered. For sure, while the EPA could be fixed, it is my belief that certain state and county agencies throughout the U.S. must be abolished, to be replaced and repopulated anew by a State Department of Environmental Quality that would have the simple and singular mission of assuring environmental quality and protecting human life, as well as all other life forms. There is no such agency in Washington State, or elsewhere, and the counties could care less, as I have observed, so long as their *Toxic Sewage Sludge* wastes are disposed in plain sight without complications.

As mentioned, when I have opportunity to discuss these contaminating and polluting matters with agency staff, whether in the office or field, they typically respond with an obviously well-tutored and well-rehearsed set of statements along the lines of: "It's safe;" "We're in compliance;" "It just goes away;" and, worst of all, they state proudly, "That's the way we do things." Totally thoughtless! You see, we can't threaten to dump some of it in their own yards to see how they like it, because they already do that to themselves by themselves as a show of blind and ignorant confidence, comprehending not the concept of long-term exposure to low levels of toxic wastes that eventually incite chronic diseases, not necessarily now, but perhaps in years or even generations beyond. Either they just don't care, or perhaps they simply don't understand or do not want to get it.

One of the great illustrative moments on this topic took place at the Fremont Farmers Market in Seattle, which also includes lots of non-farmer items, tables, and displays. We were strolling through the market on a Sunday, and there was a table featuring the King County Wastewater Treatment Division (WTD) and their horrid *Toxic Sewage Sludge* product, Loop®. The display was directly adjacent to countless booths of fresh organic veggies, brought directly from local organic farms—how contradictory! How oxymoronic. How moronic. What an opportunity!

I approached the three women at the booth and asked what their message was about. The leader began her pitch about the attributes of their *Toxic Sewage Sludge* product when applied to (or dumped upon) organic gardens and farms. I asked politely, "What's in it? Is it safe?"

I pressed on and after about ten minutes of stammering and fencing, she finally stated with conviction, "It has to go somewhere!" Gotcha! Nothing they ever say makes any sense—it is toxic waste, no matter the engendered and flowery storyline of the day.

That brief fifteen-minute encounter told me all about the peculiar *Toxic Sewage Sludge* practices, procedures, philosophies, attitudes, and convictions of King County and its wastewater treatment division that I needed to know. Such horrible and archaic ideas and practices must have come down from above, but were they from the WTD, the county government at large, the King County Executive or from the Washington State Department of Ecology? Or was it a pure dictum from the on-the-ground managers whose job description was to: "Just get rid of it?" After all, as she stated so clearly, "It has to go somewhere!" That statement alone was a severe kick in the crotch, thereby demeaning the policies and philosophies of the county, the state, and of their countless minions who are in thoughtless and unrevealed control of human and environmental health.

Toxins (poisons of natural origin, such as venoms) and toxicants (poisons of human origin, such as toxic heavy metals or chemicals), or collectively, "toxics," are poisons that incite nearly all human chronic diseases, and often death. Toxics are easily and routinely isolated from nearly any and all domestic, medical, and industrial sewage sludges, plus from wastewater effluents, landfill leachates, and agricultural and stormwater runoff, which collectively contribute to greenhouse gas emissions and climate change. These are the primary sources of toxics in our living environment, and as odd as it may seem, toxics do kill things, now or later.

Note that all toxic waste can be managed properly by the development of modern infrastructure. All it takes is the political will to do so.

Toxics are toxic, no matter if they may be released or disposed on land, in foods, in waters, or in the sky. So just who makes these decisions and authorizes their release into our living environment, non-stop, as witnessed and authorized by issued agency permits? That is a question worthy of investigation throughout this country. It is the "who" that is the criminal, in the case of open toxic waste dumping, not the "what," so just who is the Who?

Sewage sludge contains all that goes down any toilet or drain, including everything that flows in any and all industrial, domestic, and medical waste sewer systems, stormwater runoff, and landfill leachates, thereafter, to be defined as combined sewer flows. The consequences of releasing these toxics-bearing materials to the environment are impaired water quality, toxicity to wildlife, soil degradation, food-borne diseases, risk and imminent threat to public health and safety, and sometimes death, as well as environmental or ecosystem disruption, lost biodiversity, lack of sustainability, and compromised recreational opportunities, plus a diminished quality of life. They simply dump it out there and see what lives or dies, and that's it. "It's safe." "We're in compliance." "It just goes away." "That's the way we do things." "It has to go somewhere!" But no matter their contrived storyline, the end product of their immoral behavior is environmental degradation.

So, is it environmental degradation or is it ecocide that we see? Environmental degradation is deterioration of the environment through depletion of resources such as air, water, and soil, or the destruction of ecosystems and the extinction of wildlife, defined as any change or disturbance to the environment perceived to be deleterious or undesirable. Ecocide, however, as reported by the Independent Expert Panel for the Legal Definition of Ecocide, June 2021, means unlawful or wanton acts committed with knowledge that there is or may be a substantial likelihood of severe, widespread or long-term damage to the environment by such acts.[2]

Put all of this in the context of an individual citizen who seeks a moment of peace, solitude or recreation in our federal, state or county parks and forests, or by permit in the private forests of the lumber industry, and we need to appreciate the motivations of all parties. (Of note, by my years of experience, I have found the lumber industry to be most sensitive, cooperative, and empathetic to the concerns of citizens, who are their customers by default.)

One particular evening I was presenting at yet another public town hall meeting in King County, depicting once more the horrors of land-disposed *Toxic Sewage Sludge* in our forests, when, mid-speech, an unsteady voice emerged from the back of the room. A soft-voiced woman dressed in black held up her hand politely, trying to get my attention, and finally got through. "Why are they doing this?" she asked of the crowd in a shaky voice. "After all these years," she continued, "we finally moved into our new house on the edge of the Snoqualmie Forest, hoping to enjoy our retirement years in nature, but look what happened." The voice now evolved to trembling. The room went silent. "Now we can't even go for a walk near our new home. The forests are filled with their awful sewage sludge stuff, and the fumes alone make us sick. We can't even go outside. When will this stop? Who did this? What can we do?" Next came the tears, and total silence.

This story is repeated far too often. Unbeknownst to many, *Toxic Sewage Sludge* is even disposed in public parks, playgrounds, schools, and our own back yards. King County is on a roll, trying to sell their toxic wastes to the unwitting public as a fertilizer replacement or soil conditioner, for use on lawns and shrubs around your home, or in your veggie garden, all with the encouragement and blessing of their university and agency drummers, commercial purveyors, and the industry lobbyists who encourage such practices as a commercial enterprise. No right-minded person would ever consider *Toxic Sewage Sludge* as being good for us or for the environment.

"When will this stop?" "Who did this?" and "What can we do?" are the tunes most played by the citizen victims of this horrific practice. They no longer want to hear the pre-programmed and well-taught refrain of: "It's safe," "We're in compliance," "It just goes away," and "That's the way we do things." It's *Toxic Sewage Sludge*, and it's toxic to any and all living things, as I shall repeat here again and again.

There have been numerous days in the public and private forests of King and Snohomish counties when I wondered if I would see the end of the day—alive. The dangers there are of many kinds and forms, but who would have thought that the most serious danger to be faced during more than a decade of tracking *Toxic Sewage Sludge* dumping in those forests would be the cumulative risk, threat, and harm waiting to be encountered by a hiker or mountain biker as a direct result of King County's *Toxic Sewage Sludge* disposal practices.

I photograph and video much of what I see out there, careful to note the time, date, and location of each event, while focusing especially on the obvious violations of the Clean Water Act (CWA), the Endangered Species Act (ESA), the Resource Conservation and Recovery Act (RCRA), and more. It's common practice by the *Toxic Sewage Sludge* originators, permitters, planners, haulers, dumpers, and spreaders to dump when and where it is most convenient, with or without regard for the human or environmental consequences, and all under the authorization, guidance, and direction of our state and county agencies.

To calm down the highly distressed citizen victim that evening, I added to her story with one of my own recent experiences in the same area of the forest. I had left early in the morning to bike into a recently sludged area near Duvall, quite close to the site of this particular town hall meeting. Ah, ha! Bingo! No need for long distance biking today, we were into it already, with "It" being a freshly sludged forest, just out of the bounds of a collection of homes. Maybe this is where the citizen victim retiree now lived, among these hills that all drain, with every rainfall, to the streams and wetlands below, and to that same Snoqualmie River that flows to Puget Sound.

The site was proving to be especially fruitful right away. There were dead things everywhere, mostly lifeless earth worms, flies, moths, mosquitos, nematodes, and various insects and other invertebrates, plus other natural forest biome life forms that had wandered across the enticing fumes of human waste, and of all other things that run through a sewage treatment plant.

In this especially bountiful forest, the rains had recently subsided, and there amongst the trees were potholes and depressions in the ground, now filled with wet, slurpy *Toxic Sewage Sludge*, covered with a full layer of dead flies, bees, wasps, moths, insect larvae, beetles, gnats, and worst of all, dead nematodes. Lifeless rodents, lizards, snakes, and birds were sometimes also encountered adjacent to the toxic sludge dumpings, all victims of the sludge-borne toxics. If the dual mission is easy cash and easy disposal, why would anyone worry about dead things, after all?

Nematodes are small, often microscopic roundworms, white or colorless, that may be detrimental to crop plants in the case of intensive agriculture, but that enable many plant species to take up nutrients from the soil by their practice of colonizing the root hairs of higher plants in a mutually beneficial relationship among the equally essential root-inhabiting mycorrhizae, i.e., the mycorrhizal fungi.[3]

Nematodes are especially sensitive to many toxics, and therefore their demise can leave the host plants, such as forest trees and shrub species, suffering from nutrient deficiencies. I have many images of sludged forest puddles covered with a white layer of dead nematodes, wholly unrecognized by any agency staff or management who seem to view nature as a vast void to be filled with toxic waste.

I have encountered and imaged *Toxic Sewage Sludge* runoff pools of water amongst forest trees that had a milky appearance, only to realize that these massive white blotches in the sludge-stained, brown waters were the remains of countless billions or trillions of dead nematodes. Having studied nematology in graduate school, I had no doubt about the adverse impact the sludge-killed symbiont nematodes likely had on the overall higher plant forms in the forest.

Of note, mycorrhizal fungi also play a critical role in plant nutrition, as well as in soil microbiology, biology, and chemistry, only to be killed, in this instance, by an absence of knowledge or by total apathy about the critical role and the intricate symbiotic workings of soil microbes and other valued species with their forest plant hosts. It's difficult to imagine that after decades of investigations of the adverse consequences of toxic sludges, insecticides, and commercial fertilizers on the critical soil microbiome and on the greater soil microbiota that the agencies remain oblivious to the detrimental effects of their preplanned and intentionally executed toxic behaviors.

The puddles of water, and the very deep, wet layer of sludge on the forest floor covered with dead insects that day, represented many genera and species, now lost to the contaminated forest ecosystem. This was wholesale murder of natural populations by the unmanaged dumping of *Toxic Sewage Sludge* in this forest.

Fixated on taking macro images of the vast array of dead life forms, including the so many kinds of insects, I seemed to lose track of my own situation. It is situational awareness that equates with survival in these remote forests, even adjacent to luxury retirement homes. The *Toxic Sewage Sludge* fumes that had obviously killed these vast populations of nematodes, insects, and worms, had also left me a bit dizzy, or so I imagined. Uh, Oh . . . !

My next recollection was waking up at the edge of the forest access trail from which the *Toxic Sewage Sludge* had been dumped. My brain wasn't working, and I couldn't get up, so I reclined in the wet muck, waiting for my head to clear. Luckily, somehow I had rolled into the forest access road and not into one of those deep, fetid sludge puddles before passing out. A while later I did get up, but couldn't find my legs or my mountain bike, and had no idea where I was, or which way was out. My exposed skin was covered with insect bites, many of which later developed into bright red infections.

We can only imagine the infectious agents that accumulate in those toxic waste sludge piles, considering that their origin is from perhaps a million or more humans, their homes, medical facilities or industry, and who provide their liquid and solid bodily wastes into this toxic mix every day, all day, all year. The vast array of carcinogenic and immune-suppressive chemicals in this stinking goo alone is enough to give you the creeps.

Then I saw the frame of my bright red mountain bike leaning on a tree in the far distance. It rested upon a pile of freshly dumped sludge, next to a telltale sign announcing that the adjacent lands and wetlands were the newly acquired property of a 501c3 not-for-profit alleged environmental foundation. The charitable foundation solicits donor funds to purchase and protect forest lands, as they profess, but they also encourage toxic waste dumping as advised by their university and agency colleagues. I stumbled over and pushed my bike away from the continuing fumigation effects of the off-gassing of the toxic matter covering the forest floor. The earth worms, insects, nematodes, and other life forms never had a chance, but I was getting out of there. Fortunately, I had managed to record the scenes digitally, including in macro, and with some video. Ah, if only that mountain bike could speak, as well.

I revealed that story to the crowd that evening, and to the distressed citizen victim, with the hope of assuring her that she was right, that she was in fact absolutely right. This horrid and widespread abuse is common practice, because, as I reminded the audience, the messaging is: “That’s the way we do things.” My enhancement to their message is, “You also kill so many things in the process, human and otherwise, that you have no sense of, or an appreciation for, the value of life.”

The citizen victim who had just moved into her dream retirement home was not impressed by the agency and environmental groups’ non-response to her passionate outburst, so she headed for the door, sobbing and in anger. It may be that she hoped or wished that I could help her, but the oft-repeated dirty deed was already done. Neither could the state or county employees in attendance, nor the alleged environmental group members who listened in silence. They just stared as if there were something wrong with her, that she just didn’t understand that all of the stinking death she saw around her new retirement home was "beneficial reuse of a valued resource," or some other such mindless prater.

Makes you wonder just who made up these catchy phrases for public consumption: Did it come from the very top? Or was it some person at a local university who encourages us to pour *Toxic Sewage Sludge* in our yards to nourish our veggies? Could it be the county managers at the wastewater treatment division who are driven to sell this crap to the public? Or would it have been the state DOE managers who permit and oversee the destruction of our forests and some citizens' private homes by exposure to toxic wastes?

It may be that neither the King County Executive nor any of the councilmembers had ever been informed of the extreme toxicity within the contents of the county's *Toxic Sewage Sludge* product, or that by their wanton land-disposal practices they were, in fact, distributing toxic wastes throughout the state's forests, farms, and parks, thereby adversely affecting our air, foods, soils, and waters, or so we shall assume. However, the case just may be that the elected officials and their cronies are all headed for retirement, and care not to look back in fear that someone may blow the whistle, because this is highly toxic waste that they sell to us and dump in our forests and on our farms, without toxicological analysis or safety monitoring. The idea of monitoring is broken, and therefore toxicology testing has never been done on these toxic wastes, simply because it is costly and inconvenient, and no one will ever know—maybe.

They're all in it together—alleged environmentalists included—and none of them will ever challenge their own agencies on this sick practice of toxic waste dumping on our forest soils and in our sacred and diminishing waters, or on our farms, for it is these same agencies that are often the providers of the grants and lawsuit settlements that keep them all going. Will it ever end? Never, so long as we continue to accept that it's the way they do things.

Everywhere I go I see new material for our "pandemic of toxic waste causes disease," story. A major issue is that all county, state, and federal agencies understand clearly that no WWTP or POTW works, which is to say, none actually "treat" raw sewage, and as a result, we are under-infrastructured, and the practical default outcome is our loss, so get used to it. The cost for each of the 400 or so new WWTPs and POTWs needed in Washington State alone would likely be about \$1 - 5 billion each, if we seek actual "treatment" for a change, so an early disease-ridden death is instead the anticipated consequence, and cannot be avoided, because the attitude is that the alternative would cost too much. The county's and the state's cooperative Toxic Waste Relocation Project reigns supreme.

The primary challenge is to end politically driven support for the segments of commerce that yield the ever-growing waste streams and the resulting human exposures. We are being eaten and invaded by more kinds of microbes and cancers than ever before as the *de facto* outcome. It is the political-commercial consortium that feeds the lethal beast, always increasing, non-stop, primarily with government support.

The deadly combination of i) an increasing human population, along with ii) a parallel increase of the open disposal of untreated toxic wastes, all in the presence of iii) a warming climate that provides iv) optimal incubation conditions for enhanced microbial activity, and with v) an increasing transition of usually saprophytic microbes into human and plant pathogens, together with vi) spurious chemical reactions amidst ever-accumulating concentrations of toxics disposed into our living environment, all of which then present vii) the formula for our collective demise. Lest we forget, the toxics-incited immune suppression triggered by the long-term cumulative effects of exposure to low levels of environmental contaminants and pollutants results in a growing incidence and prevalence of chronic diseases, including cancer.

Lung, breast, and colorectal cancers are among the most prominent forms of cancer globally, with certain geographical exceptions, and they are all incited by long-term exposure to cigarette smoke and to a broad spectrum of other environmental contaminants and pollutants, as experienced throughout the course of life, and by those progenitors before us and of us. It is likely, as well, that the cumulative exposure to any one or more of these inciting agents, plus the additive effect of the concentration of each agent, results in mutations or in non-mutation epigenetic modifications of gene expression that trigger the initial malignant transformation event that is defined then as the origin of cancer. Once that mutation event or epigenetic modification of gene expression occurs, the progression to cell proliferation, and then to metastasis, describes the sequence of events from such exposures that likely result in cancer, disease, and death.

"Worldwide, an estimated 19.3 million new cancer cases and almost 10.0 million cancer deaths occurred in 2020. Female breast cancer has surpassed lung cancer as the most commonly diagnosed cancer, with an estimated 2.3 million new cases, followed by lung, colorectal, prostate and stomach cancers. Lung cancer remained the leading cause of cancer death, with an estimated 1.8 million deaths, followed by colorectal, liver, stomach and female breast cancers. The global cancer burden is expected to be 28.4 million cases in 2040, a 47% rise from 2020," states Hyuna Sung, et al. in *Global Cancer Statistics 2020: GLOBOCAN* under the title: "Estimates of Incidence and Mortality Worldwide for 36 Cancers in 185 Countries."[4]

Just as we are not yet finished with unrestrained polluting, unrestrained cancer is not yet done with us.

Chapter 11

Is Science For Sale?

Quite possibly, then, this era, which so congratulates itself on its self-awareness, will come to be known as the time of the Great Derangement.

Amitav Ghosh, *The Great Derangement,* 2016.[1]

A few years ago, some First Nations friends in British Columbia, Canada, with whom I was discussing their *Toxic Sewage Sludge* problems, stated that no matter how clearly they communicated to authorities in Vancouver, they could not gain traction with regard to their plight as victims of the government's thoughtless, polluting, contaminating, and dangerous toxic waste disposal behavior. There were no listeners. The Vancouver Government had assured the First Nations People that Vancouver's *Toxic Sewage Sludge* was safe and beneficial, and that it was to everyone's benefit to haul it to Kamloops and dump it on sacred First Nations' lands. Kamloops is about 100 miles from the U.S.-Canada border, due north of the Okanagan Valley in Central Washington.

Notably, Kamloops sits at the confluence of the North and South Thompson Rivers, which then flow southwest as one river to join the Fraser River on its way to Vancouver and on through the center of the city to enter the Salish Sea. The point being that all *Toxic Sewage Sludges* hauled to Kamloops from Vancouver for reckless and thoughtless land disposal soon all flow right back to the city of Vancouver from whence they came, i.e., to the same people who generated the said *Toxic Sewage Sludge* in the first place. Many interconnecting channels and waterways, including the Strait of Georgia, the Strait of Juan de Fuca, and Puget Sound in Washington State, are also recipients and victims of the great BC flushing.

Considering that the Washington State Department of Ecology, the King County Wastewater Treatment Division, and the University of Washington, Seattle, are prime advisors to the city of Vancouver, BC, on matters of managing their wastewater products, we no doubt deserve to be the recipients of the *Toxic Sewage Sludge* run-off and the associated toxic wastewater effluents that flow back to us from Vancouver and from Vancouver Island. What if they were to even think about the adverse consequences of such aberrant behaviors, and what if they were to think of a progressive, extant, and functional solution for the safe disposal of their toxic wastes, which would save lives and protect our living environment?

I suggested to the First Nations People that they may want to arrange a conference and invite researchers, activists, and other experts on the subject. Not a one-day conference, but perhaps a five-day event. "We could engage outside experts from different fields and conduct a meeting for the benefit of the First Nations People with presenters from several countries," I advised.

"Great!" Funds were secured immediately and preparations begun. The objective was to present the case for either the benefit or the adverse consequences of land-disposed *Toxic Sewage Sludge*. Dr. David Lewis, along with others knowledgeable in the field, were flown to the location, while I drove up from Seattle.

The meeting was held in a magnificent conference room in-the-round at a very nice Kamloops university in February, 2017. The First Nations hosts conducted the meeting. We each gave presentations, talking about our sludge experiences and the challenges we had faced with local, state, and federal agencies. The point was to provide a broader perspective to the First Nations People, and to determine how best to proceed in addressing their challenge.

The conference was a success, from the viewpoint of being informative and revealing the atrocities experienced by the people living in this beautiful, Central BC, alpine area. We too learned a lot from the shared experiences of the many participants, while agency representatives hunkered-down in the back, grumbling about how "outsiders" were interfering with their waste disposal plans and business objectives. Once again, the core question boiled down to: "Is this a matter of a business opportunity, or should we be primarily concerned about human and environmental health?"

The apparent and singular objective of the agencies was to present the case that, while the land disposal of *Toxic Sewage Sludge* around Vancouver was a bad idea and shall not be done in proximity to their citizenry, it was considered a highly beneficial practice on First Nations' sacred lands. The city of Vancouver worked to defeat the First Nations' opposition to the idea by inviting pro-land-disposed *Toxic Sewage Sludge* advocates from Tacoma, Washington, where, for example, the city promotes the use of *Toxic Sewage Sludge* products on food crops and elsewhere. There was a lot for them to gain by their swaying the trend and the outcome of this conference to the specific benefit of "Big City" interests. This became yet another contest between public and environmental health versus commerce, which in this case, became a contest between the city of Vancouver together with its BC government, and with support from the city of Tacoma and Washington State, versus the best interests of the First Nations People and of any and all people, anywhere and everywhere upon the Earth.

A dedicated *Toxic Sewage Sludge* dumping advocate from Tacoma, who worked for the city's wastewater treatment plant, and also for an entity that made and sold *Toxic Sewage Sludge* products, was also there. He was quiet, no doubt because he knew the city of Vancouver was contaminating and polluting the central alpine regions of British Columbia, forevermore, with the city's *Toxic Sewage Sludge*. He is known, as well, as being a contributor to the mythology that it's safe and in compliance, and that it represents beneficial reuse of toxic waste, no matter the adverse human and environmental consequences.

In turn, presenters shared information about their experiences and findings, as well as their in-depth reviews of the literature. Heavily discussed were the kinds of toxics and infectious agents in *Toxic Sewage Sludge* already being leached into the waters of the beautiful wildlands near the university venue.

First Nations leaders wisely brought in students from local middle and high school grades—future leaders within their community—who listened carefully and took notes. As officials from Vancouver gave presentations on how well they treat their *Toxic Sewage Sludge* before dumping it on the sacred lands, First Nations leaders responded with concerns about the consequences of toxic waste dumping, including its impact on their revered and solemn traditions.

One specific waterway they described on First Nations' lands had received Vancouver sludge without anyone granting permission. They had simply dumped it on a site located at the confluence of two rivers where, at puberty, First Nations youth transitioning to adulthood, were bathed in a ritual water ceremony that began long before anyone could remember, and is still practiced today. That sacred ceremony is now conducted in contaminated, *Toxic Sewage Sludge*-tainted waters. These may be considered modern times, but stories of the toxic waste disposal atrocities committed on First Nations' sacred lands offered that day were not dissimilar from the horror tales of centuries gone by, yet that continue even today, wholly uncivilized.

The outcome of the conference was that the First Nations People of Central British Columbia were now better informed because, until then they had only been preached to by the Vancouver Wastewater Treatment Division and the University of Washington pro-sludgers, who also work with the King County Wastewater Treatment Division, all of whom have implied, and without the benefit of any data, that *Toxic Sewage Sludge* is safe. The University of Washington and the King County Wastewater Treatment Division people also work with the Washington State Department of Ecology, an unholy alliance that seeks only to achieve lowest cost, most convenient methods of *Toxic Sewage Sludge* disposal on any lands, adjacent to any waters, no matter the people, the culture, or the consequences. The result of the conference is that the First Nations People are now more aware of the dangers of land-disposed *Toxic Sewage Sludge* than they had been before.

Dr. Lewis and I first met in person at that memorable conference. We had plenty of time to interact and, as a result, we put together a group of scientists from Vancouver Island, the First Nations, Dr. Lewis, and myself. After the daily sessions we would have dinner and continue our conversations on how to create a strategy to overcome Vancouver's *Toxic Sewage Sludge* dumping on First Nations' lands—a work that continues on today.

Dr. Lewis is a clear victim of his experience working with the Environmental Protection Agency, an organization driven by forces both commercial and political. His entire government career was as a laboratory scientist, or as a manager of laboratory science, where he fostered rigorous research that revealed and evaluated the toxicity of the constituents of *Toxic Sewage Sludge*. He is a serious, dedicated scientist.

At the time, he was developing an extensive database on the kinds and toxicities of *Toxic Sewage Sludge*-borne toxics. He knew people were becoming ill from interacting with these sludges. From his research, he had hoped the EPA would reveal the true toxicities of the components of wastewater effluents and of *Toxic Sewage Sludge*. He sought a broader knowledge base that could be provided to city, county, and state agencies responsible for monitoring, regulating, and permitting the land disposal of *Toxic Sewage Sludge*. He advocated for a thorough evaluation of the *Toxic Sewage Sludge*-borne toxics and their toxicities, and that the EPA would at least look toward implementing functional solutions.[2]

Citizen victims don't have the luxury of a well-funded laboratory operation, and neither do environmental activists. We do not have the luxury of spending millions of dollars on analysis and testing so we must use other means. We look for different criteria. We read the science from Dr. Lewis and other responsible researchers, and we translate that information hopefully into positive action to prevent or modify the disposal behaviors of agencies that release toxic wastewater effluents and *Toxic Sewage Sludge* into our living environment.

To this day, almost nothing useful has come out of the effort. The general response to pleas for greater awareness of the dangers of land-disposed *Toxic Sewage Sludge* has been increased resistance by municipal, county, state, and federal agencies to effectively treat raw sewage. Instead of improvement, we see an increase in the well-rehearsed storylines about the safety and quality of the products of wastewater treatment plants that favor only commercial interests and the agencies.

This self-serving viewpoint prevails because the U.S. Government refuses to adopt the European and UK models, which are based on the concept that in the EU each country or EU member has a growing population, but with finite land area. They have nowhere to dispose of their *Toxic Sewage Sludge* other than in waterways, or to create thermal decomposition plants to convert the toxic material into renewable clean energy, along with an ash byproduct that can be safely land disposed, or incorporated into concrete products—they adopted the latter alternative. Throughout the EU and Scandinavia, they have taken up the challenge, invested the funds, and converted the products of their wastewater treatment plants into useful energy and byproducts. In the U.S., we will not subscribe to that approach in large part because it insults the land-disposal industry, including the *Toxic Sewage Sludge* users, haulers, and spreaders, as well as the railroads that carry *Toxic Sewage Sludge* all the way from New York through Pennsylvania and Ohio, and on to Oklahoma.

As discussed, substantial commercial benefit is derived from the transportation and disposal of *Toxic Sewage Sludge* in the U.S. The toxic material gets moved around regardless of the consequences, and the license to do so is based upon false assumptions with regard to safety, along with a stout resistance to any form of safety/ toxicology testing.

A major challenge is that the land disposal of *Toxic Sewage Sludge* on farms, "treated" or not, is a megabillion-dollar industry in the U.S., paid for in large part by taxpayers. Supposedly, farmers benefit from land-disposed *Toxic Sewage Sludge* because the sludge contains nitrogen, potassium, and phosphate, the key ingredients in commercial fertilizer. Rather than purchase high-cost commercial fertilizer, less can be spent to have *Toxic Sewage Sludge* dumped on their farms and plowed into the soil, but at what ultimate cost, and do we actually favor short-term gains leading to longer-term adverse consequences, including compromised human health and environmental degradation?

This adverse and highly controversial activity is an ongoing tragic story about the motivations and dangers of the land disposal of *Toxic Sewage Sludge* versus its marginal benefits. The practice represents a questionable behavior pattern fostered primarily by municipal, county, state, and federal agencies, with commerce seemingly the single political driving force. Shouldn't we, as an aware and responsible citizenry, convince our leaders to stop selling us short, and to resolve this planetary insult with the better interests of Earth and its human and other life forms at top of mind?

Chapter 12

What's Really In It?

The data strongly support that applying sewage sludge to land is not safe, and if things continue as they are, the long-term consequences to human health and the environment have yet to be felt.

Jennifer M. J. Mathney, *New Solutions,* 2011.[1]

An important question generally avoided by any and all agencies, firms, and persons involved in the open dumping of land-disposed *Toxic Sewage Sludge* is, "What's in it?" No one knows! Of course not.

One approach to forcing the answer is to pose the question to all who may be encountered in the field or within agencies, as well as to members of the various municipal and county councils or boards. The question is also presented to academics and contractors who work in the field of *Toxic Sewage Sludge*, as well as to any and all persons who may be involved in or associated with authorizing, permitting, contracting, hauling, spreading or otherwise managing, supplying or receiving *Toxic Sewage Sludge* on their land or upon the lands of others. Citizens stare in wonder when this question is posed, implying that someone or some agency must know the answer—not!

How could that be? As stated so clearly by Dr. David Lewis, "It's unknowable." End of story.

I have surveyed audiences and individuals, public and private alike, on this enigma for years, without hearing a single useful answer. It's either under wraps, or wholly unknown. As stated by Gwyneth Paltrow in *Shakespeare in Love*, "It's a mystery."[2] Both the pernicious content of toxic waste and the idea of love remain as incalculable mysteries.

Land-disposed *Toxic Sewage Sludge* presents to us parallel environmental and public health crises, as well as being in and of itself a serious contributing factor to a looming and life-limiting climate crisis. From my perspective, if we're not concerned about these three directly interactive and synergistic crises, then we're part of the problem, not of the solution.

As private local citizens among a greater population base, we are confronted with a failure of basic and assumed infrastructure, a key contributing factor to compromised health and personal well-being. Clearly, this failed infrastructure-based crisis is adversely affecting our air, food, soil, and water quality, both surface and groundwaters, including our critical underlying aquifers. Historically, failed or unfulfilled infrastructure integrity has precipitated epidemics and pandemics of catastrophic magnitude. We may note them as being historical, but they are among us here and now.

All land-disposed *Toxic Sewage Sludge* and water-disposed wastewater effluents, plus leachates from both extant and legacy landfills, impair our lives and our living environment. Repaired, enhanced or renewed infrastructure by current standards will not assure the safety and reliability of our critical air, food, soil, and water resources. The basic underlying model for infrastructure must change, with a singular and clear objective of eliminating toxic wastes from our living environment, not simply moving toxic wastes around, thereby to be hidden or covered up with farmlands, forests, soils or waters, far from public view, until released yet again. The collective agencies' Toxic Waste Relocation Program simply does not work.

The EPA tells us there are three classes of biosolids, aka *Toxic Sewage Sludge*, sorted as Class B, Class A, and Class A EQ (Exceptional Quality). Class A is defined as: "Reduced pathogens to non-detectable levels, and compliance with standards for metals, odors and Vector Attraction Reduction (VAR). Class A can be used as fertilizer on farms and vegetable gardens and sold as compost or fertilizer. Class A EQ must meet and exceed Class A pathogen reduction, plus meet requirements for metals and VAR. Class B may contain higher levels of pathogens and may require an EPA permit."[3] No valid information here. As so well stated by Granny in Robert Daley's movie, *The Outlaw Josey Wales*, "I say that big talk is doodly squat."[4] If peer-reviewed scientific journal publications fail to make the case, go to culture-based logic as the best reference.

Not one word of the EPA's "biosolids" definition is true, realistic, or based on any segment or interpretation of actual science, and cannot be so, much less as may be based on any consideration of public or environmental health. The criteria for testing, and the tests themselves, are selected only as self-serving forms of altered reality, as a means to enable and substantiate the lowest cost and most convenient disposal methods for toxic wastes, and as a reward to farmers who are relieved of substantial costs from the use of conventional commercial fertilizers.

In plain English, that means the farmers get paid to take it off the agencies' hands. The agencies and their commercial partners employ false reasoning that may seem plausible on a superficial level, but that is generally unsound and intended only to deceive. By taking these toxic wastes to douse on our food crops, above our aquifers, and along and within our surface and groundwaters, simply because human and environmental health may have lesser perceived value than does agricultural success in the eyes of agencies, legislators, and industry, the disposers thereof have done their unholy deeds—"righteously."

Basically, it cannot be managed properly if you don't know what it is. Rather than sort toxics into arbitrary classes, the right thing to do would be to perform analytical chemistry investigations and toxicological studies on the isolated and identified chemistries, singly and in multiple combinations, and then at least attempt to report to the public exactly, "What's really in it," for purposes of general health and safety. It's an implied obligation of all governments, everywhere.

An informed public is a citizenry more equipped to make better decisions about their own and their environment's health, as well as to decide exactly who to elect or re-elect to office, or not. Elected officials and agencies are fully politicized on matters of waste management, and are thus rendered as being dysfunctional and unreliable. Improvements can certainly be made, and must, for the sake of us all.

Agencies and organizations are conscripted by political lobbying and marketing organizations to assure that *Toxic Sewage Sludge* is land disposed in forests, on farms, and on rangelands, despite the extensive and well-known adverse impacts on people and the environment, notwithstanding the compulsory, yet totally artificial and dysfunctional, "Class" designations for the said toxic wastes.

Abolishing land-disposed *Toxic Sewage Sludge* presents substantial political risk, and will provoke drastic change in commerce, but it must be accomplished for the singular purpose of preserving our greater ecosphere, redefined herein as the worldwide sum of all ecosystems. The ecosphere is the collective zone of life on Earth, and it is at risk of harm by the whims and greed of the disposers thereupon.

How strange, you may be inclined to wonder, yet it's a political fact of life: When public health and environmental quality challenge commerce, it's politics, commerce, and agencies that win, no matter the residual stench. And sometimes it's the fabricated public relations efforts that win the day for the agencies.

I recall one afternoon when I was on my mountain bike, deep in a Department of Natural Resources forest northeast of Duvall, and there appeared ahead of me on the track a lone mountain biker. We stopped to chat, as remote, solo bikers often do. He immediately inquired if I was noticing the thick stench, typical of human-originated sludge. "Yes, for sure," said I, "I know it well."

"What is it?" he inquired.

"More importantly," I responded with an insider's chuckle, "Can you even imagine what's in it?"

As usual, I cherished the opportunity and offered up a brief intro into the world of land-disposed *Toxic Sewage Sludge* that occurs on a regular basis in these exact forests. He listened, unable to respond, likely shocked into disbelief, as so often happens with previously unaware citizens. For his further enlightenment, I recounted the recent episode of the mountain biker who was killed and partially eaten by a cougar not far from where we stood.[5]

He gaped, wide-eyed, then burst out with, "I heard about that! The Department of Ecology reported that he was in treatment with steroids for his transgender work, and that it was the steroids that attracted the cougar."

"Total bullshit!" I blurted out in response. "Not at all. The cougar was emaciated from eating prey reduced to feeding on plants grown in *Toxic Sewage Sludge*, and that cat was nothing but skin and bones. It was very sick."

You see, I knew that cat well. It had followed me often when I was tracking and imaging sludge dumping in that same area. I had thrown a few rocks to keep him at bay while I did my work only a month or so before the biker-eating incident. All he wanted was some easy prey to eat, such as me, a hiker, a dog, or a biker; something easier to catch than a deer. The deer were now mostly gone from the area, which is not as it used to be prior to the sludge dumping. The county coroner had parked in my favorite spot when he and the sheriff investigated the gruesome scene at the Gate 10-11 Cutoff.

"I know another cat in the area that is equally ill," I added, as I biked off to go deeper and farther uphill before it got dark. When I glanced back he still stood there, eyes wide open, now another informed and wary forest interloper, but now more alert to such agency-induced dangers.

When we enter public or private forests as hikers, bikers, dog-walkers, mushroom gatherers, hunters or fishers, or simply to enjoy the solitude of a forest, we are the unsuspecting victims of the toxic waste disposal behaviors and practices of our municipal, county, state, and federal agencies, for by such toxic waste disposal works, the agencies set the stage for converting us from being enjoyers of the forests to now becoming the prey victims of the planned and incited toxicity imposed on the wildlife therein, rendering them the impaired predators of us. In this specific case, it would have been very useful and helpful to have had the guilty parties at least indicted for involuntary manslaughter, which would have opened the deadly toxic sludge disposal practices to the daylight of a serious public investigation. At least the practice and the consequences would have had to bear public scrutiny in the media, with the agencies and their leadership forced to rationalize their bad behaviors. In any case, the situation was buried and the agencies continue to carry on, as their crimes become shrouded in convention once again.

Toxics that enter wastewater treatment plants or publicly owned treatment works in raw sewage, including toxics derived from the accompanying solid wastes or physical inerts, precipitate and become a part of the solids fraction that in turn become part of the *Toxic Sewage Sludge* product and thereafter of the environment. By means of natural physical-chemistry processes, toxics are extracted by the waters in raw sewage inflows, and from the embodied manufactured or physical inerts. The toxics extracted from the wastewater inflows then migrate into and through the WWTP or POTW. Those inerts, laden with their residual full monte of any and all adhered toxics then become the toxics that inevitably, when land or water disposed, impact those higher up the food chain, such as the cougar and its unfortunate human prey.

Toxics never get away; they always get to us in oh so many ways. The raw sewage, the physical inerts screened from the raw sewage, the sludge solids or precipitates from the raw sewage, and the resulting wastewater effluents, *Toxic Sewage Sludge,* and landfill leachate products are all sources of toxics. Ordinary rainfall, snow melt, and waste-generated waters of decomposition within landfills, cause the resulting leachates to flow to ground and surface waters, with some part being piped back to the WWTP or POTW where they are once again combined with the *Toxic Sewage Sludge* from the new raw sewage or wastewater inflows, further rendering such sludges as being even more toxic as the cycles are repeated, *Toxicus ad Infinitum.*

The land disposal or landfilling of any toxic wastes, including municipal, industrial or medical wastes, non-recyclable plastics and papers, artificial turf or construction wastes, *Toxic Sewage Sludges* and septage, wastewater treatment plant grit, petroleum refining sludges or factory farm manures, plus any and all surplus, out-of-date or used chemicals and solvents, in any form, including discarded or used asphalt, manufacturing adhesives and forbidden pesticides, is not advocated by the authors, yet is invited and encouraged by the agencies, without limitation. These are toxic wastes that must be processed by thermal decomposition or by other thermolytic means to render such wastes and their residual toxics as being less-toxic or non-toxic by the best available means. In this modern era, these objectives can be met. As mentioned, all that is required is the political will to do so, plus an open recognition of the fully anticipated adverse consequences to human and environmental health by a failure to perform.

According to the Puget Sound Nutrient Reduction Project from the Washington State Department of Ecology, there are currently over 4.5 million people living in the Puget Sound region. The Puget Sound Regional Council estimates that about 1.8 million more people will move to the area by 2050.[6] This likely means that more than a 40 percent increase of nutrients to Puget Sound will be contributed just by these population increases. How will this further impact salmon, orcas, and many other species, including ourselves, if those responsible for monitoring waste disposal in Washington State, as only one example regarding this nationwide concern, do not confront the issue? By doing business as usual, what legacy will WA government entities pass on to future generations should our fish populations, by 2050, implode? A government report describing conditions in Puget Sound in 2019, indicates the following:

- Puget Sound health is critical to recovery of salmon, orca, and other marine life.
- Oxygen levels in Puget Sound and the Salish Sea are below levels needed for fish and other marine life to thrive.
- Nutrient pollution from human activities is worsening the low oxygen levels.
- Excessive nutrients in rivers and from point sources flowing into the Sound [such as from municipal wastewater treatment plants] deplete the dissolved oxygen below the minimum water quality standards.[7]

As stated, no one knows what's in *Toxic Sewage Sludge* or what new toxics are formed in it, or how it flows in soils, wetlands, streams, rivers, lakes or the ocean, or how it adversely affects forests, farms, rangelands, wildlife or human health, or how it accumulates in water, or how it bio-assimilates and bio-accumulates in foods, including in animals used for food, or what may be its fate in the environment. And what then are the air-borne toxics that are disseminated around the region chosen for waste disposal? If extreme toxics are found commonly in *Toxic Sewage Sludges*, fertilizers, pesticides, and from industrial and transportation wastes, such as and including from automobile, truck, train, ship, and aircraft sources, as well as from power plants and the Hanford Nuclear Reservation, including their water discharges, leakages, and air-borne dusts, what then must be in those air-borne and otherwise wind-blown deadly dusts from all these sites, which can loft for miles with the prevailing winds?

The answer is clear: What's in it is nearly everything, and that toxic everything is nearly everywhere.

"Externalities" are described as an economic analysis that compares the costs of premature deaths from pollution versus the value of the economic activity that caused the pollution. In the case of land-disposed *Toxic Sewage Sludge*, obvious externalities are being ignored. These are the costs that are not accounted for yet paid by society. For example, fine particulates, as wind-borne contaminants and pollutants from air blown across endless mounds of *Toxic Sewage Sludge* that spread across the vast farmlands of Washington, may lead to stroke, heart disease, lung ailments, cancer, etc. Or vectors such as flies and mosquitoes may carry serious diseases. Not forgetting that SARS-CoV-2 has been identified in all wastewater effluents and *Toxic*

Sewage Sludges measured since the very onslaught of the pandemic.[8] What could be that unseen cost to unsuspecting citizens? Is it in the billions of dollars, or has it already reached the trillions of dollars, and that's just in the United States?

If our key questions are: "What's in it?" and, "What is its fate in the environment?" then we cannot discover the answer without public and agency concern, plus a functional infrastructure that does not fear the outcome of a critical and objective analysis. In the case of *Toxic Sewage Sludge*, we now know, for example, that there is never a chemical analysis, never a microbiological analysis, and never a toxicological evaluation performed. The answers to the fundamental questions: "What is it?" "What's in it?" and "What does it do?" remain unrevealed, because the questions are never asked, and therefore the answers are never provided!

Toxic Sewage Sludge is solid waste, simply renamed by altering the definitions. Fecal matter and other organics in *Toxic Sewage Sludge* are reduced by aerobic and anaerobic digestion in the WWTP or POTW process, leaving behind chemical cocktails and solid wastes, such as and including plastics, glass, and metals, along with poorly soluble toxic chemicals, plus new toxic chemicals formed *in situ*, as well as toxin-tolerant and anti-microbial agent-resistant microbes, also formed *in situ* during the wastewater processing procedure.

What we know is there is an implied conflict between commerce and public health, near-term, as a matter of public health and safety, and longer-term, as a key inciting agent of climate change. We also know that while agencies may be good at permitting, they are wholly inept at even the idea of innovation, change, and improvement toward public health, or even a consideration of the environment as being integral to an affected biome with physical and chemical determinants, or to beneficial outcomes or progress. Those issues are left to academics and the public, often as environmental or health activists, seldom with but scant support. Therefore, these additional unanswered questions remain:

- What are the human and environmental health consequences of land-disposed *Toxic Sewage Sludge*?
- Why is there conflict between commerce and public health?
- Do we have an implied healthcare imperative?
- Who authorizes this destructive behavior?

And further:

- Considering today's well-published common knowledge about the direct associations between toxic exposure and human disease, and that we are experiencing a global explosion of cancer and other chronic diseases, why then would anyone or any agency knowingly dispose of *Toxic Sewage Sludge* or other toxic wastes on rangelands and farms, and in our forests?
- What are the political and financial benefits, and who exactly benefits?
- Why would anyone or any agency dispose of toxic wastes on alluvial flood plains or atop aquifers, or adjacent to wetlands, streams, lakes, and rivers?
- Why are the ESA, CAA, CWA, and RCRA packed with "exceptions," such that they cannot be enforced, most especially with regard to the toxic products of wastewater treatment plants, landfills or land-disposed *Toxic Sewage Sludge*?
- Who will build the first ever "Safe and Effective" sewage sludge processing plant?

To summarize:

There is an implied conflict between commerce and public health – Nearly all chronic diseases are incited by the cumulative effects of long-term exposure to low levels of environmental contaminants and pollutants, a statement heralded far and wide for decades, albeit to unreceptive ears. Air, food, soil, and water are essential resources that assure the sustainability of human and other life forms and must be protected at all cost; they are matters of national security.

The Challenge – Toxics are not safe! Toxins and toxicants, collectively toxics, are poisons that incite nearly all human chronic diseases. Domestic, industrial, commercial, and medical wastes, plus stormwater runoff and landfill leachates are the primary sources of toxics in our living environment. All that is required for their proper management is political will. Primary sources of toxics that can be managed effectively originate from domestic and industrial *Toxic Sewage Sludges*, septage, wastewater effluents, landfill leachates, and agricultural wastes, all of which contribute to water contamination and pollution, greenhouse gas emissions, and climate change.

It's Unknowable – The chemistry of *Toxic Sewage Sludge* remains unknown and is largely unknowable. Overall, *Toxic Sewage Sludge* contains all that goes down any toilet or drain, or off any roadway or parking lot, including that which flows in industrial, domestic, and medical wastes, stormwater runoff, and landfill leachates, all then to accumulate collectively as combined sewage inflows or raw sewage.

So, "What's REALLY in it?" – NEARLY EVERYTHING!

- **Biochemicals**: Chemicals characterized, produced or involving chemical reactions in living organisms (some foods, as well as some pharmaceutical and personal care products).
- **Chemicals**: Substances, elements or compounds obtained by a chemical process or used for producing a chemical effect (some pharmaceutical and personal care products, plus some industrial and commercial products).
- **Metals**: Heavy metals, ionic or salt forms of metals, or conjugates of metals, within the groups of alkali metals, alkaline earth metals, transition metals, basic metals, rare earth elements, and actinides.
- **Plastics**: Synthetic or semi-synthetic organic polymers derived commonly from petrochemicals, with variants made from renewable materials, such as polylactic acid from corn, or cellulosics from cotton linters.
- **Microorganisms**: Microscopic organisms that may include multiple species, forms, or types of viruses, bacteria, archaea, protists, oomycetes, and fungi, and that may or may not include infectious or pathogenic forms, with some forms that may or may not be resistant to antimicrobial agents or medicines, or that may produce microbial toxins or other virulence factors.
- **The principal ingredients of *Toxic Sewage Sludge* are:** Oil, fat, silt, grit, hair, bone, sand, fungi, feces, fibers, plastic, grease, viruses, bacteria, oomycetes, parasites, pesticides, chemicals, plant parts, metal parts, biochemicals, heavy metals, radionuclides, produce stickers or other industrial inerts, and more.

- **From another perspective:** *Toxic Sewage Sludge* contains chemicals that volatilize to generate greenhouse gases, endocrine disrupting chemicals, organic and inorganic particles, partially decomposed feces, nanobeads or microbeads, pharmaceutical products, pharmaceutical product excipients, personal care products, persistent pesticides, pathogenic microbes, toxic heavy metals, toxic chemicals, domestic waste, industrial waste, medical waste, immunotoxins, mesoplastics, microplastics, nanoplastics, mutagens, carcinogens, neurotoxins, hormones, hormone disruptors, epitoxins, and more.
- **Key categories of toxics in *Toxic Sewage Sludge* include:** Persistent organic pollutants (POPs), chemicals of emerging concern (CECs), and persistent bioaccumulative toxics (PBTs). Tens of thousands of synthetic POPs, CECs, and PBTs are known to accumulate in sewage sludge, including, PFAS, 1,4-Dioxane, dioxins & furans, nonylphenols (NPs), polychlorinated biphenyls (PCBs), brominated flame retardants (BFRs), and polybrominated diphenyl ethers (PBDEs).
- **Resulting in:** More than 70% of toxic chemicals detected in sewage sludge are also detected in humans, with the key characteristic that they concentrate in *Toxic Sewage Sludge*, are persistent in the environment, are extremely bio-accumulative in fatty tissues, and include degradation products that may have increased toxicity or other toxicities, and further, that they are synthetic and rarely occur as single compounds.
- **Which is to say:** *Toxic Sewage Sludge* may include any and all waste materials in any form that may enter any drain that flows to any wastewater or sewer system, including all municipal, industrial, and manufacturing wastes that include metals, chemicals, landfill leachates, pharmaceutical and personal care products, medical wastes, stormwater runoff, fertilizers, pesticides, infectious agents or biological materials, whether manmade or natural, plus new chemicals and new forms of microbes created *in situ* within the sewage conveyance system or within a wastewater treatment plant or publicly owned treatment works, or that is a discarded waste product, plus any or all of the derivatives thereof.

So what do we know?

- Nearly all chronic diseases are incited by the cumulative effects of long-term exposure to low levels of environmental contaminants and pollutants.
- Wastewater treatment plants do not "treat" wastewater; they process raw sewage, including landfill leachates, into toxic wastewater effluents and *Toxic Sewage Sludge*.
- Discharge of toxic wastewater effluents into our waterways must end.
- Land disposal (and landfilling) of *Toxic Sewage Sludge* must end.
- Environmental issues are public health and national security issues.
- Land-disposed *Toxic Sewage Sludge* generates greenhouse gases through the sewage sludge volatilome that incite and fuel climate change.
- Land-applied *Toxic Sewage Sludge* may be the most pollutant-rich material created in the 21st century.
- The Federal Clean Water Act defines sewage sludge as a pollutant.
- Land-disposed *Toxic Sewage Sludge* from industrialized urban centers contains nitrates, phosphates, toxic metals, and chemicals that flow through the food chain and persist in the environment.
- Wastewater treatment plants process wastewater, which yields: Solid waste (physical inerts), toxic wastewater effluents, *Toxic Sewage Sludge*, and gases (CO_2, CH_4, H_2S, N_2O_1, etc.), none of which are safe for human exposure.

The accelerated generation of wastes is the outcome not only of population growth, but of increasing affluence and commerce, as well, with greater impact than could ever have been imagined. Thus, toxic wastes, including *Toxic Sewage Sludge*, which are not amenable to beneficial reuse, must be processed by thermolytic decomposition into renewable clean energy and useful by-products, such as and including heat for the generation of clean energy and valuable ash, phosphates for fertilizer, and construction materials. By current standards and methods, we are experiencing a dangerous form of strategic institutionalized pollution. It is the case whereby industry, agencies, legislators, environmental groups, environmental law firms, academics, and activists deny that toxic wastes are toxic or dangerous, but rather advocate that they are safe and beneficial.

Why is that? I'll assume there are those who are paid to do so. Therefore, be bold, "SPEAK TRUTH TO POWER,"[9] stand up for what's right, or go home!

CHAPTER 13

SMOKING GUN

The world is a fine place and worth fighting for...

Ernest Hemingway, *For Whom The Bell Tolls,* 1940.[1]

The law firm's environmental consultant stood in the shallow forest wetland wearing waders, with the assisting lab tech in high boots alongside. Together they looked at the deeply sludged, sickeningly smelly waters draining from the wetlands into a pair of small streams, one headed east, one west. The consultant was now hand-motioning to the rest of us, two environmental attorneys, the executive director of a local environmental organization, and me, without saying a word.

He was looking at the heaps of saturated *Toxic Sewage Sludge* leaching, who knows what, into the slight downslope of the wetlands before us. His right hand was poised with index finger sticking forward, his thumb raised as if cocking the hammer of a revolver. He flexed his trigger finger a few times, silently gesturing to the rest of us the common sign for, "Smoking Gun," or "Gotcha!"

We were staring collectively at a forest wetland filled with *Toxic Sewage Sludge* and watching the colorful, toxic leachates flow along the natural contours of the forest and into two streams. One outflow of contaminated water winds its way for about a mile east through a series of culverts under the county access road and into the north fork of the Snoqualmie River, before tumbling over enchanting Snoqualmie Falls several miles downstream. The other putrid outflow snakes west across the wetlands for about 1,000 feet, into a culvert under the main haul road and on to Ten Creek, then through additional culverts into Tokul Creek, thereafter through a state fish hatchery and on to the same Snoqualmie River, but this time below the falls.

The point of his excited gesture was that this single isolated scene had all the markings of clear violations of the Clean Water Act

(CWA), the Endangered Species Act (ESA), and quite possibly the Resource Conservation and Recovery Act (RCRA), most notably because both wetland outflow streams drained their toxic residuals through rivulets, ditches, and culverts under numerous roadways on their way to the Snoqualmie River and Puget Sound, with all waters betwixt and between considered sacred salmon sanctuaries. Ultimately, all of these thoughts of clear violations would be ignored, for convenient, agency-led toxic waste disposal rules the day, excepted from any codes, regulations or laws, all by plan and intent.

As we all looked on together at the thick, black, fly-covered sludge slowly draining into the receiving wetlands and exit stream waters, the scene was shocking, the stench horrendous, and the culprits clearly identifiable. I was there on-site during several days of *Toxic Sewage Sludge* dumping, so I bear direct witness thereto.

This typical, filthy, and toxic scene is repeated many times throughout the Snoqualmie Forest. It is a planned and intended direct consequence of the county's scheduled land disposal of their *Toxic Sewage Sludge*, as permitted, guided, encouraged, and approved by the DOE and the King County Wastewater Treatment Division. The EPA says it's ok, as well, as do several alleged environmental groups that seek agency nods of approval and likely some grant funds in exchange for adopting agency miscreance. Local environmental groups elect to remain oblivious to the fact, or abstain from involvement or the knowledge thereof. Two alleged environmental groups actually participate in *Toxic Sewage Sludge* dumping practices in these forests, while also hosting business leaders and state and county officials as partners and board members.

The law firm's principals and consultants all knew very well that this one site, and this singular event, along with the many additional sites and events nearby that we visited that day, were all in violation, but with little recourse, and with even less of a chance to garner any statutory or courtroom attention or remediation. "That's the way we do things," as the county tells us. "It's safe." Please go tell that to the dead things in the forest. And by the way, this exact site is within a hundred yards or so of where the sludge-sickened cougar attacked the mountain biker.

That day, and to their exhaustion, I guided the shocked and depressed group to several additional deeply sludged, flowing water, wetland sites in that same forest tract. It was important to make the point that this was not a "one off," or an "isolated incident," as proclaimed so often by the agency offenders, but that this occurrence has been the typical norm and a common practice for *Toxic Sewage Sludge* disposal in these and so many other forests, from planning through execution, for decades. Some of the sludge heaps were measured at a few inches or so, but in other areas, at depths of 12 to 20 inches, all as a stinking environmental disaster. Not so good for forests, water or wildlife, or for human and environmental health overall, nor for the people directly downstream.

Not to worry, "It just goes away," we are told. The problem is, some of the dump sites in these same forests have been visited by me personally for as long as twelve years, and it's still there. Visually, it's diminished to a great degree, I'll admit, but it's still there. It doesn't just go away. Toxins don't just disappear. That's a fictional marketing ploy fabricated by the agencies and their industry-lobbying environmental organizations. So just what is the "it" that goes away?

Next we drove north along a county access road for several miles, then headed on foot down a short side trail for a three mile trek up the main haul road to another recently and heavily sludged location near Mudd Lake. The point was to further demonstrate just what was going on in the Snoqualmie forests every year, nearly all year, as a common practice and plan, so long as the weather and forest access roads permit. Without bad weather blocking the activity from time to time, there would be no controls on their rampant *Toxic Sewage Sludge* dumping in the forests at all.

On another day, and within this same general polluted forest location, this time with a local mayor and staff in tow, along with county councilpersons, a different set of environmental consultants, King County Wastewater Treatment Division personnel, and State Department of Ecology staff, we all headed up an adjoining forest access road, again to demonstrate the toxic insult against these precious forests.

Oh, what a day! As I began to lead the group left into one of the many forest access side trails, we heard a troubled shout from arrear: "No! Not left; we go right here on this other trail."

"No!" I asserted in kind, "We go left here. It's a shorter trip, and it better represents just what's going on out here." They all agreed, so left we went. Scowls abounded from the "No" person and her agency friends at the rear.

The reason for the long faces on the agency members was that to the right, on their preferred path, was what I refer to as a "set-up fake forest land-disposal site," which, and whereupon, they like to tell us, "had a fine dusting of dried 'Loop®' on the surface of the forest soil, far from any waters." Huh? A fine dusting, my ass! It's an obvious set-up! It's fake! Since when is a foot or a foot and a half of wet, stinking, *Toxic Sewage Sludge*, within and adjacent to slow-flowing waters, a fine dusting? Not by my reckoning!

They had been using that "demonstration site" to cover their tracks for years, but not this time! We veered left, including all dozen or so guests who did not really want to be there for any reason. Within a quarter mile, the stench was more than could be imagined. There before us was yet another slowly flowing wetland, fully dumped with *Toxic Sewage Sludge* which, as measured with my broken yardstick from the local hardware store, was anywhere from a few inches to as much as 21 inches deep. Again I said with glee, "A fine dusting, my ass!"

I had them follow me into the stinking mess of a previously uncontaminated forest where we wandered about taking images, measuring depths, viewing dead things, shooshing mosquitos, flies, and gnats, and wanting desperately to get the heck out of there. And, by the way, the very presence of the said mosquitos, flies, and gnats is in and of itself a violation, for the sludge generating people are required to assure "vector control," which declares that insects and other vectors of disease will be eliminated or otherwise controlled, so as to preclude or at least manage the presence of insect vectors of disease. The usual next part was totally anticipated, well-rehearsed, and could never be resisted by me. I only had to wait until it erupted, and it did, just as expected. Wait for it . . .

The agency "No" person blurted out, "This is just an isolated incident!"

"If this is an isolated incident," I instinctively responded with my well-rehearsed script perfectly on que, "then hike along with me right now and I'll take you to a dozen more such 'isolated incidents' nearby." And, as had also been rehearsed so many times before, I added, "So, how many isolated incidents does it take until such incidents are no longer considered as being 'isolated,' but are in fact considered to be the norm?" No one wanted to see, hear or think about it anymore, or go any further. Case closed! They filed out of the sickening forest scene and went home. All was quiet. Almost always I was ignored, for the truth and the facts, when witnessed, usually hurt.

I repeated the same fully anticipated and fully choreographed sequence of events multiple times over the years, each with a different audience; sometimes with members of environmental groups and foundations from California and other states, all of whom gaped in disbelief, and sometimes with clutches of citizens, other times with representatives of law firms, and sometimes with friends or family who just wanted to see for themselves. Some were total skeptics beforehand, but who thereafter became converts.

A particularly interesting and instructive visit to these forests was attended by some neighbors who were activists with several local environmental groups. They had endured more than one of my local town hall presentations and wanted to know more. As we walked into the forest, I could hear them behind me wondering aloud: "What's this?" "Look at that!" "I know what this is." "Look, here's a veggie label, and a label from an orange. It's from Mexico." "Here's a Band-Aid, and this looks like gauze from a bandage." "This is a wrapper off a soup box," etc., etc.

They soon stopped and stared at me in disbelief, mouths and eyes agape, recognizing that what they were seeing and smelling in real time, on-site that day was exactly what they had seen during my PowerPoint presentations to the community. Now, however, they were firsthand witnesses to actual forest-disposed sludges, seeing the defilement *in situ*, up close and personal, and now it really hurt. "Why do they dump so much plastic trash in the forest?" they queried.

When we filed back out of the smelly scene, one of them asked, as we passed a large pile of brown stuff, "What's that? It stinks, too!"

"That," I replied, "Is a dump truck load of *Toxic Sewage Sludge* from the local wastewater treatment plant not far from here. They told me directly that they generally just drive out here when needed and dump their load at any convenient site."

I had seen and heard it all before, over so many years, while simply hoping people will finally grasp the significance of the bare fact that anything and everything that flushes or drains from any municipal toilet, sink, floor drain, laundry machine, garbage disposal, laboratory, shower or storm drain, all flows to a wastewater treatment plant—whether from industrial, highway, medical or domestic sources—thereafter to break down many times along the way into various chemicals, materials, parts, and pieces, all laden with nearly any and every toxic substance and infectious agent known or as yet unbeknownst to anyone, and all by the intent and planning of our agencies. And there it was for them to see all over the forest floor, firsthand, right there in the Snoqualmie Forest, in freshly dumped *Toxic Sewage Sludge* loaded into a wetland that flows from there to thereafter, easily with enough plastic pieces to fill a ship.

Unfortunately, the end result is nearly always the same. No one really seems to care, once out of sight of the sick or dead animals, crispy plants, contaminated and polluted soils and waters, missing microbial life, and the like, unless, of course, you may be well-versed in any of the disciplines of ecology, ecosystems, hydrogeology, sanitation, public health, infrastructure, infectious diseases, cancer, immunosuppression disorders, endocrine disruption, toxicology, or the like. If people get sick and die from exposure to toxics, or from the myriad of chronic diseases incited thereby and thereafter, including today with the potential of a highly contagious virus, those things happen way after the fact, so they must be someone else's problem.

As well, it must be ok if our agencies and the lawyers and the large environmental groups don't mind, so who are we to complain? It must be safe. They're in compliance, and you can't prove it to be otherwise. It must just go away, as they say. But can we prove it? Yes. And does it just go away? No, it does not, it simply flows downstream to do its dirty work elsewhere and upon unsuspecting others.

Let it be well understood, the infectious disease pathogen load of *Toxic Sewage Sludge* has global impact. It is a well-known source of danger, and it is a public health hazard that can only lead to a greater public health crisis of global proportions, so long as there is no agency remedy or mitigation in sight.

Life as we know it is in a precarious situation, whether positioned by our failures at waste management, or by our tinkering with matters affecting global climate. In all cases, these insidious issues that pose risk, threat, and harm are upon us, and we must take a stand against the *status quo*, if we are to see a favorable outcome. The task is in our hands now. The bare facts are on the table for all to see.

Chapter 14

A World Smothered In Plastic

Plastic waste is now so ubiquitous in the natural environment that scientists have even suggested it could serve as a geological indicator of the Anthropocene era.

UN Environment Programme, 2018.[1]

If we're not thinking and talking about local, regional, and global plastic pollution, plus contamination of just about everything with micro- and nanoplastics, or the open disposal of mixed waste plastics by landfilling, along with the universal failure of plastic recycling, then we're not taking into account some of the key matters of public health that threaten us all today. As predicted by the World Economic Forum, it is estimated there will be 1.3 billion tonnes of plastic waste present in the environment by 2040.[2] In light of the equal prediction that global plastics production and use are anticipated to accelerate over the next few decades, where do we go from here?

Waste plastic contamination and pollution are growing globally, and at a rate far greater than human population growth, with no abatement in sight. A key problem is that most plastic items contain a suite (or a plethora) of toxic chemical additives—bisphenol-A (BPA) or polybrominated diphenyl ether flame retardants (PBDE), are two examples—which can cause serious health concerns. Endocrine disruption is just one of the negative outcomes of a world being smothered by plastics. Without question, exposure to toxics in plastics adversely affects human and environmental health.

As of this writing, the recycling rate for plastics globally is reported to have declined to about 5% – 6%, primarily as a result of the resistance of some nations to accept the adverse consequences. "The decline coincides with a sharp drop in plastic waste exports, which had counted as recycled plastic. China and Turkey have since implemented plastic import bans and other countries set plastic waste contamination limits under the Basel Convention Plastic Waste Amendments, which the United States did not ratify in 2019," states an article in Reuters, May 4, 2022.[3] Good news for those countries but unless we find a way to clean it all up, it's bad news for the U.S.

Under the heading: "Unnerving Plastic Waste Stats & Facts," the blog *SeedScientific* provides the following 2022 plastics update:

- Nearly 300 million tonnes of plastic wastes are produced every year.
- Around 70,000 microplastics are consumed by an average person each year.
- 1 million marine animals die due to plastic pollution each year.
- It takes around 500-1,000 years for plastics to decompose.
- The U.S. is the world's top generator of plastic waste.
- Plastic wastage is growing at an annual rate of 9%.
- 73% of all litter on beaches worldwide is plastic.
- 75% of all plastic produced has become waste.
- About 91% of plastic is not recycled.[4]

Microplastics and nanoplastics, generated by the decomposition of waste plastic through oxidation, UV irradiation, chemical decomposition, and microbial degradation, have become dominant contaminants of our air, food, soil, and water, as well as of our very cells, tissues, and flesh, and a topic of intense conversation. The bare fact that agencies managing our solid and liquid wastes have responsibility for our air and water quality, but seem to take little action on the matter of plastic pollution, is cause for serious concern. The heavy lifting is left to academics, concerned citizens, and a few environmental groups.

From the plastics production side, industry has stepped-up to a certain extent, for their survival depends heavily upon the plastics continuum, from fossil fuel pumping to the discarded plastic straw and its fate thereafter. The public is aware and watching and, as we all know, the first step in addressing any problem is recognition that the problem exists. The flaw between the heavy lifting and the public is the agencies, and they continue to be the light lifting factor that denies any recognition of the challenge, at all.

We have known about this matter of plastics pollution for decades, yet all energy seems to have gone into supporting industry objectives. A key step by any agency would be to end the land disposal of *Toxic Sewage Sludge*, a significant source of so much of the contaminating and polluting plastics of any and every chemical makeup or size. Sludge-borne plastics, be they macroplastics, microplastics, nanoplastics or even smaller, compromise our medical well-being, and they invade all life forms to cause havoc as the nano-transporters of toxic chemicals, a field as yet less studied.

But wait, let's go back for a moment to my forest adventures where field discussions with agency employees usually went along the same lines. There we were, either in a recently sludged forest, or in an office, or in their truck, or at a meeting or presentation, when I would politely broach the "delicate subject."

"What do you think may be the adverse consequences on wildlife and on the environmental life-support system out here from all of the plastics, macro-, micro-, and nano- that you dump in our forests and waters, with much of it bearing saturated toxic loads, in combination with all of the toxins in the sludge itself?" I'd ask.

The usual response was sung like a hymn, entitled, "That's the way we do things," as could have been crooned by Al Capone: "*They can't collect legal taxes from illegal money.*"[5] "It's not toxic, it's safe. It's not *Toxic Sewage Sludge*, it's biosolids. It's a fertilizer replacement. It's a soil amendment. It's beneficial reuse. It just goes away. We do not dump in the forest, it's land-applied. It does not flow to ground or surface waters. It does not become wind-borne as toxic dust. It does not harm people or wildlife. It does not harm plants. It builds soil structure. It remediates glacier-stripped soils. It does not contain harmful microbes or infectious agents. It supports the life of soil microbes. It does not contribute to climate crisis, it captures and sequesters atmospheric and sludge-generated carbon." And of course, "We're in compliance," as if read off a teleprompter by the boss.

It all starts at the local level, where I would show a few images of small birds in freshly dumped sludge, unable to fly, or already dead, and then a few snapshots of dead snakes, moles, shrews, rabbits, worms, nematodes, insects, and the like, for emphasis, all within a sludged forest. "One dead rabbit doesn't make for a disaster," was one noteworthy response from a dedicated state agency person, soon thereafter to become a King County sludge agency person.

"The deer are sick, and the emaciated cougars are not acting themselves, and the mosses, lichens, ferns, and such, are now brown," I would add.

"Oh, we applied herbicides along here, and that killed the plants," they sang. "They'll come back." That was in 2014. I returned there recently, and the dead plants have not recovered, and traces of *Toxic Sewage Sludge* persist to this day—it's there forever.

"Are you noticing how the birds collect the veggie stickers and the shredded and whole plastic packaging materials, and the millions of plastic fibers and hairs in the sludge, most likely for nest building?" I would ask. No answers. "It's only plastic," was the mood. Yet I have watched birds collect veggie stickers (produce labels) from sludge piles, likely flying off with their finds to make nests—toxic nests.

"Wipes are really common in this sludge," I would continue, "and the bazillions of unidentifiable plastic fragments and container parts and Band-Aids that are nearly beyond recognition."

On an expedition sometime later, I was told by one wastewater treatment plant employee that I had cost them about $3.5 million to replace their screens with a finer mesh to catch more plastics and other larger debris that usually go to the forests and farms. "Physical Inerts" or "Manufactured Inerts," they are called, and if captured, they are sent to landfill—not an improvement, just another relocation of toxic waste. The millions of tons of plastic that WWTPs/POTWs send to forest and farm, and ultimately to ground and surface waters, as well as to the sky as lethal nanoplastic dusts, are all subsequently targeted to human and other life forms, notably to our collective detriment and potential demise.

The solution is simple, build one WWTP that actually works. Build it as a demonstration project, regardless of cost, and not by the whims of the lowest bidder. Show that you are serious about matters of human and environmental health, and that functional infrastructure can be achieved. Set plans thereafter to build 100 more fully functional WWTPs that also work.

Public health is the objective, not to be compromised as a simple matter of the budget. The logical challenge is nothing more than a declaration of the basic need for fully functional infrastructure, and that can be done. Do leaders in government who are the enactors of such actions wish to be seen as taking infrastructure seriously as a matter of public health, or will they be remembered as the orchestrators of operational failure—to be recorded so in history?

In the local case, at state, county, and most municipal levels, it appears that basic infrastructure is at the bottom of the list of priorities. So long as state and federal agencies allow for the contamination and pollution of our air, food, soil, and water, why would anyone want to waste good taxpayers' money on safe, sane, hygienic, and proper treatment of our toxic raw sewage wastes in an elaborate, modern WWTP or POTW? Public health appears to be little more than a distant, superficial concern, too often ignored, and with nary a thought towards prevention of disease or protection of our living environment.

"Just as one dead rabbit does not make for a disaster," I recall thinking to myself, similarly, "one replaced sludge screen does not make for an improved WWTP." With a silent shrug I kept it to myself that day, as I wondered just who back at the office dreamed up this handy slogan.

From my perspective, if the local cougar ate the sludge-sickened rabbit and became sick and emaciated, and was therefore rendered unable to hunt its usual prey, and then decided to eat an unwary human biker, how do the agencies get to the storyline that the said biker was going through a hormone-induced transformation and thereby enticed the cat to eat him? This is not intended to be funny, just offhandedly sardonic. How many dead people will it take to force agency action to end the land disposal of *Toxic Sewage Sludge*? I hope, at least, that the State Attorney General is checking the bank accounts of the agency officials managing the business of land-disposed *Toxic Sewage Sludge*.

Prevention of disease is not on any agency's list of priorities, and so the better interests of public health get lost along the way. Agencies love to state after the fact that they, "Will be sure it never happens again." However, and while "An ounce of prevention may be worth a pound of cure," as popularized by Benjamin Franklin so long ago, that idea appears to hold little meaning. Where is that ounce when we are literally destroying ourselves with our own toxic wastes, and by the hundreds of thousands of tons?

The global plastics challenge to which we are confronted today, and which may not be reversible, remediable or curable within the lifetime of any living creature, human or otherwise, is of our own doing, and we seem to like it. We are the direct consequence of us.

The hypothetical, theoretical or sensible solution to this dilemma is that we either put on our big boy pants and decide as a global community to end the production, manufacture, and use of any and all conventional plastics and plastic products (not going to happen), or we are obliged to pour more energy, resources, and thoughtfulness into the discovery and use of wholly biodegradable, non-fossil fuel-originated, and non-toxic, plastic-like materials, which does appear to be happening in industry, academia, and federal labs; bravo for them, albeit all too slowly.

When people dug holes in the ground and discovered oil for commerce in Pennsylvania and Texas way back when, cars appeared and fossil fuel use became an issue, as in automobile, truck, train, plane, and shipping fuels. Shipping is the biggest user of fuel, not personal transportation. It's the shipping industry around the globe that represents the greatest fossil fuel and climate emergency challenges. With the largest manufacturing operations in the world located in China and India, we now ship more by sea than by air, even though aircraft are becoming more efficient. Electric vehicles (EVs) are becoming more common, as well, so fossil fuel consumption should be going down. Oddly though, crude oil production is increasing, and that's because today there is more emphasis on petrochemicals. In any case, we're not running out of crude oil, diesel fuel, aviation fuel, gasoline or coal. We have at least another ten to twenty thousand years' worth sitting below us. Who on Earth could possibly dig or pump it all? Big Oil is trying.

Big Oil does not sell gasoline for a living, they pump oil. And, now that they're pumping more, where is it going? To petrochemicals for yet more plastics. "How can that be? How can we want, need or have more?"

"Because we're not going to mine minerals anymore; we're going to replace metals with plastics, you see?"

"Well, shouldn't we recycle?"

"No, it's cheaper to pump anew from the ground and make more plastic than it is to recycle."

"Are you nuts?"

"Recycling is expensive. We should just burn it."

Big petrochemicals is on a roll, and no one dare impede them. With a global decline in transportation, coupled with the parallel discovery of new fossil fuel resources, the exploration for and exploitation of said new fossil fuel resources will continue for many decades to come, thanks to the endless current and anticipated new uses for plastics. Some quarters even see a decline of metals mining, smelting, and use, in favor of newer plastics. Therefore, we will continue to see a stream of research projects, science and popular articles, legislative threats, and implied consequences of anything that disrespects plastic.

This matter of plastic pollution is real; take a quick glance around the home or along supermarket shelves and one cannot deny our growing plastic dependency. Even the recent global pandemic has impacted the seriousness of the plastics pollution problem that cannot be whitewashed or hidden. As recently noted in *Science Advances*: ". . . the economic, public health and behavioral responses to the global COVID-19 pandemic have altered plastic waste generation, composition, and treatment practices in as-yet-unquantified ways in the United States and abroad, highlighting the urgency with which the United States must ensure proper treatment of its waste."[6]

The positive side to the story is that new plastics are planned to be less toxic and more biodegradable. Such plastics hold promise as construction materials of the future. As wisely stated in the article "51+ Disturbing Plastic Waste Statistics [2022]," published by *SeedScientific:* "Undoubtedly, plastics have helped make a vast array of commodities more affordable, adding vitality to many economies. The surge in plastic production over the decades has contributed to the prosperity of many people." Yet in the same article we are warned, "Unfortunately, the world's plastic appreciation has turned into an addiction."[7]

Chapter 15

Particulate Plastics In Poo

Unfortunately, the world's plastic appreciation has turned into an addiction. What was once instrumental in conserving nature is now destroying the planet.

Branka Vuleta, *SeedScientific*, 2022.[1]

Particulate plastics are disposed on land, on agricultural soils, and in forests via *Toxic Sewage Sludge*. What's more, the toxic consequences of ingestion, inhalation, bioassimilation, and bio-accumulation of the sludge-originated particulate plastics in the form of nanoplastics, microplastics, and mesoplastics, and that end up in our air, food, soil, and water, remain generally unknown. In any case, we are victims of plastic pollution in many forms, particulate plastic or other, no matter the source or cause, or our failure to contain.

Particles and fibers smaller than about five millimeters are referred to as microplastics. Microplastics were revealed in the 1970s when small particles of plastic were discovered floating in the North Atlantic Ocean. Richard Thompson later coined the term "microplastics" to refer to these small pieces of plastic debris.[2] A distinction can be made between primary microplastics, meaning microplastics as produced, and secondary microplastics that occur either during the use of a product, or that may have originated from larger plastic particles by weathering or by other natural degradation processes. Once plastic particles enter the environment, they pose a threat to ecosystems, as well as to human health.

One major challenge we face today is that critical pathways by which microplastics adversely affect biota remain largely unknown. Therefore, a key objective going forward is to categorize microplastics by size, shape, and chemistry, identifying those that are of the greatest biological concern. (Note: nanoplastics are 1 nm to 1,000 nm;

microplastics 1,000 nm (0.001 mm) to 5.0 mm; coarse microplastics 2.0 to 5.0 mm; and mesoplastics [macroplastics] > 5.0 mm.)

Next is to identify critical thresholds by which the revealed adverse biological effects may manifest. *Toxic Sewage Sludge* is most often disposed on agricultural lands where the practice pollutes our soils with many forms, chemistries, and toxicities of plastics by themselves, as well as with the adhered or adsorbed synthetic chemicals, heavy metals, personal care products, pharmaceuticals, and engineered specialty nanoparticles bonded thereto.

The plastics industry is estimated to be worth over $4 trillion, generating more than 400 million tons of plastic a year globally.[3] About half of the plastics produced are destined for single-use items only, meaning they will become trash nearly immediately. Without many functional and safe alternatives for reuse, recycling or repurposing, the United States now burns about six times more plastic than it recycles, with the combustion products therefrom presenting unique toxicity characteristics.[4]

It is clear that widespread land disposal of *Toxic Sewage Sludge* from municipal wastewater treatment plants onto farmlands likely represents a significant source of particulate plastics disposed in agricultural soils, with wholly unknown consequences for human health, sustainability, and food security. The degradation of larger plastic particulates into smaller, highly persistent bits, which are well within the micrometer and nanometer ranges, intensifies the problem.

Plastic particulate contaminants in sludges are carried along through wastewater treatment systems and processes, thereafter, to be precipitated in the total suspended solids that are generated within any WWTP. The functional result is that the larger portion of particulate plastics will be captured in the total suspended solids, which then precipitate into the sewage sludge, with lesser amounts being diverted to the wastewater effluents.

Particulate plastics may enter any WWTP from any source of plastic contamination. In the case of commercially produced, sludge-sourced soil amendments and fertilizers, yet another portion of the toxics-laden particulate plastics is provided to us through our food sources, either via our home gardens or from grocery stores. Such sludge-based products may be promoted as beneficial mulches, soil amendments, potting mixes or other, no matter if claimed "organic" or otherwise.

It is easy to view particulate plastics in *Toxic Sewage Sludge*, as I have documented in thousands of digital images and videos. They are seen easily as plastic-like, colored or white particles or fibers directly in any sludge sample, along with larger plastic debris, such as and including produce stickers, product labels, plastic caps, implantable contraceptive devices, catamenial devices, and all manner of plastic debris that is or may be at any stage of degradation. Larger pieces of plastic that enter the waste stream in any WWTP soon become macro-, meso-, micro- or nanoplastics, simply as a function of time and exposure, by the effects of mechanical, chemical, and microbial decomposition, plus by exposure to sunlight and oxidation.

The smallest of microplastics (MPs), and nanoplastics (NPs) for sure, are revealed most clearly when harvested and separated from *Toxic Sewage Sludge* as a colorless, oily like water emulsion in a glass lab vial, while the microplastics and larger fractions are revealed as white or light-colored specs and fibers, including when separated from dusts collected from schools and other public spaces in communities directly adjacent to sludge disposal, as we have done. The anticipated analytical chemistry results will be revelatory, for sure. And, while the micro- and nanoplastics can be toxic by themselves when inhaled into the respiratory tract, it is the adsorbed toxic chemicals that may wield the final toxic blow.

The vast array of possible and plausible synergistic toxicities resulting from the cumulative toxicities therefrom remain unimaginable. In the case of public spaces, and more so in rural areas, we can count on the cumulative adverse effects (toxicities) of all of the active toxic ingredients of land-disposed sludges, plus those from fertilizers and pesticides, all in concert with the inherent physical, biological, and chemical toxicities of the particulate plastics, as predisposing factors to a vast array of diseases. Lest we forget, the combustion of carbonaceous fuels, be they wood, fossil fuel or other, add to the pathology incited by what we breathe.

All plastics degrade and morph at will in any environment, always transforming to a lesser size with yet more proportionate surface area to which toxic chemicals may adsorb. It's the way of physics and chemistry. All plastics of any size that are flushed, drained, washed or otherwise discarded into any drain or sewer, will always accumulate in a WWTP, to be released thereafter to the environment at the whim of the laws of physics. Basic analytical chemistry challenges, and a strong reluctance to disclose, plus difficulties with sampling and separation methods, all hamper reliable quantification and analysis of the flow of particulate plastics through any WWTP, including what may be adsorbed or adhered thereto. This is especially true for nanoplastic and microplastic particles and fibers, both of which have distinct physical and chemical properties, as a direct function of their small size and extreme spectrum of aspect ratios.

Wastewater treatment plants are the modern-day interface between engineered and natural systems, whereby large amounts of particulate plastics from domestic inflows may include pharmaceutical and personal care products, macroplastic wastes, and tire and brake debris from stormwater runoff, all of which are destined to reach surface waters or agricultural lands that are the targets of land-disposed *Toxic Sewage Sludge*. These particulate plastics can impact soil ecosystems, crops, and livestock, either directly or through toxic endocrine-disrupting substances added during manufacturing. The substances contained therein or thereupon include short or medium-chain chlorinated paraffins and various plasticizers, which may represent as much as 70% of the weight of the plastic products or plastic debris.[5]

Plastics are polymers (materials or chemicals composed of many repeating subunits that form macromolecules) that may attract and accumulate harmful pollutants from the environment by adsorption or by another form of a chemical bond. Adsorbed or bonded chemicals, such as persistent, bioaccumulative, and toxic chemicals, include polychlorinated biphenyls (PCBs) and polycyclic aromatic hydrocarbons (PAHs). PCBs and PAHs represent a large group of manmade toxic chemicals that have broad industrial applications. They are highly stable chemicals that are resistant to extremes of temperature and pressure.[6]

The ecotoxicological characteristics of these ubiquitous chemicals are well known and reveal their threat potential, with emphasis on their bioaccumulation and toxicity in living systems. Other critical characteristics of their toxicity include toxicokinetics, which is to say, the kinetics of chemical uptake or adsorption of the toxics, as well as the kinetics of distribution, metabolism, and elimination of the said toxics, known collectively in the field of toxicology as ADME; toxicodynamics (their potency as toxic chemicals); their potential for accumulation in the environment; and their biotransformation into other compounds that may or may not display toxicity characteristics.

Photoactivation of PAHs within *Toxic Sewage Sludges*, for example, as when exposed to sunlight and the open environment via land disposal, adds further to their potential for increased toxicity, with much yet to be explored and determined. In any case, none of these products, or their physical, chemical or biological inclusions, can ever be proclaimed as being safe.

Of importance, the transfer of particulate plastics and the broad array of hazardous substances bound thereto from WWTP products, and thereafter transported to farm and table, has never been evaluated in depth at all, for that work alone could end the pernicious practice of land-disposed *Toxic Sewage Sludge* forever. Such assessment of the contamination of agricultural soils by particulate plastics and their chemical loads cannot be completed easily, even though agricultural soils represent one of the greatest environmental reservoirs of toxic chemicals known. Forget not, as well, that soil security is a matter of national security, for food security is a direct function of both soil and water security. Evaluation of particulate plastics in dusts blown on communities from agricultural soils, and most especially in the dusts that include particulate plastics and the raft of toxic chemicals and infectious agents adsorbed thereto, remain as a near untouched area of environmental health sciences, yet of growing national security and public health concern.

What's more, the mere presence of particulate plastics in wastewater effluent streams suggests they pose a serious threat to critical ecosystems. Studies investigating the removal of microplastics from WWTPs show that perhaps as much as 90% of plastics from raw sewage inflows may be sequestered within the wastewater biomass and removed from the effluent stream by precipitation into the total suspended solids; thereafter, to be land disposed as *Toxic Sewage Sludge.*[7]

A key reason for striving to combust, not incinerate, *Toxic Sewage Sludge*, so as to eliminate land disposal in any form, is to eliminate the full menu of nanoplastics, microplastics, and macroplastics and their bound toxics from the environment. The detection, isolation, and analysis of any of the many sizes and forms of particulate or fibrous plastics found in any waste stream from any WWTP raw sewage inflows to their *Toxic Sewage Sludge* and toxic wastewater effluent products, and on to the analytical lab, remain complex, costly, and difficult tasks.

Bon Appétit, nano-marinade, *et al.*

Alarms should sound with a 2021 report from *Science* writers Matt Simon and Katie Palmer where they describe how microplastics are being found today even in infant feces:

> Whenever a plastic bag or bottle degrades, it breaks into ever smaller pieces that work their way into nooks in the environment. When you wash synthetic fabrics, tiny plastic fibers break loose and flow out to sea. When you drive, plastic bits fly off your tires and brakes. That's why literally everywhere scientists look, they're finding microplastics—specks of synthetic material that measure less than 5 millimeters long. They're on the most remote mountaintops and in the deepest oceans. They're blowing vast distances in the wind to sully once pristine regions like the Arctic. In 11 protected areas in the western U.S., the equivalent of 120 million ground-up plastic bottles are falling out of the sky each year. And now, microplastics are coming out of babies. In a pilot study published today, scientists describe sifting through infants' dirty diapers and finding an average of 36,000 nanograms of polyethylene terephthalate (PET) per gram of feces, 10 times the amount they found in adult feces. They even found it in newborns' first feces.[8]

Generally, what we discard cannot be reused, recycled, upcycled, downcycled or repurposed; it's either dumped, buried, burned or hidden from view. Plastic recycling is in large part an aspiration, except for specialty plastics that are mostly pure and recyclable. Of the millions of tons of plastic waste collected globally, less than a third is recycled. In the United States, that amount may be closer to just ten percent, or much less, with about 25 percent going to landfill, and about 43 percent incinerated, and the remainder discarded in the landscape.

Microplastics and nanoplastics, or any MPs or NPs that may host adsorbed contaminants of any biology, chemistry or dimension, (i.e., contaminated microplastics (cMPs)), are major sources of global pollution, with known adverse impacts on living things, including on plants, most notably food crops. The absorption, uptake, and transport of cMPs released into agricultural systems adversely affect plant growth and crop quality as a function of the physical and chemical properties of the cMPs, as well as of the biological characteristics of the plant species in question, all of which are highly complex and controversial. Contaminated microplastics also adversely affect soil properties and the critical soil microbiome community, thereby further affecting soil health, plant nutrition, plant growth, and food crop quality.[9]

Contaminated microplastics pollution has emerged as one of the world's most serious environmental issues, with the potential for grave consequences for soil and plant ecosystems. That's because the interaction between plastic particles in soils and plants incites oxidative stress, and also induces adverse impacts on photosynthesis, metabolism, gene expression, and other critical plant growth parameters. The ecotoxicity of cMPs also adversely impacts critical soil characteristics, such as soil structure, moisture holding capacity (MHC), and soil rhizosphere pH.

Contaminated microplastics may also provide a significant habitat and transport medium for microbes. And, as a potential pollutant vector, cMPs can be transported directly into the food chain. During their transport to the environment, cMPs may undergo mechanical breakdown, photochemical degradation, oxidation, hydrolysis, microbial biodegradation, biofilm disruption, and particle agglutination, leading to variability in density, size, shape, and surface forms that may subsequently be found in soils, sediments, and waters; ground or surface waters alike.

Contaminated microplastics are known to be accumulative, persistent, and toxic in the environment and in biological systems, and they are known also to react with water-borne contaminants and pollutants to cause acute toxicity in a vast spectrum of organisms. Therefore, the source, detection, toxicity, fate, and health risks to humans from cMPs must be investigated for each affected environment and as a defense mechanism against their contribution to the extreme toxicity of toxic wastewater effluents, *Toxic Sewage Sludges*, *Toxic Sewage Sludge* leachates, and toxic landfill leachates.

The development of quantification techniques for the analysis of contaminated microplastics, the definition of their degradation products, and the methods of their transport in aquatic and marine ecosystems, are essential to providing an understanding of the adverse consequences of releasing contaminated microplastics into our living environment, a serious work not done.

In addition, and from the human health perspective, cMPs also play a critical role as vectors to carry bacterial threats, such as infectious and antibiotic resistant bacteria, as well as gene sequences that code for antimicrobial resistance or virulence, plus deadly viruses and other microbial forms, whether by natural mechanisms for delivery or as may be realized by their environmental association, or specifically as may be intended for use as biological weapons of mass destruction. Lest we forget, nanoparticles of specific design and chemistry are engaged globally as delivery vehicles for many pharmaceutical and personal care products, specifically because of their effectiveness in delivering drugs, chemicals or biologicals to specific sites at specific rates to achieve an intended biological reaction or effect. This is settled science, not conjecture.

Membrane bioreactor (MBR) systems installed in line with the wastewater treatment process, and most especially in wastewater effluent outflows of WWTPs or POTWs, along with a mandated thermolytic decomposition system to achieve thermal decomposition of all sludges and all other solid waste products to an ash product, can be an effective means for reducing or eliminating the contamination of receiving waters by cMPs. In any case, the MBR system must be coupled with a continuous flow monitoring system, plus regular analysis of the resulting waste products to ensure the elimination of the collective contaminated and contaminating microplastics that are introduced into wastewater flows by industry and by infrastructure systems and their managers.

With growing public concern about plastics, we may be approaching a serious revision of how plastic products are manufactured, used, and disposed. Waste plastic is a major global toxic pollution issue that cannot be ignored, with land disposal of *Toxic Sewage Sludge* clearly adding to the many known challenges involved in sustaining human and environmental health.

CHAPTER 16

PERSISTENCE PAYS

Memories are wonderful things
if you don't have to deal with the past.

Celine, *Before Sunset*, 2004.[1]

As attested, in Washington State and throughout the U.S., we suffer the adverse consequences of past and current policies and practices with the polluting and contamination of our living environment by the land disposal of *Toxic Sewage Sludge* and release of toxic wastewater effluents. Adding to the challenge are the extant and legacy landfills producing never-ending sources of landfill leachates into our surface and groundwaters, plus the menu of greenhouse gases and other noxious emissions released to the sky. Marginal improvements are of no measurable benefit at all. All who come after us will bear witness to what we have done, and they will suffer the consequences. No, it does not just go away, and yes, we did it, and yes, we did know about the adverse consequences beforehand. Political expediency won the day.

Agencies know well that wastewater treatment plants do not treat wastewater, and that landfills represent some of the greatest and longest-lasting sources of contamination, pollution, and climate-changing greenhouse gas emissions on the planet. They know as well that the cost of providing cleanup or restoration, or any replacements or alternatives thereto, would likely cause current national, state, and county agencies to go bankrupt, and therefore are ignored by plan and intent. The job is just too great to tackle. Bad ideas and practices from the past are the devil. The ideas of cleanup and restoration are just too big to even think about, from which then originated the harmful policy of Toxic Waste Relocation as the favored alternative.

Based solely upon my personal observations and experience, there has been no single or standard tactic, strategy or protocol for how citizen victims or communities should confront or otherwise challenge these archaic methods of *Toxic Sewage Sludge* disposal, or of any other toxic waste relocation or disposal process or strategy, near, adjacent to or upon our neighborhoods, towns or greater communities. The few such battles that have been won appear to have been successful solely as a result of the persistence of local citizens, and for no other cause or reason. Citizens have persisted because they knew well or learned that the agencies and the sludge and garbage industries were powerful, persistent, and forceful agents against them, but if they were to continue the good fight for long enough they could perhaps become the greater threat, and thereby win the battle.

Generating public emotion and opinion against the land or water disposal of toxic wastes, as permitted, regulated, and enabled by agencies and their allied commercial firms who do this devil's work, then becomes the success-focused strategy. Matters of life or death, or toxic insult, or the law, or the adverse consequences therefrom, have relatively little force or influence, for commercial interests always win. Commerce always wins, except when confronted by an active populace. Public emotion has power, for the better, and is generally correct.

In the case of Union, WA, on the lower west side of Hood Canal, citizens just gave up because they had far too many people with opposing viewpoints and few with personal concern and focus. Larger environmental groups then came in and simply undermined everything. They offered counter statements, such as: "Toxins would get purified in the soil," for example. So, how do you purify toxic heavy metals and persistent toxic organic chemicals out of the soil and water? Do you not expect them to leach and flow down into groundwaters and beyond to be diluted, and then to become part of the community drinking water supply? Or to drain into the lower depths of the Hood Canal and there to form anoxic zones (i.e., deep zones in the Canal depleted of dissolved oxygen)? Do you fantasize that some magical exotic microbes may then render the toxics as being non-toxic? Is your own survival thereby also dependent upon some ethereal magical microbes?

Note well that the agencies claim that anoxic regions of the lower Hood Canal directly adjacent to the hamlet of Union result from "natural phenomena," whatever that may be, rather than from induced eutrophication (the dense growth of microbial life in response to the introduction of excess nutrients that deplete oxygen and result in the death of resident animal and other life forms), as incited by runoff and contamination from the adjacent lands. Our suggestion here is that contaminated runoff from highly polluted sites on lands above Union provide a steady supply of excess nutrients into Hood Canal directly below, as one obvious contributing factor.

Note also that the southern regions of Hood Canal host some of the highest rainfalls in Washington State. The streams and surface waters that flow from the forests and fields on the plateaus above the eutrophic waters of the south end of Hood Canal are the exact ground-zero locations and sources of the land-disposed toxic wastes of interest, as provided by agency-approved facilities, among other sources.

Hood Canal is ringed with WWTPs and POTWs that empty their toxic wastewater effluents into the Canal, and as well, there are many forested areas around the Canal that host current and historical sites for forest-disposed *Toxic Sewage Sludges*. No, these are not "natural phenomena," no matter the ongoing agency and environmental group cover-up chatter.

Eventually, the landowners of the *Toxic Sewage Sludge* disposal fields, which sat atop the town's water supply, and only as a direct result of continued public pressure, changed their minds. They weren't worried about lawsuits, disease or illness, or about environmental contamination and degradation, but they were worried about negative public opinion and their reputation. "My God, I live here, and all my neighbors are accusing me of poisoning their water supply, etc.," was said to have been heard. And so they stopped, although hundreds of thousands of tons of toxic waste had been dumped upon those watershed lands by this activity over the years, with full agency encouragement and approval. Municipal, county, and state agencies stood by silently, never taking sides with the public against the polluters, never performing the required water testing, and never standing up to offer an opinion.

It is irresponsible county and state policy that supports the lowest cost and most convenient methods for toxic waste disposal, even directly over a community drinking water supply. A state senator attended one Union town hall meeting, and along with representatives of a larger environmental group, they supported the agency's position, an all-too-common behavior pattern. The greater cause for protecting human and environmental health is so often of little concern, most especially in the face of clearly stated commercial interests.

In other similar locations, what happened over time was that they simply moved their *Toxic Sewage Sludge* dumping work to other forests or fields nearby, or just out of reach, or just out of sight, but only after citizens complained. Those citizens didn't really stop anything because their efforts only caused relocation of the toxic stuff elsewhere close by, as in nearly all cases. They just keep moving it around. It's a given that county, state, and federal agencies fully intend to dispose of toxic wastes upon our lands and above or in our waters, so long as directly affected citizens fail to challenge. If citizens driving by don't see it and complain, then it's ok, is the working philosophy.

Whether in Ellensburg, Port Angeles, Sequim, Bothell, Winlock, Kenmore, Toppenish or Toledo, or on a Reservation, agency representatives quietly attend the town hall meetings, sitting in their preferred back row section, so as to hide within the threatening atmosphere. Ellensburg presented a special reaction opportunity for the state. We conducted two town halls there in response to citizen requests for help. Tens of thousands of tons of *Toxic Sewage Sludge* from King County were accumulating on the land surface, including in the fields of farmers who produced hay for foreign markets, and which were alleged to be organic by having never been exposed to *Toxic Sewage Sludge*—not! The groundwaters below Ellensburg supposedly represent clean waters flowing eastward from the Cascades, even though the center of the flow still holds a massive unrestored landfill with its leachate accumulation ponds, and with an endless supply of landfill leachates flowing downward into the underlying groundwaters, continuing then east to the Columbia River.

The hall was packed for the two events, with citizens anxious to learn about possible adverse consequences from the developing *Toxic Sewage Sludge* dumping plans, as simply a continuation of the long series of planned dumping events for their community. The local "biosolids" manager for the Washington State Department of Ecology was there, and being an agency representative and a land-sludging advocate, and as agency representatives in support of commerce often do, he embedded several local farmers around and amongst the hundreds of citizens in the audience. Each embedded farmer had been coached on the best questions to ask, so as to create doubt amongst the attendees about the ideas or data provided in our presentation. They asked, or stated, after politely raising their hands, something along the lines of, "Ah, I'm skeptical about what you say that there are toxic things in sewage sludge."

The audience of around 250 groaned and someone suggested I ignore him and his friends. I did, and the farmer-sludgers waited until the end and then wanted to talk. I had spoken with each of them beforehand and had observed their pre-presentation coaching by the DOE manager. They knew they were violating their contractual agreements with their buyers, for sure, and they did not want to create any friction with their buyers publicly. I offered to visit their hay farming operations the next day to observe their practices firsthand, but they were not interested in having me see anything. Free fertilizer replacement material in the form of *Toxic Sewage Sludge*, as provided by the county and state, with then a greater bottom line, was just too good to be interrupted by an outsider such as myself—people, public health, and the environment be damned.

A third town hall in Ellensburg was scheduled by the DOE, not by me, with the DOE and their University of Washington sewage sludge celebrity figure expecting to present their side of the story. Citizens called me with urgency, "What shall we do? They are going to try to paint a rosy story about the wonderful benefits of being buried in *Toxic Sewage Sludge*." The answer or resolution I offered was to suggest that the citizens of Ellensburg simply not show up, and they did not. Five concerned citizens who did not get the message comprised the entire audience, and they exited the near empty auditorium very early.

Subsequently, and once again, the state of Washington DOE and the King County Wastewater Treatment Division overruled the better interests of the directly affected and justifiably concerned citizens, and of environmental health. The King County Wastewater Treatment Division and the DOE forced their *Toxic Sewage Sludge* dumping practices on the lands of the citizens of Ellensburg. Note again that the ground and surface waters of the sludge-dumped areas all flow east, directly to the Columbia River, a sensitive and protected waterway, contaminated and polluted continuously under the advisement and consent of county, state, and federal agencies—"Roll On, Columbia," be damned!

At a Port Angeles community presentation, an agency person stood in the back row in the dark with his digital camera mounted on a tripod. He took an image of each of my PowerPoint slides for later reporting to the agency. After the presentation, I wandered over to where he was positioned in the shadows and offered, "If you would like a copy of my presentation, I'll be glad to email it to you," to which there was no reply, just a blank stare after which he exited the venue. "If you don't have anything to hide, then why hide?" I wondered, as I went to engage with others still lingering in the room.

My Canadian experiences were particularly enlightening. The city of Vancouver delighted in hauling and disposing their *Toxic Sewage Sludge* on First Nations' lands, all of which bad behavior resulted, as told earlier, in a multi-day conference supported by the First Nations. Washington State agencies were there, as well, snapping images of several days of presentations for everyone back at the office to review.

Nothing is hidden. All of my observations and findings are available for any and all agencies or their representatives to review. I have sent complete copies of my PowerPoint slides to agencies that have made requests, and I have met with county councilmembers and presented the clearly imaged evidence of inappropriate land disposal of *Toxic Sewage Sludge* without restraint. As well, I have met with the State Attorney General's office, the Commissioner of Public Lands, and with multiple county councils and their executives, always with a mind to full disclosure of the observations and findings, and with a clear and open intent to establish a complete record. I image and video all that I witness, and I present to elected officials and agency members the evidence that has been shared with the public, including visual evidence of the possible county, state, or federal violations, rarely with a response.

The purpose of societal infrastructure is to provide a functional set of facilities and systems to support the sustainable functions of households, services, and industry, locally, nationwide, and globally. Subverting the central and intended functionality of society's basic infrastructure foreshadows the end of society as we know it. Functional infrastructure is destroyed by agencies that subvert its intended use and function by giving in to commercial interests over public need.

There's a legacy landfill in Washington State, just off the I-5 near Highway 510, as one bad example, that's now a dog park. Leachates and the surrounding runoff will be there passively doing their toxic drainage work for thousands of years. The site is constantly settling because the landfill was constructed in the earlier, common style where there's just a big depression in the land that was filled with garbage. It has no side or bottom lining; it just leaches and drains to the adjacent lands and streams, and to the wetlands, estuaries, and waters of Puget Sound. Only in more recent times were some basically ineffective liners installed around the top.

This legacy landfill will leach into Puget Sound and into the directly affected orca (killer whale) birthing grounds, and also into the surrounding vast and ancient salmon habitat, for the next couple of thousand years. So, they put a dog park and a garbage transfer station on top of it—safe and concealing. Now they're trying to lease these legacy landfill surface lands as industrial sites. The thinking must be that if you wait long enough, people will forget where the highly toxic landfills were situated, and that developers can now get away with placing schools, playgrounds, sports fields, housing, strip malls, dog parks, and other people-inhabiting spaces atop the forever-persistent, forever-leaking, toxic accumulations. Whatever you put there is just going to sink, and stink, and leach, and pollute, for a very long time to come.

"They should put a wastewater treatment plant out there," someone suggested to me at a community meeting.

"Yeah, sure," I responded, knowing that would only add highly toxic wastewater effluent outflows and even more legacy landfill leachates that would enter the sensitive salmon migration and orca birthing waters of South Puget Sound. Bad idea!

"Are you kidding?" I added. "Putting that much weight on that endlessly leaching, decomposing, gas-burping, and undulating garbage pile would do no more than simply make it sink and leak faster. What's more, the mix of toxic gases seeping out of and up into the sky from that massive decomposing hulk will be yet another major contributor to greenhouse gas emissions for generations to come." This legacy landfill is one of so many in Washington State that must be mined down to its basement layers and restored to as natural a state as may be feasible.

They all leak and leach toxics, including CHRL, right into the orca birthing grounds of Southern Puget Sound, to the sickness or death of these magnificent, endangered creatures. And despite the fact that all orca populations are protected under the Marine Mammal Protection Act (MMPA), and with two populations receiving special protection under the Southern Resident Orca Distinct Population Segment (DPS).[2]

As mentioned, in the earlier days of Washington State, the theory was that all estuary lands and other wetlands were wastelands that must be filled with solid waste, so as to make them more useful. For the most part, that sick idea is now gone, but not the still lingering ideas that the microbes in wetlands rapidly decompose any wastes dumped therein, toxic or otherwise, much to the chagrin of federal agencies that do focus on the health of oceans, fisheries, forests, lands, and air, in opposition to the opinions of the EPA or the agricultural agencies. Their thinking must be that oceans and their life forms are just too big to kill.

That landfill/dog park site is adjacent to the outflows of the Nisqually River and McAllister Creek, favored natural salmon breeding grounds, both of which drain into the renowned Billy Frank Jr. Nisqually National Wildlife Refuge. The Nisqually River and McAllister Creek also carry toxic wastewater effluents from the wastewater treatment plants of towns they pass along the way, plus the *Toxic Sewage Sludge* runoff from the contaminated lands along the riverbanks in their meanderings. It's national and state policy, not simply local bad behaviors that pollute and otherwise corrupt these precious waters that originate from the clean snowmelt and rainwaters of Mount Rainier and adjacent regions of the Cascades. Beginning as pristine waters, within but a few miles they become sewage and wastewater effluent-tainted, agricultural runoff waters for that greater region entering Puget Sound.

Just imagine, for a moment, the cumulative toxic consequences of the combined stormwater runoff from roads, highways, and impermeable surfaces of all parking lots, plus the industrial wastewaters, and the medical and domestic wastes, as well as drainages from land-disposed *Toxic Sewage Sludges*, toxic wastewater effluents, fertilizers, and pesticides, and the runoff and leachates from the many extant and legacy landfills, all combining their accumulated toxics to do their synergistic toxicity bad deeds to the life forms of

Puget Sound. And yet a similar predicament can be described nearly all over the United States, even though that fact does not make things at home any better.

No, it doesn't just go away, it accumulates in the waters, sediments, and aerosols along the way, then into the cells and tissues of the life forms at the end of the toxicity chain. We can't survive us, just as all of the human-affected life forms along the water flows are destined to suffer the same ill health consequences as does our unfortunate and ignorant selves. Multiple science publications continue to report the broad distribution of highly toxic Perfluoroalkyl and Polyfluoroalkyl (PFAS) chemicals in the ground, surface, and rainwaters of the world.

As with nearly every place I've presented, the most affected people are usually citizens who live in smaller towns, or the local farmers who worry about things like water, soil, and air when they buy land or a home. Almost no one projects the problem beyond their domicile, neighborhood or town. I wander around before and after the presentations and engage in conversations with everyone I can. I want to know why they are there and what they may be trying to get out of the presentations, and mostly, why they put up with agency-imposed contamination, pollution, and disease. "But we live a healthy lifestyle," they persist, unaware that an invisible toxic lifestyle is brought to them by courtesy of overseeing agencies and their partners in commerce.

The city of Yelm, WA, was different. There I encountered an interesting and engaged community that attended the several town halls convened for the informational benefit of local residents. I didn't have to go out of my way to meet anyone. The citizens would engage with me when I was outside or over in a corner of the room. That was great, because when I give a presentation I always wonder: "How is the audience receiving this? Am I wasting anyone's time? Is it a good story? Is it going to be helpful to them? Are they learning what they may need to help defend their air, foods, lands, and waters from the agencies and their commercial associates?"

In Yelm there was a lot of feedback. I found more people wanting to be actively engaged in the solution or outcome than almost anywhere else. Usually, people just sit there for an hour or two then go home. That's the way it is. It doesn't go anywhere. Yelm was different from that perspective. People wanted to be engaged. They weren't questioning me. They weren't asking: "Can you show me a publication on that subject, or can you prove this?" There wasn't any hostility at all, for they were experiencing directly in real time the adverse effects of agency exploitation. It was about a deeper perception of the problem, and now I was assisting them in realizing that what they were worried about was really something worth worrying about. I provided validation. I had no idea what they did before, but they were much more of a positive group.

The typical situation and attitude in so many other communities is that they are victims of the state, the county, the community, and the commercial sludgers. Yelm, on the other hand, wanted information in order to fight back. The case was similar among Tribal and First Nations communities throughout the United States and Canada. Citizen attendees were often there for a first time introduction to the topic of toxic waste disposal, and to discover why and how they were being exploited, and what to do about it.

In Yelm, they did get the state to move the sludge elsewhere, so from a greater perspective, the citizens achieved local relief. However, toxic waste is always going to be generated, and unless someone destroys it, the problem isn't going away, it's just moved, as has been experienced so many times before. That's because county, state, and federal agencies prefer to shuffle, relocate or hide toxic wastes, rather than seek resolution for the benefit of citizens; something that could take many years.

The owner/operator of the contracted sludging operations intended for Yelm died recently of pancreatic cancer. Similarly, it's interesting to note that a company that owned the *Toxic Sewage Sludge* spreading operations in the Snoqualmie forests, a father and son team, both died of liver cancer, not far apart in time, and also just a few years ago. The latest similar story is not yet fully clear, more anecdotal at this point, but a surviving former sludge truck driver from Central Southern Washington State reported at a similar town hall meeting that he may be the lone survivor of about ten sludge truck drivers from his location. He implored the audience to end the idea of *Toxic Sewage Sludge* dumping on the virgin, clean soils in their area.

Toxic waste is toxic, no matter the business incentives or attitudes. The deceased sludge contractors drove the spreaders after King County dumped the sludge in the forests. They had a millions of dollars contract to spread it. No one was monitoring what they were doing, so they made more money in a lesser amount of time by dumping it to greater depths to their own convenience and satisfaction.

Curiously, while the state issues permits for *Toxic Sewage Sludge* dumping sites, and is responsible for overseeing the actual on-site dumping process (rarely, as I have observed), and while they inspect the end result after the fact (also rarely, and then by driving by in a pickup truck and looking out the window, as I have observed), it remains quite interesting that both county officials and the state profess that the land-disposed *Toxic Sewage Sludge* was applied "at agronomic rates." How so when it is often dumped on steep hillsides above protected salmon waterways, and at depths that may be from 4 – 18 or more inches? And all the while professing responsible oversite to assure "compliance." It's simply not true! I've been there. I go there. I observe them. They make it up as they go!

Yet the contractors who manage the toxic sludge dumping and spreading operations in the Snoqualmie forests are generally nice people, based on my personal experience. One day I came down a rocky slope far too fast and blew out both tires on my mountain bike—front tire and rim non-repairable, left forearm broken. It was a nine-mile hike back to my car. There I was in the middle of nowhere and I couldn't repair the damage, and along came the repair truck that serviced the sludge loading and spreading equipment. He stopped and gave me a ride out to my car. He and his partner are now both dead from cancer.

Another time I had a flat tire far into the forest. It was deathly cold, everything was frozen and I couldn't use my fingers to fix the mangled tire. I was freezing. Along came the same truck and the driver again asked if he could help. Same guy, both times, stopped to help me while I was trying to reveal them for supporting the sludgers in the forest. I've had experiences where some of these guys really wanted to kill me because sludge spreading is their job, and they have orders from King County that if I show up they are to turn off all the engines, pack up, and leave. They wouldn't get paid for that day, so they weren't at all happy.

Sludge tracking has been a difficult path to follow because I still have not exactly learned how to present the core message as an important public health principle for the benefit of us all. I've presented the information to a number of people and organizations, such as county councils, state legislators, state department heads, and so on, and they just look at each other and walk off. At the end of the day, they just don't want to think about it. Toxic wastes don't vote well for them, and the public knows nearly nothing about the dangers, so why get involved, most especially when the remedies will cost at least many tens of billions of dollars for state-wide fixes.

Toxic wastes don't vote for them, and salmon don't pay taxes, but citizens do both, yet so often communities appear more prepared to die from an uninvestigated cause than see tax rates increase. It's the way self-destructive systems work.

At a recent county council meeting, directly after the chairman had been carted off to hospital and later died from COVID-19, all of a sudden it seemed important to hear what I had to say. Everyone in this particular meeting had recently been in close proximity to the chairman. They had been sitting together around small tables. Now they were on the edge of their chairs listening to me talk about how many of the known toxics in *Toxic Sewage Sludge* may incite immune suppression disorders, and about the raft of SARS-CoV-2 and so many other viruses and infectious agents that are being detected and isolated from raw sewage, *Toxic Sewage Sludge*, and toxic wastewater effluents from across the U.S. Now they wondered if they were going to be safe.

Note well that the United States Government monitors *Toxic Sewage Sludge* from many wastewater treatment plants throughout the country on a regular basis for the presence of SARs-CoV-2, the inciting agent of COVID-19, to evaluate the prevalence of the virus across the broader population. SARs-CoV-2 is an excellent analyte to detect and count because it is persistent in raw sewage wastes, just as it is in processed sewage in the form of wastewater effluents and *Toxic Sewage Sludges* that are released unabated to the environment.

Although municipal, county, and state agencies wish this were not the case, it is, and with each drop of effluent, and with each lump of disposed *Toxic Sewage Sludge* into our living environment, we and all of nature are dosed with SARs-CoV-2 and a greater raft of infectious agents, including a broad array of viruses, bacteria, fungi, parasites, prions, and more. While some infectious agents may be diminished in the great outdoors rapidly, others persist and are disseminated wholesale to our collective detriment.

By their practices of disposing WWTP/POTW effluents and sludges into the environment, the agencies and their contractors are performing biowarfare, bar no other greater form worldwide. SARS-CoV-2 is detectable and has been detected in any and all wastewater treatment plant effluents and sludges tested, and that's just the way it is, along with so many other viruses, including those considered to be potential or actual biowarfare agents. What's more, by the similar unrestrained release of the toxic chemical constituents of such effluents and sludges, the thought may be expanded to also include toxic chemical agents as a form of chemical warfare against our citizens and nature.

Each human body responds differently to almost any kind of infectious agent. Some people can actually be carriers and have no symptoms. It's rare but true and that can go up the whole gradation from a cold to flu to COVID-19. The problem is they are all still infectious disease agents. That only really matters after you are about thirty-five or forty years of age, because that's around the age where your immune system is in decline. Human beings were only designed to live to be about thirty-six, or approximately three generations. We have extended that to a more than doubled lifespan with modern technology, mostly by advancing our systems of transportation, infrastructure, shelter, food, sanitation, and medicine.

When someone was twelve in primitive times they were pregnant and having babies, then after twelve years of age the whole function of the oldest person was based on grandma taking care of the grandkids while mom did the work. That's how human evolution progressed—three times twelve. By the end of the third generation, they went off to die and that was the end of it. In our most recent 50 years, we have extended our forecasted lifespan from around 60 years or so to now about 80 years or more and expanding. The limitations today are more a function of lifestyle choices and less about finding a mate, grubbing about for food, and finding a suitable rock for shelter.

Industrialization and technology have extended our potential to live a longer, more productive life. On the other side of that—as I was explaining to the councilmembers while drawing an upward curving line of global human population increase and inserting a second line to show increasing pollution—it is clear that pollution is increasing at a rate greater than population growth. Each person today creates more pollution than someone did ten years ago, or 100 or 50,000 years ago. That's because of things like consumerism, with less focus on the requirements of daily life, such as accessing a mate, food, and shelter. As a direct result of that behavior pattern, we now have a greatly impaired and toxic planet, overly filled with humanity's toxic wastes, and with real consequences yet to unfold.

There was a day when the United States Government, meaning in this case the Environmental Protection Agency, relegated these issues to the states, and all of a sudden, the rules became local and individual, and states and municipalities could do whatever they wanted. That's the nut of the problem. The federal government turned over these national issues to states who then gave in to the counties and municipalities who had their own agendas and interests in getting paid, rather than concerns about human or environmental health. The original intent of our environmental laws was given up by the federal government to untrustworthy states and counties who have their own personal interests. That's the problem.

In Washington State, the Department of Ecology's solution is to move *Toxic Sewage Sludge* further away from towns to sites where people can't see it. Then the problem goes away. But it doesn't. No matter what part of the country we may investigate, it is this same public and environmental health abuse matter that we must suffer. There's a solution that's functional, and that I have worked on for years. You can't keep moving it. You have to get rid of it, and the "It" of which I speak is the highly toxic, *Toxic Sewage Sludge* and the overly abundant consumer waste products we all create.

CHAPTER 17

SPEAKING TRUTH TO POWER

The way to right wrongs is to turn the light of truth upon them.

Ida B. Wells, *The Light of Truth: Writings of an Anti-Lynching Crusader.*[1]

Speaking truth to power is a non-violent political tactic, employed by dissidents against the perceived wisdom or propaganda of governments they regard as oppressive or authoritarian. The phrase originated with a pamphlet, *Speak Truth to Power: A Quaker Search for an Alternative to Violence*, published by the American Friends Service Committee in 1955.[2] "Speak Truth to Power" is also the title of a global human rights initiative under the auspices of Robert F. Kennedy Human Rights. In any case, speaking truth to power means to stand up for what's right.

Which shall prevail in a democracy: i) A citizen's right to be free from state-imposed disease and death, or, ii) State rights to dispose of toxic waste or other toxic materials in a manner that exposes citizens to toxin-incited or imposed disease and death? A gunshot, for example, is loud and deadly, and one dies. Toxics are without immediate alert, yet many die. Death by toxics represents a failure of government to provide functional infrastructure and safety.

Toxics from wastewater treatment plants—be they in the form of grit, *Toxic Sewage Sludge* or wastewater effluents, or from landfill leachates or stormwater runoff—do what toxics do, they are toxic, they make us ill. Therefore, governments at all levels, from municipalities, to counties, states, and federal departments, should be held accountable for their agency's toxin-induced illnesses and death. Citizen victims cannot fend for themselves in the face of agency-asserted, agency-released toxics.

The Constitution establishes a federal democratic republic form of government. That is, we have an indivisible union of 50 sovereign states. It is a democracy because people govern themselves. It is representative because people choose elected officials by free and secret ballot. It is a republic because the government derives its power from the people.

By the underlying tenets of democracy, citizens also have the unalienable right to health by their own measures and decisions, not to be alienated by the serendipitous, politically driven decisions of ineffectual governments. Be it governor, county executive, department head or division manager, hauler or spreader, the land disposal or landfilling of toxic wastes represents the gravest threat to human health and survival, now and forever. We are both the source and the victim of waste production, disposal, and exposure, the primary limiting factor to the health of us all.

The Ninth Amendment to the Constitution is intended to assert the principle that the already enumerated rights of citizens are not exhaustive and final, and that the current listing of certain rights does not deny or disparage the existence of other rights, revealed or not, explicitly and including our unalienable right to health by our own measures and decisions. Governments and their agencies must be proactive and protective of their citizen constituents, or they have failed by default.

Policy is critical to achieving vital public health advances. Policy that includes legislation, rulemaking, and public campaigns offers a set of tools for improving health at the population level. There are five steps for using public policy to produce gains in population health:

- *Define the Problem:* Quantifying the impact on morbidity or mortality is the key starting point.
- *Elevate the Audience:* Seek to capture and sustain the attention of the public and policymakers.
- *Select an Approach:* Employ a strategy that will achieve the greatest gains in population health.
- *Enact a Policy:* Foster policy that can be implemented at scale. Policy change is plodding and incremental, and most legislation is never enacted. Litigation can serve policy when legislative action is insufficient or infeasible but is costly and often fails to prevent future harm.
- *Employ Analysis:* Apply ongoing analysis and evaluation to confirm that the selected policy is leading to the desired outcomes and to monitor for unintended consequences.

So, who buys all that becomes waste, and is it needed, and can it be avoided or prevented, and if not, can it be disposed of safely to ensure public health?

The underlying and resulting fear of agencies and their partners in commerce regarding safe and effective management of toxic wastes is two-fold: i) Retribution as a result of lost commerce, and ii) Adverse reactions of taxpayers in response to increased water and sewer costs, even when such proper management is for their personal health benefit. Complacency is the killer, in and of itself. Political and agency leadership are essential, yet largely not available.

We generally ask logical questions about what is in it (i.e., in toxic wastes), yet we seldom ask the equally important question, what does it do? We must know what is in it and what it does because our inherent spectrum of thought for survival is fashioned by our cumulative life experiences, no matter the spurious events that may send us asunder, at least for a while. In the case of toxics from toxic wastes, there is the axiom that, "Once exposed, twice injured, no way back."

So, just who gives license to agencies to expose us to endless toxic-laden wastes that generally yield acute or chronic illness and death? Is that just the way it is, or is there a way out of this? Must we be citizen victims, simply because we are here? Can politics and commerce be driven to protect us, rather than drive us to illness and demise? Does anyone really care? We seek discovery of the compassionate person of leadership, and we must take pause to think about this.

One approach to citizen safety is to consider the open land disposal or landfilling of toxic wastes from the perspective of the pharmaceutical industry. We can ask the simple question: Why would we not regulate the administration of toxic wastes to the public—by way of air, food, soil or water—as equally as we would regulate the administration of pharmaceutical products by any route of administration?

Although the regulatory path to achieve the approval of a pharmaceutical product may be time-consuming, costly, and arduous, there is good cause and reason for such complications and travails, and that is simply the matter of safety, regardless of the benefits or not of the resulting new drug product. And yet, in this example, we need not look at the full spectrum of challenges along the way to regulatory approval of any new pharmaceutical drug, but to simply review one quick step, and that is the requirement to evaluate the therapeutic index (TI) of the intended product.

The TI is a quantitative measurement of the relative safety of a drug (or substance). It is a comparison of the amount of a therapeutic agent or substance that causes the therapeutic effect to the amount that causes toxicity, or as may be restated, the therapeutic index of a drug is the ratio of the dose that produces toxicity to the dose that produces a clinically desired or effective response. The TI of a drug, of a substance or of another material may be determined in animal or cell systems as the lethal dose of a drug for 50% of the population (LD50) divided by the minimum effective dose for 50% of the population (ED50), i.e., the therapeutic index = LD50/ED50. A higher TI is preferable to a lower TI, which is to say, a patient would have to take a much higher dose of such a drug to reach the toxic threshold than the dose taken to elicit the sought after therapeutic effect.

There are no such criteria applied to the land disposal, landfilling, water disposal or air disposal of chemical, drug or biological wastes, or to any other toxics, which are clear agency responsibilities. No agency seems to accept such burden, despite the adverse consequences, so long as the commercial and political equilibria are achieved and maintained.

Chapter 18

Just Like My Shoes

With one of the richest, most diverse and beautiful landscapes in the world, we have an important legacy to protect.
The health of our environment must be protected, as it is essential to our quality of life and to the strength of our economy.

Chris Gregoire, Washington State Governor, 2012.[1]

Way back when I was a kid, we were a family of meager means, to say the least. My father, after years of hand digging the first-ever sewer ditches in Lower Manhattan, became a prison officer and that was the single source of family income for five of us. I mentioned to him one day that my little toe was sticking out of my olive drab, Army surplus tennis shoe (actually, they were laceless Army prison shoes). My wardrobe consisted of previously used, prison-recycled, former WWII U.S. Army surplus things, including leftover shoes returned from the European Theater. All of this was my very cool stuff, not to be mocked.

The response from the big man was always a well-intended and stumbling statement, overly stated, oh so many times, "Well, when I was your age," he would say, "I didn't have any shoes," implying that I was lucky to have even one shoe, little toe inside or not.

Answer back, "… but my feet are growing."

The hardworking pragmatist, who labored at Rahway State Prison for 35 years and worked six ten-hour days for $100 per week (4 p.m. to 2 a.m., plus a two-week unpaid time-off), responded, "No, your feet aren't growing, your shoes are shrinking!"

"Time off for good behavior," we always barked in return, a sick insider's joke among prison worker families. In any case, the next day, yet another pair of very old, laceless Army surplus shoes appeared, always well worn, and much appreciated.

My father exited the U.S. Army in Texas in 1919, following a few years participation in an unrecorded war of some sort in Mexico. He was a highly technical, experienced horseman and mule skinner who moved wagons and materials around the dry outback of Northern Central Mexico, no doubt with some plan and intent, never to be revealed. They all exited the Army and the war back then with tuberculosis, and his obstinate cough persisted until death. He passed at 93, happy to have had a job since he was ten, and proud that he had attended school from "about the forth to sixth grade," as he put it. The point is that his sphere of concern extended from home to work and back, and everything else just took its own course.

Along those very same lines, and just like those shoes, we don't fit the Earth anymore. Of course, it's not because we're exploiting and consuming the remaining finite resources far too fast as a result of population growth, coupled with rampant consumerism, it's that the Earth must be shrinking, just like my shoes. Makes sense, but my very special and much appreciated shoes had been recycled, repurposed, refitted, and reused until there was nearly nothing left to discard, but they always worked. Today, nearly all goods find their near-time fate in the trash.

We consume Earth's non-renewable or marginally renewable resources faster than the air, soils, and waters can be remediated or replenished, knowing well that many essential resources cannot be recovered at all, not now or ever. Today, we consume faster than nature can provide or restore. Nature by its own means endeavors to balance some essential resources, while we are no more than woeful passive observers revealing limited means of assistance. We exploit; we produce; we consume; we pollute; we deplete; we over-exploit and over-produce; we over-consume; we become polluted and depleted; then exhausted nature takes over; no more room for us. The Earth is in fact shrinking by our very creative handiwork; it needs to be refreshed.

The plain fact is that population growth is crowding us out, as uncontrolled mineral exploitation operations search for new resources to feed the potbellied pig of malignant consumerism. It's all a matter of population growth-incited resource exploitation and the concomitant land degradation and deforestation resulting therefrom, along with the consequential air, soil, and water pollution, all of which result then collectively in yet further soil degradation and decline, and all of which also colors our surface waters, thereby to absorb yet more sunlight, resulting in yet more non-dissipated heat—CO_2 be damned. How do you win this one? Is the enemy heat, of and by itself, or is it the heat being trapped and stored and accelerated by us that is prescribing our potential fate? It's both. In any case, it does look like we are way more of the problem than of the solution.

Anthropogenic activity is impacting Earth's climate characteristics and its dependent ecosystems in ways that are clearly dangerous, disruptive, self-limiting, and ultimately unsustainable for humans and other living things. Greenhouse gases are accumulating in the atmosphere at alarming rates, with expectations that the outcomes could be experienced by our descendants for but a few added centuries beyond today, and not much longer thereafter. Our time here as a species is limited, the literature tells us. Perhaps in just 100 years the remaining human population may be only ten percent of what it is today. Who then will be left to complain?

We are experiencing firsthand the self-prescribed adverse consequences of our past, present, and continuing contaminating and polluting behaviors, along with the continuous and accelerating outpouring of toxic gases and chemicals into our living environment that present a rate of pollution no longer able to be degraded, absorbed or slowed, much less reversed ahead of some approaching "point of no return" (PoNR), and that leaves us in intellectual and political paralysis. We're not quite sure when that PoNR may be revealed, or upon us, but it will likely be sooner rather than later, and it may be just about here. This is not an intended tale of impending gloom and doom, but it is to what we bear witness right now.

Let us not forget the fundamental phrases of our hymn: "We are in endless conflict between commerce and public health, and commerce always wins;" while "Nearly all chronic diseases are incited by the cumulative effects of long-term exposure to low levels of environmental contaminants and pollutants;" and that, "Agency Toxic Waste Relocation Programs are not working;" plus, "Prevention of disease is always more beneficial than treatment." What's more, we have reduced this bane of our existence to but a few relevant and substantive phrases, including: "Farmers want to produce crops for available markets, at any cost;" and, "Agencies fail to evaluate, monitor or regulate just about anything and everything, all in support of commerce, never for the benefit or health of the people, or of the environment."

What's more, the Earth no longer meets our resource requirements on any level, simply because it's either too small, or there are way too many of us, or we don't plan to manage our natural resources properly. We expect to populate, exploit, consume, and pollute more, while the Earth keeps shrinking, just like my shoes.

Past forecasts and projections of the seemingly elusive PoNR were simply the feeble beginnings of a critical and generalized realization. More precision is now required in the analysis if we're intending to slow the resource exploitation or consumerism trains from going off the tracks ahead of obliteration, which is to say, going off the steep cliff of failed sustainability. Serious matters of resource and population imbalance, deficit-driven food production, and climate-forced human migrations are arising, now well in advance of any useful or actionable predictions, raising serious questions about the continued habitability of certain regions of the Earth, or eventually any of it at all.

We appear to be in a late stage act of the performance. Can that train be slowed, stopped or turned around? We don't know the answer, while seemingly few care to find out, so long as there may be a cost. Hard questions are hard to digest, much less resolve this far into the game, and with full and planned ignorance of the blatant and calculable externalities—those side-effects or consequences of activities that may affect others without being reflected in the incurred costs of the associated goods or services.

Certain agencies and academics divide the risks and threats to the Earth into a few major categories with the objective of understanding the nature of a predicted or predictable event, and with a mind toward changing the course of things to our benefit, just in time. Their work centers not on a last-minute rescue attempt, but rather on a "No Surprises" concept, such that we may be able to predict and avoid disaster, and not have to simply clean up the mess later if we were still here and able. These categories generally include those threats to global and national security that are within our means to avoid or deflect, and those for which we may be able to detect and prepare for very well, then brace ourselves. The idea of "No Surprises" is a part of my former life, and holds firm even more so now than back then.

Giant objects from space, for example, present foregone conclusions, which is to say, brace yourself and pray. Smaller objects from out there headed our way may be intercepted and shot down before they hit home—maybe. We have agencies that track meteors and meteorites hurtling toward Earth, and who then make predictions about if and when they're going to hit, how big the damage may be, and whether we could send up a missile to blow it up or change its course. Those objects present real danger, and some are so big there won't be an Earth left. But in those cases, there isn't anything we can do about it, other than track them and wait.

"But what if the Sun goes out?"

"Then pray some more, and get on with it."

We have several alternate spectra of possibilities to look at here. On Earth, there are those impending seismic events, predictable or not, that may do us in by way of a crushing quake or eruption, or by the ensuing giant waves and floods from the aftermath. Either way, the "Big Ones" are real and imminent (so why is it that wealthy people live in beach houses, and what do they know that we don't?).

As an example, the Pacific subduction zone runs right off our coast from Northern California through Oregon and Washington, and on to British Columbia, and we're now 350 years into a 200-year cycle, meaning that long ago, a subduction event was supposed to have happened—very difficult to predict. A tsunami from that event will likely be experienced first as a nine-point something thud, followed by a wave that will wipe out most of the low lying regions of Northern California, Washington, Oregon, and British Columbia, and will probably take out most of the Hawaiian Islands, except for the tops of a few volcanos. It's also going to have a major adverse impact on Alaska, and as far away as Japan. But what can you or I do about it? You can't pour cement over it or build a large wall. It's a big deal and a real deal awaiting some imminent or future timeframe.

So, we have things from space that we can't do much about and we have things in the Earth we can't stop, so what is there we can do much about? We can do a lot about domestic and foreign terrorism, which are major threats, and we can do a lot to improve how our waste is processed.

USAMRIID, the U.S. Army Medical Research Institute of Infectious Diseases, is the largest organization in the world that studies infectious agents everywhere on the planet. Everyone's in agreement that epidemics and pandemics are such serious threats that you can't let politics, cultures or borders get in the way. The Director of National Intelligence has a program called, "No Surprises" and that's just what it's for, and that's why the human waste from every international airport in the United States is monitored on a regular basis, looking for transmissible infectious agents.[2] It's that big of a deal. We also do research, which is enough to curdle your blood, on creating new kinds of infectious agents so that we can practice developing treatments and vaccines against them in anticipation that nature, or a domestic or foreign entity, may try and do such a dastardly deed. Cyber surprises are of yet another sort, not to be ignored.

All of these threatened or actual blights on our survival are created mostly by natural, exploitative or evolutionary processes, such as and including, for example, the domestication of animals, and not necessarily by scientists in secret labs. Nature loves generous food resources, as found in dense populations of anything—animal or plant, human or other. The nature of life has but two ambitions: nutrition and reproduction, the essential elements of survival. Infectious agents of plants or animals seek nothing more than a densely populated receptive host range that is willing to provide adequate nutrition and a favorable location for rapid and perpetual reproduction, until a host immune system attempts to interfere, then to move on to the next set of hosts, locally or globally, yet once again.

Often, and as witnessed by our long history of epidemics and pandemics, infectious agents wreak havoc upon us, accelerating their infectious agent populations until the immune system of the ill-fated host either stakes out its territory successfully, or the pathogen runs out of receptive and nutritious hosts, which is to say, until the target population is decimated beyond its reproductive capacity, never to return. We have a functional immune system for no reason other than to support our varied attempts at self-perpetuity, merited or not.

On the same note, one of the key flaws, and as well one of the key milestones, in the development of growing human populations is the recent advent of animal agriculture. The domestication of food animals brought us into intimate contact with an array of non-human species for many reasons, at extreme and intimate densities, and with a high cultural cost. The outcome is zoonotic diseases as human infectious diseases that originate in non-human species, mostly in domesticated herd animals, that then "jump" to humans as alternate hosts, thereafter, to become the infectious diseases of us.[3]

Zoonoses originate primarily in our domesticated herding food animals, generally with either favorable or stressful environmental conditions, complicated even more by a changing climate that helps the host-switching tactic of infectious agents, which is realized as a host-exploiting maneuver of the animal-infecting pathogens and parasites. Herds of mass-produced food animals, many suffering unfavorable conditions in over-populated feedlots, coupled with an expanding need for vast amounts of animal feed materials and clean water, along with a requirement for vast amounts of fertilizers and pesticides to support production of the animal feed required, together with the polluting effects of fossil fuels, as needed for power and transportation, represent one of the greatest sources for new human infectious agents and of polluting greenhouse gases ever encountered, all within a vast arena of global commerce based primarily on the ever-expanding need for human food.

Today, possibly half of the world's animal feed crops are grown in *Toxic Sewage Sludge*. Added to that, the most abundant animal agricultural product is chicken meat, with nearly all of those chickens grown on a strict diet of plant-based agricultural crops that are likely grown in *Toxic Sewage Sludge*.

Herd animal production is wholly dependent upon the production of plant-based foods that then represent the primary source of human food, and of toxic waste pollution. While the primary sources of plant-based foods used for animal husbandry are generated by farming corn, wheat, soybeans, sorghum, canola, hominy, barley, field peas, rice, and a few others, and as well that vast amounts of fish meal and bone meal are also added to animal feeds to meet their nutritional requirements, who then asks the question about what toxic chemicals are bioassimilated and bioaccumulated in animal feeds from plant-based agriculture, and at what concentration? Are those same toxics then taken up and bioaccumulated within the meat products that we eat for dinner? The question remains in large part unasked, with the critical answers therefore nearly wholly unknown.

Corn appears to be the most widely produced animal feed grain material in the U.S., with soybean-based animal feeds representing the world's largest source of animal feed protein. Wheat is the second, most produced cereal grain behind corn, with the global trade of wheat now greater than that of all other crops combined. China, India, Russia, and the U.S. represent the four largest individual wheat producers in the world, with the combined UK and EU wheat production just behind China.[4]

If in fact half or more of all of the world's supply of animal feed is grown in *Toxic Sewage Sludge*, and by the common biological phenomena of bioassimilation and bioaccumulation of toxics into animal feed plants being a reality, then that most-consumed agricultural product of all, globally, i.e., chicken meat, must in turn be laced with any or some or most of the most common and most likely toxins that are taken-up from the chicken feeds to the most cherished animal food for people of all, and that must be chicken meat.

Are our collective toxic wastes in fact being relocated from contaminated plant-derived animal feeds into our most prized and consumed meat products? And are those said toxics originated from the combined plant-derived and animal-derived products that were grown in the cumulative and synergistic toxic stew or brew of fertilizers, pesticides, and *Toxic Sewage Sludges*? Most likely, YES! That then becomes the question that represents the basis of our survival, regardless of epidemic, pandemic, tsunami, earthquake, object from space, war, or other.

Making matters worse is our gene-driven innate need for socialization, which presents us with the concept of contagious diseases among us and by us and providing opportunity for a mix and match set of community transmissible diseases by which we develop accommodation or resistance, or not. In any case, the evolved infectious disease etiological agents are here to stay, so long as we are available to provide hospitable living quarters. Which one will go rogue next? Or by the hands of whom, intentional or not?

Other key threats to our survival include our genetic predisposition to competition for resources, territory, mates, and mere existence. Competition may be either a favorable heritable trait or a doomsday predisposing factor for self-destruction. We have had it both ways. By this concept of competition, we are destroying ourselves over matters of food, water, land, commerce, procreation, and the like, for the sake of survival, well beyond the risk and threat of infectious diseases to ourselves or to our animal food stocks. What a circus! It's fun, but we do have to pay—it's a one-way pass without a return ticket if we don't wake up soon. Mind the gap!

Consumerism is not just about buying products and creating plastics and all of that, or about forcing the petrochemical industry to make yet more plastics and gas and oil, but it's also about real consumerism from the simple perspective that we're eating the Earth bare. We can no longer grow enough food to feed the population we have. In our affluent societies we don't quite see it on the surface, but globally, we have major food challenges in the face of actual Earth changes.

A climate emergency reveals there is too much water in one place, not enough in another, and often of the wrong kind. As a result, human- and climate change-incited desertification processes yield drier lands, with former highly productive agricultural lands now in fact becoming deserts, and unable to support any crop production. As well, many soils are now degraded by decades of woeful applications of toxic wastes, including *Toxic Sewage Sludge* and toxic wastewater effluents, plus landfill leachates and produced waters, as well as commercial fertilizers and pesticides, all of which result in soil degradation, with diminished remaining soil structure and function. The observed, imagined, and fantasized short-term gains in exchange for immediate commercial rewards always result in longer-term disasters, generally irreversible.

We foretell of our out-of-control consumerism, buying material things we do not need and wasting too much food, simply because we appear to want to have more less-nutritional foods and "things." These are the very same issues that drive the petrochemical industry to make ever-so-much-more plastic, resulting in more greenhouse gases, and with a compelling need for more transportation by air, ship, truck, and train, simply to move this stuff around the planet. We're actually becoming ecocidal—spreading population-based, global, environmental ecocide—and not necessarily by the fault of Big Oil, coal companies, cattle producers or rice fields, but because we refuse to believe in or accommodate the need for sustainability or growth management. We're like the proverbial hoard of locusts that eats everything on Earth, until it's gone. We can't seem to stop the self-destructive and aberrant behaviors, and it's not like us to fix anything when we can just buy anew!

We now see mass migrations across the Earth. People are leaving Eastern Europe, Africa, Latin America, Asia, and the Middle East, not only because of war or civil unrest, but because too much of the marginal land that used to support food growth is now dry, without new water in sight. The waters of the world are in the wrong place, and therefore the food is in the wrong place, and therefore people are migrating to other places. We happen to be blessed in Washington State and Oregon and in parts of Idaho and Montana that still have enough snowfall to create surface and groundwaters to supply most rivers. It's a diminishing situation, even in these areas. Clean water is a priceless commodity. To irrigate certain desert areas of the western U.S., for example, there have been proposals to build pipelines from the Great Lakes or from the Mississippi River to Arizona, but that's not a workable solution—at least not yet.

Now that we've gotten ourselves into this mess, there is a nearly insatiable desire and need for a supportive infrastructure, generally in the form of shelter, warmth, nutrition, water, social structure, and the need to accommodate our wastes, with the latter being generally by stealth of disposal, rather than by a healthy plan, no matter the consequences. As a result, and well beyond space rocks, earthquakes, tsunamis, volcanoes, wars, epidemics, pandemics, and depleted resources, we seem to have left a solid path of destruction in our wake. We have fouled the lands and waters with our unsanitary wastes, depleted and contaminated the soils, and polluted our environment with total disregard, no matter the destructive or diminished outcomes. As such, it is not necessarily growing populations that wreak such havoc upon us, but the unregulated permissiveness of the alleged regulatory agencies, at all levels, from top federal echelons down to local communities, as failures to govern, and primarily without public health, environmental health, and survival at the front of mind.

Can we now go back up the chain of responsibility to assign cause, or is that just too slippery a slope to navigate? Perhaps that massive space chunk from out there may be the kindest and most forgiving answer—after all, it would provide a restart, albeit with no one here to receive the message.

Depleted lands made unsuitable for agriculture, fouled air and waters left unsafe, and foods now unfit for consumption: Are these some of the gloomy features we are not willing to discuss? As sentient people we have sought to understand the core of our life-limiting challenges. We divide the issues into challenges for which we have no remedies, and push those aside, but thereafter, we do seem to have divided ourselves into those who deny any and all science and its findings and predictions, including those who somehow profess survival instincts, those who withhold viable information for whatever insane purpose, and those who seek logical remedies, albeit with outcomes unknown. A climate emergency didn't just happen, we are part of why it's happening, and way ahead of schedule.

I mention a few things about which we are hapless, helpless, and nearly hopeless, with the rest representing things we can do. We're just not good at it. Therefore, our current situation is a direct result of a reckless approach to the weakly attempted maintenance of our critical living environment as we persist in disposing of toxic wastes ever so blatantly.

Just how did land disposal, water disposal, landfills, and garbage incineration begin, anyway, and by whom, and with the permission of whom or what, and why is it perpetuated to this day, now with full knowledge of the grim and catastrophic consequences? And who ever dreamed up wastewater treatment plants that do no more than return our toxic wastes to our precious lands and waterways? They do not treat wastewater at all. Madness, for sure. Yet we default to government agencies at all levels for appropriate remedies, which in turn default only to the better interests of commerce. Thereafter, it all falls back upon us to defeat the selfish interests of the vast and clueless business and government consortia, hopefully and selfishly to turn the tide toward our basic interests and instincts for survival, no matter the mindless clatter that ensues.

We seem to have it all wrong as a society! In fact, living in, eating, breathing, and drinking our own wastes is not really a good thing. Is it? The Romans didn't think so, and we've hardly made any substantive improvements since then.

First we discussed the idea of global warming, then we revealed our emerging, albeit dubious thoughts about climate change, and more recently, we elevated the chatter to there even being a possible or highly likely climate crisis event going on. Today we openly speak in no less terms than that we are well into a flaming climate emergency—who would have thought? The universe is actually bigger, stronger, and more predictable than are we, so we must work to fit into the grand scheme of things, not to over-exploit the given resources so as to hasten our demise. It's all about carbon gases, and this time around it's about the carbon gases being over-produced or generated primarily by us, and at most, scantily recaptured. One might argue that a single volcanic eruption can spew megatons of greenhouse gases into the atmosphere, so why bother? But should that curtail consideration of the part we play in what some like to call an end-of-the-world-as-we-know-it probable scenario?

We need a way out of this dilemma. Our inevitable wastes from municipal, medical, stormwater runoff, and industrial sources can be better managed, but only if the agencies are willing to move past their lurid affair with commercial interests. One way of eliminating wastes is by the prevention of waste generation, for sure, but that would entail gigantic changes in consumerism and public behavior—not about to happen. Another way is to end landfilling of garbage by sorting for recycle or reuse in large part, as can be done, then committing the remaining refuse to renewable waste-to-clean energy plants—unlikely to happen any time soon because that would interfere with fossil fuel exploitation. Conflicts abound and we are decidedly both responsible and irresponsible.

The core of the issue is that engineers, scientists, and technologists who manage and work in these infrastructure settings are taught that liquid and solid wastes are in fact valuable and beneficial products as presented in waste streams, and that they are intended to be a key part of commerce, as is, not that they are likely killing us wholesale, such as they do. Wastewaters can in fact be treated successfully to enhance and replenish diminishing water supplies, just as solid wastes can be recycled, at least in part, with the remainder converted into renewable clean energy and other useful, non-polluting products, with full carbon capture and utilization. That's not going to happen in a world where infrastructure is managed by political persuasion and corruption only, rather than by the better interests of human and environmental health—never.

Remedying such aberrant and malevolent behaviors (i.e., deviant behaviors by premeditation) takes some measure of thoughtful effort, but the neglect we articulate by our failure to take any purposeful action prescribes such fate. It is the practice of non-action that then becomes the single strongest possible action. It is a forcing action even greater than that of the universe. To be certain, it is our clearly decided and well-articulated non-actions that become our strongest possible and intended actions. Logically, it does seem like our agencies should do something to protect us from the evil consequences of land-disposed *Toxic Sewage Sludge* and water-disposed toxic wastewater effluents, but our failure to act is also to blame, and that said failure-to-act is an actual thing, as prompted by us.

Let's consider for a moment the consequences of the simple, common, and far too often deadly Shiga toxin-producing forms of *Escherichia coli* (STEC) that incite foodborne diseases, and that often lead to hemolytic uremic syndrome (HUS), hemorrhagic colitis (HC), and/or death. About 2.8 million acute illnesses around the world each year are attributed to STEC, with perhaps 60,000 of these experienced in the United States. Shiga toxins are the main virulence factors found in STEC and in other pathogenic bacterial species and strains, with the primary prototype toxins being designated as Shiga toxin 1a (Stx1a) and Shiga toxin 2a (Stx2a). Simply look at the vast science and health literature in the peer-reviewed press to get a quick glimpse of the fateful adverse outcomes of exposure to these bacterial-borne toxins.[5]

Shiga toxin (Stx) is a highly potent bacterial toxin found commonly in the outtake products of nearly any wastewater treatment plant and is also the main virulence factor of Shiga toxin-producing *Escherichia coli*. This deadly Stx (the toxin) originated in *Shigella dysenteriae* and is also found in some serogroups of *E. coli*, then known as Stx1. The diarrheal disease shigellosis is incited by Shiga toxin, as carried by and originated from *Shigella dysenteriae*, but it is quite often transferred to *Escherichia coli*, which results in a highly contagious and severe diarrheal disease. Some *E. coli* strains also produce another toxin, Stx2 that has the same deadly mechanism of action as Stx1, but is antigenically distinct.

While these bacterial strains, along with their toxin-coding genes, may have originated in ruminants (cattle), they are transmitted to humans in beef products or in their manures, especially in beef food products that are served to the human populace at large. Humans then, may have first encountered Stx as a consequence of infection with *S. dysenteriae* type 1, and/or along with certain serogroups of *E. coli*, such as the well-known and highly deadly strain, O157:H7, and all as released into common public wastewater systems. So then, when our agencies state they test for *E. coli* in wastewater effluents and in the other product known as *Toxic Sewage Sludge*, and when they note the diminishment of *E. coli* in these matrices by the "treatment" features of their wastewater treatment system, which *E. coli* is it that they measure and monitor? Is it the *E. coli* with the toxin-coding genes, or may they be testing for any and all *E. coli* strains therein? It's a wild crapshoot, at best, and in total darkness.

As another consideration, wastewater treatment plant worker's clothes and bodies are sprayed constantly with Stx, mec (the mecA gene encodes a methicillin-resistant *S. aureus* (MRSA)-specific penicillin-binding protein), and other virulence and antibiotic resistant bacterial genes during the course of their workday, which workers then carry to their homes and elsewhere in and upon their bodies, clothes, and shoes, and within every opening of their body. Note well that a WWTP worker and tour guide at Seattle's West Point Plant happily told us during a public tour that, "We're all now immune to infectious diseases and to the flu because we are sprayed with 'this' every day."

If that behavior and message is a typical product of King County's training program for its sewage plant workers, that action then for certain is a complete, blatant, and dangerous public health failure. Of the many tens or hundreds of thousands of infectious agents encountered within any and all raw sewage sources, wastewater effluents, and sewage sludges, any or all of which will include viral, bacterial, fungal, parasite, and other human and animal infectious agents, it remains wholly unknown and likely unknowable as to the kinds and numbers of infectious agents released to the receiving waters and lands within or from the infectious agent-laden sewage waste materials, or their infectious nature, or their host range, or their fate in the environment. You need not seek further to identify or evaluate any one or more of the vast list of infectious agents in these waste materials, for focusing on just one or two key public health pathogens or infectious disease examples reveals a tale of incompetence, neglect, and inadequacy.

All of the science and engineering ever needed to clean our wastes and divert them for beneficial reuse exists, yet politically driven commercial interests rule the day. Therefore, we are forced to accept the inevitable consequences of living and swimming in our own filth until the day arrives whereby we become big enough to confer full responsibility for human and environmental health solely upon ourselves and demand corrective action by our institutions and agencies at all levels, in the face of otherwise disastrous consequences.

By the degree of our in-action or delinquencies, as well as by policy, and by the ill-will of policy- and decision-makers, we will be held accountable for either a diminished planet, or for thriving landscapes on an abundant Earth that can truly be enjoyed by all. The immediate challenge is to identify the agencies, institutions, and/or individuals who are directly responsible for setting environmental and public health policy, which is to say, we must identify those responsible who diminish human and environmental health by their advocacy of environmental contamination, pollution or degradation in the name of convenience, commerce or self-interest.

It is clear now, just as it has been for decades, that land-disposed *Toxic Sewage Sludge* bears a heavy toxic load that does not just go away, it is not readily degraded in soils by soil microbes, and it is not simply carried away when bonded to soil organic matter. In fact, it contaminates ground and surface waters, it contributes to greenhouse gas emissions from soils so contaminated, and it does harbor viruses and other infectious agents long past the immediate days following land disposal. Soils contaminated or polluted by land-disposed *Toxic Sewage Sludge* can never again be considered as being "clean" or without contamination, they are polluted forevermore.

Toxic waste is toxic waste, no matter the misstatements or falsehoods implied by agency management and staff. To accuse, indict, and prosecute the agency miscreants who are guilty and responsible for these toxic behaviors against human and other life forms then falls upon us to reveal and disclose in the total absence of any legal or regulatory oversite or prosecution within state or federal legal systems. We have a continuing job to do, so as to protect human and environmental health, amidst rampant agency failures to do so.

Chapter 19

The Documentarian

To assume that we must resign ourselves to turning our waterways into rivers of death is to follow the counsel of despair and defeatism.

Rachel Carson, *Silent Spring,* 1962.[1]

During the process of threading this toxic waste and sludge-tracking tome together, I was contacted by a bright and artistic young naturalist with a mind to the future who was planning to create, write, direct, capture, and produce a documentary film about the subject of land-disposed *Toxic Sewage Sludge*. She is therefore hereafter, "The Documentarian."

She had already travelled the Earth, seeking truth, or at least the justification or rationale for the atrocious practice and behavior of exposing our entire population to our own filth and waste. She had traversed the U.S. from border-to-border, and coast-to-coast, inquiring of academics, environmentalists, and citizen activists alike if any of them could possibly explain the reasoning behind or the justification for such age-old, epidemic-causing, disease-inciting practices as the land disposal of *Toxic Sewage Sludge*, and on our food farms, no less.

The Documentarian soon discovered that it was local, county, state, and federal agencies, in concert with industry, entrepreneurs, and their political advocates, that support and practice these high-risk and objectionably heinous health- and life-threatening bad behaviors. The same gruesome stories emerged from across all segments of the U.S.; there was no sane rationale or justification to be found anywhere, she discovered, for these dangerous abuses. We connected right away.

The Documentarian called and we made arrangements for her to visit Seattle to tour the thickly sludged forests, farms, and waterways of Washington State, and to see (and document) the ongoing statewide devastation firsthand. We made an expedition plan based on some experienced guesswork about where we should go. The objective was to observe and record the process as practiced and performed on any typical workday, weather permitting, without annoying the involved agencies, or their sludge haulers, spreaders or contractors, or the receiving foresters or farmers. We would not be the proverbial inconspicuous flies on the wall, but we would be the iconic flies on the feces, or so to speak.

She arrived on a September evening, and the next morning we departed before dawn for Central Washington State to investigate some of the near endless sludge-contaminated wheat fields, fruit orchards, hop farms, and vineyards, and to experience the magnitude of the sludge-dumping practices on critical, human-consumed food crops. We were not to be disappointed.

The adventure that day was a lengthy, 14-hour jaunt, highly fruitful, and with a full demonstration of the toxic waste dumping procedures and crimes being committed. Out on the prairie hilltops, with camera and sound-recording device rolling, we encountered some of the worst displays imaginable in northern Central Washington, and in places that have long held a history of toxic waste dumping. In the vast wheat fields, encompassing hundreds of square miles of formerly wild, carbon-sequestering prairies and natural soils, we observed and captured what was perhaps the most inappropriate dumping of toxic wastes on food crops that anyone has ever seen, all in real time and in live action. They have been performing this act for many years, as I have observed, so the scene and the players were just as expected.

Central Washington is already deeply covered with deadly toxic dusts and other toxic materials that have been accumulating over decades from regular applications of fertilizers, pesticides, and *Toxic Sewage Sludges*, each and all of which, whether together or in part, may generate some of the deadliest toxic dusts known anywhere. The winds in the region blow nearly non-stop year round, with no one spared the toxic insult.

Worse yet, and adding to the insult, due south of the sludged region sits the Hanford Nuclear Site, with a direct and regular air flow northward across most of Central Washington, and onward farther north into Canada. Hanford encompasses more than 500 square miles of land, nearly all of which is contaminated with millions of gallons and millions of cubic yards of radiation and non-radiation forms of toxic waste, all as dumped and accumulated over the past seven or eight decades.

At Hanford, for more than 30 years, the U.S. Department of Defense and the U.S. Department of Energy produced tons of plutonium for use in the U.S. atomic weapons program, as just one bad example. Beyond the atomic weapons related projects, Hanford also produced substantial quantities of extremely hazardous chemicals, such as and including hexavalent chromium, among others, as well as additional radioactive and nonradioactive toxic materials, some of which continue to flow and leak into the adjacent lands and waters to this day, including into the Columbia River, likely forever.[2]

For many years, Hanford disposed of its radioactive and nonradioactive chemical wastes by simply dumping, pumping, burying or otherwise discharging the many classes of toxic wastes directly into or onto the adjacent lands, despite the known toxicities and their persistence in the environment. No one cared, so we must care now, for our very survival. Many native plants and animals were contaminated, *in extremis*, and they will remain that way for many millions of years ahead of us.

So, will the sins of the father then truly become the sins or the inheritance of the sons, including the burden of the never-to-be-accomplished "cleanup?"

We will not even entertain the foolish and improbable thought, as has been reported, that the U.S. Government actually undertook a massive program of collecting and burning certain plant and animal life forms from that said 500 square miles of ever-contaminated Central Washington State that included contaminated tumbleweed, rabbit poop, and certain other tainted biota. And we will not even consider that the result was the broad dissemination of the ensuing toxic ash products that likely also contribute to the toxic load of the dusts that shall forevermore blow north and elsewhere.

I will not go near there, without due cause. When I took The Documentarian on a long, casual tour around the site, once she saw it up close and personal, and the all-engulfing dust storm, she totally lost it and demanded that we immediately drive west to Moxee, and wherever beyond, just so she could avoid breathing what she was witnessing—filmmaking and storytelling be damned. It happened to be a very windy day in that desert region of Hanford, and we were covered in those dusts, simply by being passing observers.

An entire regional industry and economy has been formed around the very peculiar idea that 500 square miles of radiation and chemical contamination can be cleaned up—not! And it never will. There is a solution, but that will continue to be spurned, most likely because the economic deficit that would be realized would potentially shut down the local economy. "It" can't be cleaned up, and no one expects it to be. It's the gift from the Department of Energy and other branches of the U.S. Government that just keeps giving.

The Documentarian quickly converted in her mind the volume of soil that would be derived from a 500 square mile parcel of land, say three feet deep, into cubic yards and tons that would have to be either removed, or at least decontaminated or buried. She told me in a flat minute they should just abandon the site, evacuate all remaining souls, and fence the place off forever—so perfectly correct. Meanwhile, I was tapping rapidly on the calculator in my cell phone to get to the numbers, but only made the first tap or two before she derived the massive amounts of contaminated soils that would have to be processed. I said not a word, and we moved on.

We were also on another serious assignment that day, so we made stops at three middle schools in three different local communities that were in the direct wind paths north and east of Hanford. The point was to get a glimpse of the exposures the middle school students experienced by the position of their schools in relation to the toxic winds. We collected dust samples from parking lots at each location. The samples were field processed by collecting and sifting each separate sample through three levels of fine stainless steel meshes, then sealed in lab jars for transport.

The first stage of the project was to submit the samples to an EPA-certified lab for testing for the 12 key toxic metals (all samples tested positive), end of story for now. The next two stages of the project are to test for a series of industrial chemicals, and then for the radionuclides common to Hanford wastes. Overall, the objective is to gain a profile of the metals, chemicals, and key radionuclides to which the students at the selected middle schools are exposed, thereafter, to monitor the levels over time, and to consider the likely sources, consequences, and remedies.

The Documentarian did just what documentarians do, they document everything, and in this case, she documented my rear end, waiting for the moment when I would be busted for crawling around on my knees in school parking lots on some unexplainable mission. She laughed and filmed. I crawled, sifted, and collected, and we got away free.

Although Chernobyl, Fukushima, and Three Mile Island were nuclear accident disasters of various and extreme magnitudes, the very thought that Hanford represents a disaster by plan and intent, and without a strategy for mitigation or cleanup, makes the case well that Hanford has earned the right to be referenced as the most dangerous and contaminated site in North America—it cannot be cleaned up. The greater point is that when the combined blown dusts of Central Washington move ceaselessly across the broad landscapes, including across cities, towns, and remote farms, the combined *Toxic Sewage Sludge*, fertilizer, pesticide, chemical, and radiation-laden dusts spare no person from toxic waste exposure, not one person, ever.

And while we know the federal government spends billions of dollars per year on their purported Hanford Cleanup programs, of various sorts, it is quite apparent that these wastes cannot ever be cleaned up. The radiation and nonradioactive chemical wastes can be buried, grouted, moved, reformatted, vitrified, "treated," or the like, as part of some grand Toxic Waste Relocation Program, but the state and federal government's Toxic Waste Relocation and reshuffling programs will most likely have no positive or favorable effects at all on health outcomes for the people of the region, far and wide, and for millennia ahead. The obfuscating and creative reports and their best of intentions are no better than fiction, Pulp Fiction: "*And I will strike down upon thee with great vengeance and furious anger those who attempt to poison and destroy my brothers*."[3]

When I'm out there in the field with a like-minded witness, such as with The Documentarian, we're always wondering, "What are we doing here? We're tracking and exposing ourselves to these highly unsafe dusts and *Toxic Sewage Sludges* that actually kill people, just as they have for more than a century or so. Don't we have a normal life? What is this about? Why are we here?" "Ah, because we live here, with one another, on this planet of limited resources," we would remind ourselves.

Together we were dealing with some of the most maddening circumstances, and The Documentarian and I were on a mission to make sense of, and to provide warnings for, other people. "Maybe there's some kind of odd bonding that happens by this strange obsession in such a toxic and hostile environment," we pondered. "What a sorry thought," we concluded.

As The Documentarian and I drove toward our somewhat vague destination, I explained to her, as I have repeatedly stated to others, that the rules for these intensely dumped *Toxic Sewage Sludge* sites in Central Washington are bizarre, obscure, complex, and wildly different. That's because the U.S. Government, mainly the Department of Energy and their contractors, still show little or no meaningful progress on preventing, fixing or cleaning up any of the nuclear radiation contaminants and pollutants lost to the air, soil, and water, from way back when until now, in these regions.

The myth of "Cleanup" thrives. Although plans abound, as plans often do, there is no apparent progress. In effect, Central Washington State, and all proximal and distant air, land, soil, water, and food that are or may be directly affected downwind or downstream, have no toxicity relief in sight, there is no one in charge, and nature's fix, in these cases, can be measured in perhaps millions of years, at best.

Historically, and as presented to us first by the U.S. Atomic Energy Commission (AEC), then later by the Energy Research and Development Administration (ERDA), subsequently by the Nuclear Regulatory Commission (NRC), and now by the U.S. Department of Energy, is the endless release of historical, recent, and current radiation contamination that seems to be considered part of the built-in local dangers that we receive and perceive, and it is.

Are we to be hoodwinked more into believing that radiation exposure is benign, or that remediation must be put off until some mythical distant time? Not so! And if you think even for a moment that there may be some benign or peaceful use to be proclaimed by the toxic effects and manifestations of the combined and synergistic toxicities of *Toxic Sewage Sludge*-borne toxins, along with the toxic components of fertilizers and pesticides, then also along with the loads of historic and contemporary toxic radiation contaminants from such places as Hanford, then that is a very misguided thought. We see no evidence of remediation, reclamation, change or cleanup, at all, only a concerted continuance of past behaviors, most especially ones that include adding of the toxic wastes of the past with the toxic wastes of today, and all within wholly predictable, cumulative, and synergistic toxic formulations and formats that will inevitably harm us all.

There is no clear evidence of any effort or work product that reveals or proclaims clearly that there may be any less toxic exposure now than has been experienced over the past 80 or more years, and in fact, by adding yet more pesticides (including some that are now restricted), fertilizers, and domestic, industrial, medical, and stormwater runoff toxic wastes, to the regular menu of toxic exposures, the citizens of Washington State, and most assuredly the people of Central Washington, are guaranteed a firm opportunity for a life of inconvenient illnesses and untimely death by the hands of our local, county, state, and federal agencies that seem to have a very limited tolerance to hear citizens' concerns for human and environmental health, including of their own.

If this were not true, we would not be where we are now amongst the plethora of accumulated and ever more accumulating toxic wastes. Were it not for citizen environmental activists, often as individuals, and such as The Documentarian, we would be even farther into an ever-expanding and worsening toxic debacle than we are today, with no hope for any form of prevention or remediation. Were it not for courageous citizen activists who work to keep track of the endless pollution we see (and most often don't see), more of our precious lands would be rendered wholly unusable and uninhabitable for the rest of time, to our collective detriment.

Nodding in courteous agreement to my lengthy rants that day on such ignorant behavior, The Documentarian then told me a little about her journey. She had moved to California via Mexico and was living in Temecula, north of San Diego, south of Riverside. It used to be beautiful, wide open ranch country and now it's just beautiful ranch country with homes dotted everywhere.

I mentioned to her that Riverside is where I attended the University of California (UCR). It was during my graduate studies at the UCR agricultural campus, aka, The Citrus Experiment Station, that we reluctantly evaluated the use of sewage sludge as a soil amendment and fertilizer replacement on various crop plants to see if it enhanced the virulence or pathogenicity of plant pathogenic fungi and the resulting plant diseases, or whether it might enhance plant growth and productivity. As well, we investigated further to determine if such sludges may adversely affect agricultural crop efficiency, or if such treatments enable plants to resist plant diseases by some wholly unknown mystical strategy, or not, and the answer was always:

"No—not! Don't do it! Do not add sewage sludge to our agricultural soils and crops. There is no benefit to the soils or crops. It is toxic waste. People hate it," and so on.

Sewage sludge was always found to be too toxic to soils, soil microbes, groundwaters, and plants. Interestingly, and for relevance and comparison, while California has strict laws, codes, and regulations for the management of *Toxic Sewage Sludge* wastes, Washington State has nearly none. And, while some states revere *Toxic Sewage Sludges* as being akin to holy human shite, others do take the time and effort to evaluate and reveal the contents and adverse outcomes.

As the story goes, The Documentarian became interested in the sources and kinds of pollution that were contaminating her favorite oceans and beaches everywhere, so she started investigating what was causing it. That led her to scour the U.S. for sources of actual on-the-ground experience and information, and when she contacted me, she began with, "Everywhere I went, I met people who said, 'We can tell you about our experience and research, but you have to go to Seattle and find Richard Honour. He's the only person we know who actually chases sewage trucks around the mountains on a mountain bike.'"

That's what prompted her to come to the Pacific Northwest. Whether she was in Texas, New York, Virginia, Pennsylvania, Ohio, North Carolina, Oklahoma, or anywhere in her quest for information, she kept bumping into my name, and so she came to the source, valid or not. Anyway, she ended up calling me and said right off the bat, "I was very nervous about calling …"

We had a great phone conversation, which then led to, "I've been thinking about making a documentary about what happens with human sewage, and several people told me to stop calling it 'biosolids,' and that I had to call it '*Toxic Sewage Sludge*,' because otherwise that strange guy in Washington would get after you." That's what happened, we laughed, and here we are today.

In the process of filmmaking, The Documentarian discovered that New York, New Jersey, and most especially Manhattan sewage sludge was first sent on trains every day to New Jersey, Pennsylvania, Ohio, and Texas until that schedule changed or stopped, and now it goes to some of the same places, and Oklahoma. After her visit she went back to Oklahoma to continue documenting the sewage sludge trains traveling across the country. She and I were laughing because the whole silliness is, "People in Texas and Oklahoma are so stupid they think it's useful to take and make use of other people's shit." We had a lot in common.

We could only joke about how preposterous it all seems, and how related the economics of *Toxic Sewage Sludge* management seemed to be to the enormous sums of money spent on solid waste management and the vast global transportation business of plastic wastes, and the like. If you make things, it appears there may be more revenue generated in management of the residual wastes than there is in actually making and selling the intended products.

Just as with the Hanford debacle, no identifiable person or agency seems to want the radiation contamination and other toxic wastes actually cleaned up or otherwise removed, for so long as there is toxic waste available to manage, the economy benefits. Similarly, so long as there is municipal solid waste and *Toxic Sewage Sludge* to manage and transport, no one wants to think about the vast economic deficit that would result from preventing the production, transport, and disposal of the forever-generated waste products—it is basic economics we most likely face, and perhaps not any intended lack of consideration for human or environmental health.

I don't know how it came about, but The Documentarian revealed that she was also concerned that women are not well-recognized in the film industry, not just in conventional Hollywood film work, but also as documentary film creators or artists. They're not recognized as talent with regard to the creative aspects, as well as to the production, writing, editing, cinematography, filmmaking, and all of that.

There's one shirt she wears that's black with red writing, and it's something about women in media not being appreciated, and so says her baseball hat, worn backwards. She lives that life and is motivated to enhance the role of women in media. Her film is focused on investigating, throughout the U.S., the fate of sewage sludge: "What's in it? Where does it go?" People burn it. People use it for fertilizer. People dump it in landfills. All of this she wants to understand and document.

With regard to what's in it she says, "You're like the only one on Earth who wants people to know why it's bad, what's in it, and why it's toxic—what is it that kills things. Because the rest of the world seems to be saying this is a nutrient. It's so wonderful. It's good for your spinach and lettuce. You know, let's eat this stuff," she carries on.

She didn't tell me a lot, and she shouldn't, because people who write, make films, and create do not want to be influenced by others' opinions or by early, unsolicited feedback.

As we travelled through the mountains and across the state, she asked me to narrate the passing scenes, so for two full days I didn't shut up for a minute, and she appreciated it, or so she said. This was all new country to her. There were at least two occasions when I thought, "Oh boy! She's going to faint. I'd better shut up for a moment, at least."

On the first day driving to Eastern Washington I told her, "You know, sometimes when I'm going out, and because I need to film or image where King County's trucks go on the Central Washington prairies, I may have to make three or four hundred mile trips several times before I can find even one exemplary and especially violated location, and then it may already be a month old. The counties, farmers, and haulers, and the Department of Ecology," I told her, "all tend to keep the activity secret in collusion, as if they may be detected by actual concerned citizens, or that they may be performing criminal acts." According to laws enacted to protect citizens and the environment, the open land disposal of toxic waste is in fact a criminal act, simply protected by the exclusionary principle, and all as approved by the U.S. Congress.

So there we were, headed out on Interstate 90 over the Cascades, then north on Hwy 97 to Hwy 2, then east toward Wenatchee. We crossed the Columbia River, turned south, then drove up Badger Mountain Road, past massive, previous King County *Toxic Sewage Sludge* dump sites, and then headed north toward Waterville and Mansfield. We had driven over the Cascades and were heading east when I said on a hunch, "Oh, let's go up here today. I haven't been up this road for a while, so maybe they're dumping sludge today on the wheat fields north and east of here."

Son of a gun! As I looked in the rearview mirror, there, approaching from behind at high speed, was a King County Loop® *Toxic Sewage Sludge* double dump truck, headed toward us, belching carbon into the sky. We were doing about 60 mph on this 50 mph blacktop two lane country road, and the monster approaching from behind could have been exceeding 65 or 70 mph. So, when we came to the place where we were planning to turn, meaning, where I knew the truck was likely to turn, I went straight, and as soon as he was out of sight, I turned around and we followed him.

"Oh my Gosh!" I said, "The Great Spirit came down and blessed you today, Woman, because people just don't get to see this! They only get to see the nasty aftereffects." I was convinced this had to be divine intervention of some form.

The Documentarian never takes off her video cameras and audio equipment. She wears it all like the body armor of a SWAT Team, and so she turned around in her seat and began filming more sludge dump trucks coming from behind. Then, when they screamed ahead to pass us via the oncoming lane, she was filming them as they were going up and down the highways and hills, across the prairies, through the fields—really artistic stuff. Off in the distance the double dump trucks were weaving through the hills amongst the picturesque wheat fields, really nice visuals of toxic waste disposal in delivery mode, on the run, in real time, hundreds of miles from their wastewater treatment plants of origin way back in King County, guzzling and burning fossil fuels by the thousands of gallons in a never-ending quest to get back in time to repeat their extreme polluting invasions the next morning.

Of note, the Washington State Governor proclaims that he intends to achieve his GHG control objectives on time, I shall imagine, while continuing the sewage sludge double dump truck carbon invasion, as well. We must seek to discover if the Governor is informed or may be aware of the atrocities engaged in by his Department of Ecology and their county minions—I doubt it.

So then I realized, "I know where they're going." So, we took a short cut along another backcountry road and got there first, because I finally figured out where they were headed. We were parked along the edge of the road and here they came, as expected, straight at us, from over a hill in the distance, and I said, "You see that little dark bump or dot thing coming over the hill over there? It's probably a mile and a half out. That's the sludge truck that raced past us on the two-lane blacktop." The next hill over we watched him stop, as anticipated. He then made a turn and drove into the wheat fields.

For a better view we had to move a little through the hills, and there it was, that rare Hollywood to-die-for scene, especially if you are searching about 10,000 square miles or more of rolling hills and wheatlands looking for sewage sludge-laden trucks dumping their toxic loads, and we were able to observe (and document) the entire procedure from a public highway.

I continued to narrate, barely containing my excitement: "This is what they're gonna do. He's gonna back over there. He's gonna dump out of the back dumper of the double dump truck, i.e., the pup, and then he's gonna move forward, unhook it, and then back up the main dump truck. And then he's gonna dump that, and then he's gonna exit the cab," and I was joking, but I've seen this enacted so many times, "... and then what he's gonna do next is, he's gonna open his pants and pee, and then get back in the truck, hook up the pup and head back to Seattle." I said, "I've seen it a million times. It's the way truck drivers operate. After you've been in that truck all day for hundreds of miles, any break will do."

So, what does he do? He gets out and pees, and she's got it all on film, just as per the unwritten script.

As soon as he leaves and gets away over the hill, BOOM!, here comes another double dump truck, the dark green ones from that county far to the north of Seattle, which is contracted by King County to haul their obnoxious loads. It just doesn't happen this way! The second truck pulls in, and there's always a sign in the field, a white sign, and on one corner it reads: "WP" for West Point Plant. The other point on the sign reads, "SP" for South Plant, and the other one indicates "BW" for Brightwater Plant. When the trucks arrive, they essentially identify themselves with a wastewater treatment plant of origin by their positioning, meaning, from whence they and their toxic load originated. They back up per the guidelines and dump the sludge in the designated area.

Next came the front loaders, and then the spreaders, within a minute, with the front loaders filling the spreaders, and the spreaders driving back and forth across the extremely dust-blown landscape spreading their toxic loads right away, then on to the next sequence. In some cases, the thousands of tons of *Toxic Sewage Sludge* may be dumped in large piles and left for days or weeks, and in others, the sludge is loaded into spreaders and immediately spritzed across the once clean and beautiful, virgin prairie rangelands.

The prairies out there are massive, natural dustbowls created by historic melting glaciations during and after the last Ice Age, 15,000 to 18,000 years ago, when the ice dams burst and glacial lake floods left behind the sediments. It is a view, landscape, and scene meant to blow your socks off, and now they're trying to fill it with *Toxic Sewage Sludge*.

It's all about scheduling, so as to land dispose as much of the *Toxic Sewage Sludge* as possible across as much of the farmlands as can be done before the seasons and weather change, in order to have the next human food crops started for the next deadly cycle. As we stood there marveling at our good fortune, the sludgers spread their toxic wares on this dry earth, while pools of dust just blew up into the sky, travelling for miles to the northwest that day on the ceaseless winds.

While we observed the spreading, the loaders dug into these big piles of *Toxic Sewage Sludge* and loaded the dried sewage into the spreaders, and here we were filming them driving through wheat fields and across the sacred prairies of the original Indigenous inhabitants, shooting black human crap out into the fields. What we were seeing firsthand were plumes of brown dust coming up from the wheels as they went through the sludge, carrying the fine, sludge-tainted dust and spreading it everywhere, for hundreds of miles, and beyond.

The dusts go into the sky. You could see it blowing off in all directions. There are towns not far from there, and they must be accustomed to it. This airborne material is combined then as a mixture of the residues of windblown fertilizers, pesticides, and *Toxic Sewage Sludge*, plus the toxic chemicals and radiation dusts from Hanford, not far to the south, accumulating and blowing and going all over the place, and I told her again, "Woman, you are so lucky! It just doesn't happen like this! You are witnessing and recording the intentional process of air, food, soil, and water pollution in actual real time, by plan and intent, and with the unimaginable loads of the inciting agents of cancer and other chronic diseases on board."

As we were leaving, and just after the second truck departed, BOOM!, here comes another dark green sludge truck over the horizon from King County! It is four hundred miles from Seattle to this location in Central Washington and back, so you can imagine the fuel costs to haul tens of thousands of tons of *Toxic Sewage Sludge* to this one site alone on all available days, weather permitting.

As mentioned, *Toxic Sewage Sludge* is roughly 70% water, so the weight is essentially the weight of water—toxic, dirty water. Millions of dollars in driver, fuel, insurance, and maintenance costs alone, plus the U.S. DOT highway fees and the implied liability costs. Then King

County has to pay or otherwise arrange for the Beneficial Use Facilities (BUFs) to take it. There are many variations of the contracts, logistics, and financial arrangements between the state, the counties, the BUFs, the haulers, the spreaders, the farmers, and the facilitators, but the cash flies as cash will do, and the subject toxic wastes are spread onto the heartlands of Washington State, where they are "disappeared" forevermore.

The BUFs, who sell it to the farmers, get paid to spread it across the landscape, as well. Do you then wonder who pays for the equipment and labor costs, and the fuel bills for loading and spreading all of that *Toxic Sewage Sludge*, and also for managing such widespread operations? It's a many millions of dollars per year boondoggle industry, just to dispose of *Toxic Sewage Sludge* on our agriculture crops, and to degrade our air, food, soil, and water.

Then the third truck arrived, and he went in and repeated the same cycle all over again, and she's just about to faint. "I've been all over the U.S., but there's nothing like this! They told me I had to come to Seattle to see you actually chasing trucks around, and that's just what we did!" she carried on in astonishment.

"What must be the death toll?" we wondered aloud during the drive to the next devil's dump spot.

The Documentarian and I left the active and horrible scene, fully entranced, both disgusted and yet delighted by our good fortune, to go to another location nearby, by way of the county road. I offered, "If they're dumping here today, then I'll bet I know where else they are doing their nasty deeds right now."

From previous experience, I had some ideas about where they might also be dropping off, and just sitting on the dumped sludge piles until they could get it spread. We drove the few miles, and there they were, more mountains of fresh *Toxic Sewage Sludge*! I took pictures of The Documentarian standing along the road near the sludge. She'd never seen the real thing before, up close and personal, in all of her travels.

We had just gotten back in my car when all of a sudden we were hedged in by two pickup trucks, front and back, with one veering toward me at speed so as to run us off the road; a clear case of vehicular assault by intent, obviously fueled by road rage, and likely incited by full awareness that they had been outed while participating in the polluting policies of King County and the Washington State

Department of Ecology, and were caught. They know me, they know my car, and the one farmer boy got out.

The Documentarian was sitting there in the passenger seat, wearing her baseball hat. She won't wear an N-95 mask because she knows well the toxicity of the dusts in which we were now covered; she wears a full respirator mask because she's terrified of the toxics in the thick dusts. She left her cameras running, audio and video, and the little farmer guy came up and started swearing and screaming at us in the foulest language you may have ever heard about what they were going to do to me and to her, and this and that, and on and on, and she was trembling in fear, but still filming. She was trembling there in the front seat, but in her coolness, she had everything still running. She captured it all. They didn't catch on to that.

Others appeared from the second vehicle and pulling out their cell phones, squatted down in front and back to take pictures of my license plates. Then they came over to my window, which was open. The one farmer guy, the older one, I know very well, and he was the focus of my attention. We had testified together on opposite sides of the spectrum at a King County council meeting not long before. It was just me that day, first invited at 10:30 p.m. the evening prior, versus 18 or so representing the State's Department of Ecology, King County, their University of Washington advisor from the Department of Environmental and Forest Sciences, and their BFU contractor, and who knows who else. They had obviously had a long strategy preparation meeting ahead of my arrival to counter all that I might report, but it didn't work. They told me I had only 5 minutes to make my case, while their toxic sludge-dumping team likely had hours to plan their complicit schemes.

"What case?" I asked. "Why did you invite me here? What is this about?"

"Oh," said the council leader that day, "so you can explain your ideas about sewage sludge."

So, I took 15 minutes and focused on the obvious and well-published relationships between land-disposed *Toxic Sewage Sludge* and infectious diseases and cancer, as you have heard reported earlier in this work.

The King County councilmembers and this now belligerent farmer were very pleased with themselves that day in the chambers, and with the comments of their advisors and contractors—what a crew.

I have wondered what their stance would be once *Toxic Sewage Sludge* dumping is ended for reasons of its carcinogenicity and other toxicities. That day will come. The land disposal of *Toxic Sewage Sludge* will end, once the current elected officials are denied reelection for their support of practices that compromise public health and that incite environmental degradation, due to the greed and self-interests of their politically driven commercial adventures. This practice is not "Economic Development;" it is blatant corruption, contamination, and pollution of valued resources that are of definite national security concern.

Only weeks prior, I had faced this farmer defending his unethical livelihood practices in the middle of the wheatlands, and this time, it didn't look so good. In any case, I did know the fellow, and there he was now screaming and swearing at me and the whole interaction was recorded. Along with streams of foul and disgusting language slung at The Documentarian, to be presented, distributed, and reflected upon for a long time to come.

I will assume that behavior patterns of individuals like this enraged farmer must be a result of breathing in all of those toxic pesticides, fertilizers, radiation chemicals, and *Toxic Sewage Sludges* over a lifetime, or so I will imagine. In any case, they can be highly threatening people, as demonstrated so clearly by their degraded mouth workings and mental aberrations out there on the prairie—most likely the end-product of their own self-applied and self-induced toxicity, generation upon generation.

Finally, the younger, highly unstable, and extremely foul-mouthed of the two, and seemingly the most volatile, meaning the one in the Oshkosh b'gosh-like overalls, started going back to his truck and so I asked the BFU manager/farmer I knew, "What's going on? I'm out here on a county public road showing what you are doing with King County's *Tòxic Sewage Sludge*, and you and your pal go nuts. This is a county road."

Well, he just started swearing and screaming all over again. I spoke very calmly because I knew everything was being recorded, and I was trying to be civil, so I said, "You know . . . we should just have a civil conversation about this. I'm here on a county public road and you have no right to object to what I'm doing." So he started calming down because he realized what I was saying, and we started talking.

Anyway, the situation became more civilized, but you couldn't have made a Hollywood production of a mafia movie with more horrible, disgusting, violent threats and screaming and intimidation than what was going on with these farmers. These are passionate people regarding their rights to dispose of toxic waste on their lands, but it doesn't stay on their lands, it becomes widespread, wind-borne toxic dust, presenting high-risk exposures to all adjacent neighbors and others, end of story. Contamination and pollution of the air, food, soil, and water are of no concern to them. They don't want any kind of rules or government regulation, no matter the degree of public health compromise or environmental degradation that may be caused by what they do.

At the end of his violent display, what he said was so perfect for the film: "We don't want people like you around here, so just go back where you belong and stay there." It was perfect. I couldn't have written that script any better. All captured on film and with full sound. So, it was quite an expedition, that first and most memorable day of our filming.

I've never told anyone about the so many similar interactions I've experienced over the years. It's not that we were out there for vengeance, or for vandalism, or for some form of extreme advocacy, or that we are mad at anyone, and it's not a political agenda, but one of singular concern for public health. In any case, this specific interaction demonstrates the kind of attitudes we experience from people who receive from, and are contracted by, actual government agencies to spread this toxic filth upon our food crops! I won't tell you this, but one set of irate farmers told me that, "It didn't really matter anyway, because all of the resulting food crops were exported to Asia, Latin America, and Africa," so what does that tell you? Does that then imply Washington State thinking, attitudes, and policy? Kind of makes you wonder.

Once The Documentarian recovered from the tense interaction, we headed south to some other likely sludge-dumping farming areas to fill out our day. She wanted to see sludge on grapes because Washington wines go all over the world, and she wanted to see sludge on hops because Washington hops go all over the world for beer-making. So, we went south, and she videoed grapes and hops, because those hops from the Central Valley of Washington are used in most beers produced in the world, or so we are told. I have images of hop plantings buried in *Toxic Sewage Sludge* delivered from King County.

It got dark before we finished the trip, so we called it a day and headed back to Seattle. We rode in silence for quite a while, then pondered what the morrow would bring. "They told me it would be like this if I met with you," she mused. "So," she added with enthusiasm, "where are we going tomorrow?"

The next day she wanted to see sludge dumping in forests, so we went into the Snoqualmie Forest, just east of Seattle. She's not used to such forests with things like bears and cougars, so I took her first to a recently sludged site where a cougar had attacked a mountain biker. We got out to see the site, but, as it turned out, I got out and she locked the doors and stayed in the car. I said, "We're going over here and I'm going to show you exactly where the cougar hid the biker's body under a log," but I was only speaking to myself.

We then went to a couple of other places, and I could show her where they had dumped sludge in the wetlands and where they were now building luxury homes in a forest that just a year or two ago was full of toxic waste. She found that disgusting. These people were never going to know their luxury homes were built in sludge-dumping fields.

Farther into the forest I took her to that bluff I described earlier that's an 800-foot cliff looking out onto the wetlands below. We hiked up there. This is one of the most stunning places in the Snoqualmie Forest. You're up on these big granite outcroppings and looking at the landscape below, and there are ponds, streams, forests, and wetlands, with the leaves on the trees just turning color. Beautiful!

As we approached I said, "This is where they took their sludge spreaders and made a three point turn up against the edge of the cliff with the sludge spreader still running and they were shooting human waste off this ridge down into those wetlands and forests below."

"Oh my God," she said, "I've never seen anything like this. You're telling me they backed up here and dumped shit off this cliff?"

"Yeah," I sighed. "You can still see it down there, even after all this time, slowly leaching off to those wetlands, which flow to Elwell Creek, then north to the Skykomish River and on to Puget Sound."

"When was it?" she queried.

"I can't remember. Three or four years ago," I recalled.

"Well, it's all gone now, right?" she asked hopefully.

"Not quite. Follow me." So we started walking through the forest, and I was showing her the still thick sludge deposits on the trees and ground, and the overall sludge-disposal situation.

"You see the ground here? It's all covered with fir needles and other plant debris, and," I said quietly, "all that black stuff under the leaves is sewage sludge residuals from three to four years ago. That will be here for many years, slowly releasing its load of toxic chemicals as it decomposes."

She's filming and recording my narration, and all of a sudden the realization bomb went off and she starts hopping up and down—she finally saw the black, tar-like material that coated the surface of the forest floor. "Get me out of here. I don't want to walk on that. Now I'm all contaminated. Get me out of here! What is wrong with you?" she's hollering at me. "You made me walk in this again! I want to film it, but I don't want to get in it!"

She was dead serious. She wears those medical booty things on her feet, like when someone comes in your house to fix your heater, and they wear rubber booties. And she has gloves on and not a mask but a full, double-filter respirator, plus that backwards baseball hat and everything. She's terrified of this toxic stuff.

She said she's afraid of it because she'd never heard of it before and then once she heard of it she really wanted to investigate it, and then the more she got into it she realized how toxic and lethal it is. Everyone told her how toxic and lethal it is, but no one does anything about it, and so on, and now I made her walk in it – Strike one.

On the way out she said she was tired and wanted to rest. "Don't they have any logs around here I can sit on to rest?" she asked.

"Of course," I responded, "But just who are the 'they' of whom you speak, and just why would you want to sit on that log over there? It's buried in *Toxic Sewage Sludge*." – Strike two. "And the toxic dusts we are breathing are in the air all around here," I added – Strike three.

More screaming and despair, and we hiked out of the contaminated forest in a hurry, and without speaking another word. So, needless to say, we had a full day in forests and a full day out on the prairies and she got way too much information, and she was sure she was going to die at any moment.

I have a lot of activist women friends, and they kept calling while we were on the road or out and about, and she could hear them on the car speaker, and was curious, "Who are all of these women? Why do they call you all the time?"

"Because I work with them," I told her. "They're on the Yakima Reservation, they're in Central Washington, they're down in Prosser, they're over here, they're over there, and they all happen to be women. I don't know many men who do this thing—environmental activism. I think if I get down to the genetics of it, probably women are more caring about the adverse effects on populations and especially on children and food, and about destroying soils and waters. I do think it's almost a core genetic sensitivity." She smiled, understanding.

Before leaving the forest, The Documentarian told me, "I go all around the country and to all of these places, and so often I think these people are all just full of shit. They're just bullshitting me to try to make a better story. But you tell me this stuff, and then you take me out and you make me walk in it! This is about validation," she said. "I get it now, there's no more doubt. Let's get out of here."

The next morning she flew back to California, then off to Oklahoma again to another sludge operation, but later she called and said, "Ok, I really want to come back. Can we go and do several more days? I'm all better now. Can we go do more? We didn't finish our job."

"Yes, of course," said I. "There's much more to do."

The plan is to visit a handful of wastewater treatment plants where I can actually take her inside to see how they operate, and then out to their sludge-drying areas before it's loaded on the trucks. She wants to document that.

The million-dollar question that she wants to convey is: "How do we manage all of this human waste safely for a change, other than through incineration or putting it in landfills or trying to recycle it which isn't working?" I know the answer. I've experienced just how little of recycling works. After several decades of sludge tracking, I have chronicled everything I see and it isn't working!

As The Documentarian told me, she's been all over the United States tracking this shame, and she came here, and it was just: "Holy shit! We're out here in this and the trucks are coming by, and these people are screaming at us and we're tippy toeing through the forest. This is sludge!" And she's almost screaming and crying in fear. "No one ever sees this," is what she's saying, "and yet it's all over the country. People have to know."

I look forward to The Documentarian's now-experienced, creative interpretation of this undeniably serious, complex, and highly toxic issue. She hardly took a breath during our entire two-day adventure. We discuss her findings often, and I am supplying her with a trove of images and videos spanning a decade and a half of sludge tracking in Washington State, far more evidence of this perpetuating abuse than she had ever hoped or bargained for.

Part III

The Solution

Chapter 20

Nexus

We find ourselves poised at a fateful point.
We've diligently hauled ourselves to the top of a roller coaster
with no idea how steep and long will be the plunge that awaits.

Eugene Linden, *The Winds of Change,* 2006.[1]

"In a world without us," he asks, "would the climate still change?"

"Yes," she responds, "for sure, but we're making it happen faster."

Therein lies the heart of a thoughtful conversation. We are responsible collectively for our global state of being, and we must take action to fix things, and it can be done. On 28 July, 2022, the United Nations General Assembly recognized that a clean, healthy, and sustainable environment is a universal human right. "This refers to the three main interlinked environmental threats that humanity currently faces: climate change, pollution and biodiversity loss…"[2]

What we learn by our investigations and observations is that, yes, the climate is changing, just as it always has and just as it always will, because that's what climates do, they change. The difference is that we appear to be accelerating the rate of climate change by our polluting activities, mostly by the generation of carbonaceous and other gases. Similarly, we also learn that all life forms evolve forward to invade and colonize all possible niches and environments, simply because they can and because that's what they do.

Our role in these events should be protective, but it's not; we continue to impact all natural and other processes and events on the planet, mostly to our detriment, and without much of an appreciation for the causes or consequences. And while natural processes offer multiple and diverse settings across, above, and within the Earth, and while life forms invade and adapt in their endless pursuit of nutrients and reproductive opportunities, we provide obstructions, often dangerous obstructions, because bureaucracy overrides logic.

We must discern if we actually fit into the natural scheme of life on Earth or if we may be no more than obnoxious interlopers. We seem not to be programmed to fit into any natural environment but may be best adapted to take over and annihilate at will, not as good neighbors. So, yes, the climate is changing, and yes, life forms are invading and colonizing all possible niches, as climates and life forces do, but our self-assigned role seems to be focused on disruption, and the killing of all things along the way, and not fitting in anywhere.

More than 7 million species of plants and animals are estimated to have lived on the Earth, although revised estimates suggest that just slightly more than 2 million species may have existed here. However, if you include microbial life forms, the estimated number of species then becomes much larger, reaching perhaps 11 million, and with newer estimates proclaiming there may be more than a trillion species of microbes alone, including those not yet discovered. But what are the criteria for defining a species, and who says so, and just who's counting? Is the static number of extant species a reasonable measure of the stability of life, or do we also count those that lived before us, plus those who live now and the species that will emerge next?

The far ranging estimates of the number of animal and plant species on Earth lie somewhere within 2 to 30 million, an enormous discrepancy, but then again, who's counting? No matter the wild or more conservative estimates or projections, about 1.2 million animal species have been formally described, with the vast majority identified on land, not in the oceans.

From another quantitative perspective, that of biomass, or of biomass production, Earth's life forms are dominated by plants, revealed as providing perhaps more than 80% of the global biomass, with animals providing less than about 1% and livestock likely outweighing wild mammals by a factor of ten. Humans seem to provide no more than about 0.01%. Other estimates suggest that as many as 86% of all life forms likely reside in terrestrial environments, with the rest dwelling somewhere within the Earth or inhabiting marine environments. If humans make up 0.01% of all life forms, what then is the rest, meaning, just what is the composition of the remaining extant biosphere on Earth, and what is the resulting biomass of life within that biosphere?

In 2018, Bar-On, et al., in a study for the National Academy of Sciences, quantified life by estimating Earth's biomass as tons of carbon, the primary building block of life. They state that plants, mostly trees, dominate life on Earth, accounting for more than 82% of biomass, to be followed by microbes at about 13%, with animal life measuring about 0.4%, and humans weighing in at perhaps 0.01%.[3] Of note, they also tell us that livestock accounts for more biomass than all human and wild animal life on Earth combined. Looking at all these numbers allows us to understand more about our planet and our place within it. In any case, these estimates remain in flux.

One of the most widely cited figures regarding species comes from Camila Mora and colleagues who state in *ScienceDaily*: "About 8.7 million, plus or minus 1.3 million, is the new estimated total number of species on Earth, the most precise calculation ever offered, with 6.5 million species on land and 2.2 million in oceans."[4]

Another source, the International Union for Conservation of Nature (IUCN) *Red List of Threatened Species™*, has evolved to become the world's most comprehensive information source on global extinction risk. The Red List reveals the current status of threatened and endangered species, providing regular updates on the status or disappearance of life forms on or from Earth. They state clearly that the listed species represent "only a fraction of the actual number of species on Earth, because many have not yet been found, identified or studied."[5]

If you were to tackle the chore of sorting out the most reasonable estimate of the number of species on Earth today, you may discover, as have I, that the real and factual number is wholly beyond comprehension and thereby unknowable at present and may not become known with validation for decades. And, while new species may evolve at evolution's pace in response to environmental forces or stimuli, and with the availability of adequate nutrients in a favorable space, the rate of disappearance of extant species is being witnessed at an even greater pace, climate change or not, and by wholly unpredictable means and rates. Note that the term, "in a favorable space," is the operative here.

Scientists estimate that perhaps as many as 99% of all living organisms that ever lived on Earth are now extinct. The planet's natural ecosystem is constantly evolving, and as the environment changes certain species will begin to disappear. There have been at least five instances of mass extinction, so the threat of a sixth may not seem so unusual or distressing. However, it is estimated that the next extinction event could occur in a timeframe unlike some of the previous mass extinctions. Rather than taking millions of years, we are seeing influential changes during the course of a single human lifetime that could end abruptly as the next mass extinction event.

In the case of the human species, population increases had been declared to be a threat to our very existence, but now we see that global human populations appear to have at least stabilized, or perhaps to have even declined by a small measure. It will now be a waiting game to see if global contamination and pollution rates begin to decrease, at least at a rate of decline parallel with the stabilization or decline in global human populations. No matter the calculus related to population size, or to the concentrations of contaminants and pollutants, and no matter the orbital shifts of the Earth, or the degree of axis tilt, or the reach or direction of precession (the slow gyration of the rotation axis of a spinning body, such as the Earth, about another line intersecting it that describes a cone), human attention to the generation and management of toxic wastes will play a key role in our survival.

Planetary physics and even greater universal influences beyond our control will dictate the inevitable adverse results of major Earth changes, far out of reach of the relatively minor, yet ever so impactful, causal characteristics of carbon dioxide, methane, and nitrous oxides (CO_2, CH_4, and NOx [NO and NO_2]), or of fossil fuel exploitation. It is the mitigation element we can manage for now, so long as the geopolitics are taken off the list of causes.

Climate change is seen as an alteration in global or regional weather patterns over time, in large part attributable to the release of greenhouse gases to the atmosphere by nature or by human activity, seen then as global warming. Also, the amount of energy we receive from the sun is variable, and it too similarly impacts that thing: a warming climate and Earth changes. The Earth orbits the sun once per year, but the orbit is elliptical, not circular, and its axis is tilted, not straight, any or all of which factors change over time, thereby making the Earth cooler or warmer by modifying the amount of radiation received from the sun.

The three main forces that impact Earth's natural climate cycles are eccentricity, obliquity, and precession. Eccentricity reveals the elliptical shape of Earth's orbit around the sun, and is an important influencer of Earth's cycles, with regard to climate impact. Obliquity is the tilt of the planet's axis relative to the plane of Earth's orbit around the sun and affects the average temperature of the Earth and the prominence of the seasons. Precession is the wobble factor within Earth's axis (it does wobble, and in both directions), realized as the degree of axis tilt and the direction in which this tilt points, both of which factors may alter Earth's climate. Our planet is currently in an interglacial period (period of milder climate between ice ages). If there were no human influences on climate, scientists say, Earth's current orbital position within the Milankovitch cycles predicts that our planet should be cooling, not warming, continuing on a long-term cooling trend that began about 6,000 years ago.[6]

What we are witnessing today is climate refugees on the rise, with people losing their lands and livelihoods as a consequence of environmental degradation and changing climatic conditions resulting in natural hazards.[7] From the perspective of historical geology, this crisis or change is natural, and the environment has been evolving all through the recent geological past. However, it is the influences and effects of human activity on nature that are threatening, and that have been shaping local, regional, and global environments in less than beneficial ways. We now wallow in the quagmire of our own manufacture, and it must be turned around post haste for the trail of filth we leave in our wake will become the source of our own collective and self-imposed demise, now or later—our choice. Political choices are generally for later, when it then becomes someone else's responsibility, in another time.

Ineffectual and insensitive political and industrial influences must be subordinated to the greater interests of humanity, and in direct concert with the maintenance of a livable planet, because the additive and cumulative effects of all sources of greenhouse gases are contributing to global warming and incite, thereby, continuing changes in the climate that are altering the nature of our living environment.

Big Oil, big petrochemical, big plastic, big carbon, big emissions, and big climate crisis are all one and the same, and cumulatively, they represent the key threat to us. "The permafrost isn't going to release a catastrophic explosion of carbon that would, say, double overnight the amount of carbon dioxide in the atmosphere," says Ted Schuur, who leads a project on permafrost thaw and climate change at the University of Northern Arizona. "Instead, this carbon is going to leak out from all over the Arctic and, over time, add a substantial amount to the carbon humans have already added by burning fossil fuels."[8] Way back in 2018, a report prepared by the UN's Intergovernmental Panel on Climate Change gave humans a maximum carbon budget of 580 billion tons in order to have an even chance of limiting warming to one and a half degrees Celsius.[9]

Sustainable development requires broad collaboration and the focused efforts of multiple international organizations, national governments, cooperating non-governmental organizations, local communities, and the public to address the challenges of mitigation of environmental degradation (the deterioration of environmental quality from ambient pollutants or detrimental activities), and to solve the various and many complex and challenging issues at all levels of government and society.

Environmental degradation is the main factor that impairs sustainability, and involves key processes that include desertification, pollution, land degradation, snow ablation, glacier retreat, dams, and population resettlement, and their relationships to environmental sustainability. Environmental degradation is adversely impacting human health and well-being, leading to lost biodiversity and ozone layer depletion, and requiring high cost and effort for recovery, including restoration of green cover, soil remediation, and protection of endangered species.

Environmental degradation is fostered by increasing population density, urbanization, industrialization, poor infrastructure, and other development schemes that are exerting ever-increasing pressures on the carrying capacity of land and resources, leading to a rapid rate of deforestation, escalating desertification, and loss of species.

Physical factors, such as earthquakes, hurricanes, monsoon rains, and poverty in some countries, also accelerate the decline of environmental resources. However, human activity is interfering with the stability of global ecosystems by the use of fossil fuels, expanding agriculture and forestry, the use of nitrogen and phosphorous fertilizers, and the increasing exploitation of scant fresh water supplies for irrigation in agriculture, industry, and waste disposal. The impacts of these changes include warmer climate, decreased ozone, increased lake and stream acidification, decreased tropical forests, amplified soil deterioration, eutrophication of estuarine and coastal marine areas, and increased development of anoxic conditions in freshwater and marine systems and sediments. Collectively, water, soil, and land use are the underpinnings of the advancement of civilization and population growth, so they must be managed properly.

Cumulatively, environmental degradation is the deterioration of the environment through the consumption of assets, the destruction of ecosystems, and the eradication of wildlife and their essential habitats. Environmental degradation occurs when the Earth's natural resources are depleted, and the environment is compromised as manifested by pollution of natural resources, extinction of species, and the rapid growth in other species populations. The term also refers to lost biodiversity through depletion of air, food, soil, water, and habitat. It is any change or disturbance to the environment perceived to be deleterious or undesirable and is one of the greatest threats in the world today.

Environmental degradation is characterized by any lessening of the capacity of the Earth to meet our social and environmental requirements. The societal consequences include increased poverty, overcrowding, famine, adverse weather extremes, acute and chronic illnesses, mass migration, war and human rights abuses, and an increasingly unstable global situation that portends chaos and disaster.

A scientific assessment by the United Nations Environment Programme (UNEP) found that economic growth has come at a devastating cost to the planet, and urged governments, businesses, and people around the world to act to reverse the damage before it is too late. The result is three interlinked environmental crises: climate disruption, loss of biodiversity, and pollution that collectively threaten our viability as a species. They undermine food security, water security, and human health.[10]

A clear nexus is formed between the human-incited and intended malpractices of land-disposed *Toxic Sewage Sludge*, returned toxic wastewater effluents, landfilling, landfill leachates, deforestation, excessive use of fertilizers and pesticides, soil pollution, soil erosion, overgrazing, excessive and poorly managed extraction of minerals, drought, urbanization, industrialization, desertification, and the creation of toxic wastelands and hazardous waste disposal sites, plus the more weather-related forces of floods, typhoons, droughts, rising temperatures, and fires. These collectively are encompassed within five primary environmental concerns i) lost biodiversity; ii) water pollution; iii) deforestation; iv) general pollution; and v) climate change.[11]

Without a doubt, most global pollution is caused by the production and burning of fossil fuels, while most water pollution is caused by industrial waste, agricultural runoff, and sewage and wastewater flows, plus mining activities, marine dumping, accidental oil leakage, and leakage of sewer lines. Underlying these environmental insults in all cases and by all opinions, it is deterioration of water quality by the land disposal of *Toxic Sewage Sludge*, plus discharges of toxic wastewater effluents and petroleum industry pollution that must be remedied, posthaste.

Environmental pollution is defined as contamination of the physical and biological components of the Earth and atmosphere to such an extent that normal environmental processes are adversely affected. Of nine worldwide processes that underpin life on Earth, four have exceeded safe levels: human-driven climate change; loss of biosphere integrity; land system change; and high levels of phosphorus and nitrogen flowing into the oceans from fertilizers.[12]

Undoubtedly, the top causes of air pollution include the burning of fossil fuels, agriculture operations, industrial emissions, mining operations, indoor air pollution, wildfires, microbial decomposition of organic matter, decomposition of toxic waste, transportation, open burning of garbage, construction demolition, decomposing animals and plants, and the release of radioactive elements, lest we forget the contributions of microplastics and nanoplastics, as well as natural events.

To make matters worse, environmental degradation/pollution has become a major contributing factor to the scourge of infectious diseases globally, including antibiotic-resistant infectious diseases. The fundamental cause of antibiotic resistance, of course, is antibiotic use, including in animals. The inciting agents of infection may be any viruses, bacteria, fungi or parasites by direct or indirect contact, or through wounds, such as from insect bites, cuts or abrasions, or by contaminated foods or water.

Wastewater epidemiology began in the early 2000s, with researchers isolating the residues of illegal drugs to understand community use. But over the last two decades, wastewater-based epidemiology has been expanded to look at the remnants of pharmaceuticals, alcohol, and pathogens to identify existing substances and emerging infectious diseases to reveal the overall health of a population. What is being learned is that our wastes and wastewater systems are in fact the agents of our and nature's demise, unless we decide to end the practice of living in our own wastes and filth—simply because of the failures of our agencies, at municipal, county, state, and federal levels, to ditch the antiquated practices of sending our collective wastewaters as raw sewage into poorly engineered facilities that do not actually treat anything. They simply separate the solids from the waters, and then send each into our air, foods, soils, and waters as a botched effort to relocate our wastes.

Impaired food, water, and living conditions resulting from environmental degradation foster infectious diseases, including and most especially antibiotic-resistant infections. The six leading pathogens for deaths associated with antibiotic resistance are *Escherichia coli*, *Staphylococcus aureus*, *Klebsiella pneumoniae*, *Streptococcus pneumoniae*, *Acinetobacter baumannii,* and *Pseudomonas aeruginosa* while the "first comprehensive analysis of global impact of antimicrobial resistance (AMR) estimates resistance itself caused 1.27 million deaths in 2019—more deaths than HIV/AIDS or malaria—and that antimicrobial-resistant infections played a role in 4.95 million deaths."[13]

One single pathogen-drug combination, methicillin-resistant *Staphylococcus aureus* (MRSA), caused more than 100,000 deaths attributable to antimicrobial resistance in 2019, while six others each caused 50,000 - 100,000 deaths attributable to antimicrobial resistance: multidrug-resistant tuberculosis (MDR-TB), third-generation cephalosporin-resistant *E. coli*, carbapenem-resistant *A. baumannii*, fluoroquinolone-resistant *E. coli*, carbapenem-resistant *K. pneumoniae*, and third-generation cephalosporin-resistant *K. pneumoniae.*[14]

While these antibiotic-resistant bacterial infectious agents, and their resulting infectious diseases, are typically detected, managed, and reported simply as infectious diseases, we must become cognizant of the fact that it is environmental degradation in large part, locally and globally, that yields the resulting antibiotic-resistant infectious diseases and the subsequent deaths; it is not the infectious agents alone. In the case of nearly any and all infectious diseases, whether of an individual or of a population, the critical factors of the environment must be evaluated and considered, as well as key elements of disease etiology, management, and control.

On the other hand, the matter of climate crisis continues to pose more questions than answers. One of the key questions is centered on the cost of full carbon capture and utilization (FCCU) technologies, versus the cost of releasing more toxics and greenhouse gases to the environment. There are not as yet, in practice, any clean FCCU technologies or methods for the explicit purpose of capturing greenhouse gas emissions from *Toxic Sewage Sludge* production or use, including from its transport and land disposal, not even when considering the toxic and impairment consequences to our air, food, soil, and water. The long- and short-term lessons of exposure to *Toxic Sewage Sludges* are not yet fully learned but are emerging, as we suffer. On the brighter side, newer FCCU technologies are being shown to capture nearly all carbon released by the combustion of *Toxic Sewage Sludge* in a well-designed fluidized bed combustion (FBC) facility that intends to capture nearly all exhaust gases for beneficial reuse.

As Earth changes manifest with an impending climate crisis, a common question is: How may we lessen the rate of approach, defend against, or plan to endure the expected adverse impacts of the disaster once upon us? In doing our part, the answer of merit, in my view, is centered on the absolute need to reduce the addition of greenhouse gases released to the sky to about zero, not just a marginal reduction. Should nature contribute volcanic eruptions, spewing clouds of sulfur dioxide, thereby inherently cooling the planet, or volcanic carbon dioxide, a greenhouse gas that has the potential to further warm the planet, then there is little we can do about it.

The 2018 Intergovernmental Panel on Climate Change (IPCC) report stated that humanity had already exceeded global warming of 1C above pre-industrial levels and is set to reach 1.5C of warming between 2030 and 2052. The report follows warnings about the state of the planet from leading scientists, including a later statement in 2019 that people will face "untold suffering due to the climate crisis," unless substantial changes are introduced.[15] Capture and use of atmospheric CO_2 is not the only answer to this looming crisis, but it is where we can begin, now that climate and earthly changes are upon us. However, a fully functional and economically feasible direct air capture (DAC) system to collect the accumulating carbon emissions cannot likely be created or operated, although costly efforts are being made.

Reduction of deforestation, along with increased reforestation, have great potential to impact climate change positively by capturing greater amounts of CO_2, but that would require renewed and expanded forest preserves, globally. Also, enhancing soil quality to favor carbon-sequestering fungi and other microbes by employing altered agronomic approaches will enhance carbon capture. In both examples, total elimination of the land and water disposal of toxic wastes, such as and most especially *Toxic Sewage Sludges*, industrial wastes, toxic wastewater effluents, pesticides, and overly toxic artificial fertilizers, will be an absolute requirement. Soils, forests, and wetlands must not be considered waste disposal or toxic waste relocation sites, and extant and legacy landfills must be restored to their historical settings over time, with longer-term approaches to combating open pollution and the consequences of climate change considered a mandate.

Back in 2018, the IPCC highlighted the urgency of declaring a climate crisis and the importance of achieving net-zero greenhouse gas emissions by midcentury. In addition to aggressive decarbonization across all sectors of the landscape to prevent additional greenhouse gases from entering the atmosphere, the IPCC SR15 report also touted analyses by the International Energy Agency (IEA), Lawrence Berkeley National Lab, and the National Academies of Science, Engineering, and Medicine, and concluded that some extraordinary level of carbon dioxide removal (CDR), via carbon capture and utilization (CCU), will be required to limit global warming to 1.5 degrees Celsius and from an even more devastating 2 degree Celsius scenario, an unlikely event in this politically driven, commerce-centric environment.[16]

Early in our geological history, excess CO_2 from volcanos and comets was postulated to have been the greatest plausible natural source of greenhouse gases globally. Today, the major greenhouse gas challenge is considered to be human-generated pollution, realized as excesses of nearly everything, Through expanded resource exploitation, we are generating more carbon emissions than the atmosphere can manage, with the net result being excessive atmospheric accumulations of greenhouse gases.

The U.S. Department of the Interior states on their website: "The United States faces a profound climate crisis, and the Department of the Interior is poised to take action to avoid the most catastrophic impacts of that crisis and to meet the moment. The climate crisis is transforming where and how we live, and presents growing challenges to human health and quality of life, the economy, and the natural systems that support us. In 2020, the United States experienced

22 extreme weather and climate-related disaster events, with a cumulative price tag of nearly $100 billion."[17]

Unfortunately, the various, outdated ideas related to recycling or beneficial reuse of *Toxic Sewage Sludge* involve risks of increasing greenhouse gas emissions, nutrient leaching into the environment, and multiple adverse impacts on soil biodiversity, as presented herein. Carbon dioxide, methane, and nitrous oxide, all potent greenhouse gases, are produced during the wastewater treatment process, as well as after *Toxic Sewage Sludge* and other biological wastes are disposed on land. While newer methods for the treatment of raw sewage and other wastewaters have led to marginal improvements in the quality of wastewater effluents, they have also increased the overall volume of *Toxic Sewage Sludge*. So what does this all mean?

Although *Toxic Sewage Sludge* from any WWTP or POTW is considered by some to be a valued source of nutrients and organic matter for reuse as a fertilizer, or as a soil amendment in agriculture, or for the remediation of contaminated sites, or to rebuild depleted soils, it is considered by others, as documented, to be a major source of soil and water contamination and pollution. So to better understand the differing points of view, let's break it down.

Dewatered *Toxic Sewage Sludge* from a wastewater treatment plant still holds about 73% water. Within the solids, on average, dewatered *Toxic Sewage Sludge* contains about 50 - 70% organic matter and about 30 - 50% minerals (toxic or otherwise), including about 1 - 4% inorganic carbon, plus about 3.4 - 4.0% nitrogen (N), 0.5 - 2.5% phosphorus (P), and various amounts of other nutrients, including micronutrients that can be recovered. Extractable nutrients available in *Toxic Sewage Sludge*, such as phosphorus, are destined to become scarce or exhausted in the next 50 - 100 years; therefore, phosphorous recovery from wastewater and from *Toxic Sewage Sludge* is becoming an increasingly important endeavor, not to be ignored.

Of note: Nitrogen, phosphorus, potassium, and micronutrients can be recovered from any *Toxic Sewage Sludge* material and used in commercial fertilizers, as is done successfully in many venues worldwide. However, the health and ecotoxicological risks associated with the presence of so many contaminants in *Toxic Sewage Sludge* emphasize the overall risk, threat, and harm associated with the outdated practice of direct land disposal.

Dewatering *Toxic Sewage Sludge* for any use appears to result in the least amount of greenhouse gas emissions, while conventional incineration of sludge produces the most GHGs. Today, newer sludge-to-energy engineering solutions can reduce these emissions greatly, including to near zero, while providing safe and useful residual products for industry, plus valuable, renewable clean energy options.

Technologies utilizing enhanced thermolytic decomposition methods for the reduction of *Toxic Sewage Sludge* include combustion, pyrolysis, and plasma arc gasification. Older heating or conventional incineration methods simply generate revised forms of toxic waste that must be land disposed as reformed toxic wastes. Combustion is considered to be the most beneficial reduction method for *Toxic Sewage Sludge* because it eliminates the requirement for dangerous and contaminating landfilling and alternative land-disposal strategies.

The good news is, thermolytic decomposition of *Toxic Sewage Sludge*, such as by FBC combined with FCCU technologies, provides complete thermal reduction of all toxic chemicals and volatiles, including pathogenic microbes, while producing renewable clean energy and high-demand residual ash. Most importantly, the *Toxic Sewage Sludge* is thereby eliminated from disposal in land and water environments.

Following thermolytic decomposition, the residual *Toxic Sewage Sludge* ash product is rich in silica, alumina, calcium oxide, and iron oxide, which are of substantial value for the production of building materials such as bricks and concrete. The metals are thereby stabilized and solidified, so the potential risk of metal toxicity is reduced. Also, the addition of activated carbon to *Toxic Sewage Sludge* immobilizes the bioavailable fractions of polycyclic aromatic hydrocarbons and metal elements.

Clearly, the safe management of and functional alternatives for actual beneficial reuse of *Toxic Sewage Sludge* and other toxic wastes remain as critical and necessary objectives in the reduction of greenhouse gas emissions from wastewater treatment plant processes. Our forest and agricultural lands are precious and must be protected, preserved, and restored—no longer abused or ignored as toxic waste dump sites.

Our soil and water resources, and thereby our food resources, are already compromised by centuries of exploitation, and must be reinstated as being essential to survival, not destined to added abuse.

If the concept of sustainability applies anywhere, it must be for the assured continuity of reliable food resources—it is a matter of both national and global food security.

The engineering methods and technology exist to achieve these critical objectives in a safe and cost-effective manner. The causes and remedies are clear; all that is required is the political will to support the key global objectives of mitigating climate change and restoring public health and environmental sustainability, based on a longer view of absolute need, rather than on a shorter view of immediate and out-of-sight disposal gratification, as a means of "Toxic Waste Relocation."

The commonly stated parable of waste generation and toxic waste management yields no solace, for as we age and gain more personal experience in these matters of toxicity, we also broaden our perspective into the likely outcomes, which must not be simply a demand for even more analysis, of and for its own sake. Conventional methods and practices become the obvious errors, and the whole idea of progress then takes on new meaning. More specifically, the costs of waste generation and disposal, and of the resulting pollution, yields yet more harmful outcomes, thereby generating even greater challenges to our health and healthcare systems, often with unresolvable consequences. Progress by itself, without wisdom gained, may become more the problem, not necessarily a part of the solution.

There are measures we can take to slow the consequences of a changing climate, so as to mitigate its crisis nature, and so as to be solution providers, not spectators. As well, there are events and processes we can only observe as mere spectators of the movements of inevitable astronomical phenomena and physical geology. The rest we must address, and not make worse.

Here on Earth, for example, we can gradually eliminate the exploitation of fossil fuels as being no more than available petrochemical raw materials. More locally, we can strive to end landfills and the land disposal of *Toxic Sewage Sludge* and the water disposal of toxic wastewater effluents into our fragile living environment. Next in line is the absolute need for contained consumerism, which must be managed to ensure greater application toward local manufacturing, agriculture, and trade. The unalterable alternative may otherwise be a sudden decline of human and other life forms as we know them. The challenge is no less than in the first case devising means to conquer political and commercial driving forces that mark our emerging decline. Orbital physics cannot be altered, as our historical geology tells us so well.

The functional resolution to this conundrum is to slow the change, as it cannot be stopped, otherwise it is just so much *más de lo mismo*, with no wisdom applied at all. And we must recall that while Ben Franklin advised, “An ounce of prevention is worth a pound of cure,” Albert Einstein revealed that, “Intellectuals solve problems, geniuses prevent them.” We must prevent environmental degradation by stopping the land disposal of *Toxic Sewage Sludge* and other toxic wastes and the chilling consequences thereof, for prevention must be the rule.

CHAPTER 21

KNOW THE RULES

It is one thing to be told and even to believe we need to save the trees and other plants. It is quite another to actually do it.

Stanley A. Rice, *Green Planet,* 1957.[1]

The first to document the adverse health effects of *Toxic Sewage Sludge,* as applied under the EPA 503 rule, was Dr. David Lewis, together with Virginia Tech and Rochester Institute of Technology colleagues. Lewis reports that in 1978, a soil scientist working in the EPA Office of Water advised, "The application of some low levels of toxic substances to land for food crop production should not be prohibited." Lewis then states:

> And so began what I would argue is the worst conceivable plan EPA could have ever come up with for containing and disposing of the nation's industrial and municipal pollutants. That plan was to discharge 'pretreated' hazardous industrial wastes into sewer systems, treat the sewage sludges that settle out at wastewater treatment plants with chemical and biological processes to reduce pathogens and odors, and then market the treated sewage sludges as cheap fertilizer products.[2]

What, by Gosh, were they thinking?

In 2002, Dr. Lewis and others went on to assemble the first cases of sludge-related illnesses and deaths to be found in the peer-reviewed scientific literature. As documented in his book, *Science for Sale*, health problems experienced by individuals living within one-half mile of a disposal site included "skin rashes and burning of the eyes, throat, and lungs." According to this research, "Approximately one-fourth of fifty-four individuals exhibited *Staphylococcus aureus* infections of the skin and respiratory tract, including two mortalities from septicemia and staphylococcal pneumonia." People also complained of nausea, vomiting, stomach cramps, migraine headaches, flu-like symptoms, and sudden death.

As far back as 1998, Dr. Lewis began investigating reports from dairy farmers who had applied sludge as fertilizer and were experiencing "a precipitous drop in milk production, sick cattle, and excessive cattle mortalities."

In a scientific experiment conducted in 2000, where sludge containing lead and other hazardous materials was spread in an African American neighborhood, Lewis reports the finding that: "Small children often directly ingest lead-contaminated soils when playing in their yards, and when the contaminated soils are tracked into their homes. They also inhale dusts, and pick up dusts from handling contaminated surfaces."[3]

Other reports on the suffering associated with *Toxic Sewage Sludge* appear in the book, *Fateful Harvest*, by Duff Wilson, as revealed by a locally affected citizen, Patty Martin. The author, a reporter with the *Seattle Times*, focuses his investigation on toxic wastes being sold to farmers as fertilizer in the area of Quincy, WA. He found that "The tainted fertilizer—containing lead and dioxins—is believed to be destroying crops, sickening animals, and endangering the nation's food supply."[4] Wilson details stories of reduced hay production, severely ill and dying horses after eating the contaminated alfalfa, economic loss to farmers, and major health issues in the community. In this work, he follows Martin, Quincy's mayor at that time, as she battles to reveal the unsettling truth of what was occurring. As a result, Wilson won a Goldsmith Prize for Investigative Reporting from Harvard University.

Moving forward to today, I recently offered three town hall presentations to the small community of Toledo, WA, where, once again, citizens were in fear and uproar because the Washington State Department of Ecology, the permitting regulatory agency, had arranged for landowners to have *Toxic Sewage Sludge* spread on about 400 acres in the nearby Layton Prairie region. Neighboring residents did not know what to do. So someone told them they should call the sludge tracker, Richard Honour.

During those town halls, the first in a post-Civil War era barn, another at a local church, and the last at the city's new high school, individuals shared publicly their personal stories. One fellow now opposed to the land disposal of *Toxic Sewage Sludge,* had previously worked in the business as a sludge truck driver. He astonished the group at the first meeting when he had this to say: "There were 10 of us driving sludge dump trucks, now I'm the only one still alive." He is convinced the sludge they worked with had destroyed the lives of his friends and fellow sludge workers.[5] Others attending despaired over the significantly high incidence of cancer where they were living close to a sludge-spreading operation. Perhaps 165 residents attended the barn meeting, about 125 the meeting in the church, and 80 attended the third town hall at the high school, including multiple state and county elected officials. As of the 2020 census, Toledo sports a total population of about 631 residents, a great turnout.

So who then would it be, we ask yet again, who advocates, plans, permits, executes, reviews, approves, and continues this abhorrent sludge dumping practice, the land disposal of *Toxic Sewage Sludge*, on public and private lands, whilst even the federal permitting authorities, upon whose responsibility for the oversight and prosecution of the Federal Acts falls, forbids by its own accord such heinous acts upon their own adjacent federal lands? Might it be the case that federal agencies know well the crimes committed or the adverse consequences of such toxic waste disposal actions on the lands and in nearby precious waters? And, was a crime committed? Is this a failure or violation of code, regulation, statute, or law, or of the directly responsible state permitting agency, or of its manager or employees?

Investigation, indictment, and prosecution via federal laws that seek to protect our precious air, food, soil, and water resources seem to have run aground amidst the need to subvert the disposal of toxic wastes in favor of the better commercial interests of agricultural practices, while forevermore working to supply rationale and justification for such widespread and wholesale degradative operations upon our landscapes. What then is the value to the citizens of laws not revered or prosecuted, or of regulations cast asunder? Has public health no meaning? Is environmental health of no concern or merit? Are environmental degradation and Earth changes simply someone else's bad day at the office? Is the thought of prevention of disease too complicated? Are we here for an 80-year or so stint, and that's it? Will there be no one after us? And if so, why not? It seems like the idea of stewardship has gone south if that is the case.

In light of the revealed and blatant pattern and practice of the land disposal of toxic waste on public and private lands, we must step forward with a duty to investigate, bear witness, report, and then resolve the deadly behaviors placed before us. There is inherent danger in a person or agency's poorly monitored or wholly unevaluated pattern and practice of bad behavior. And most especially when performed consistently, including of any repeated, routine or generalized act, not merely as an isolated or sporadic procedure or instance, and that may be manifested as a custom or act affecting, or potentially affecting, more than a single individual.

So, where do we go from here? Do we as citizens continue to suffer toxic insult, as orchestrated, blessed, and allowed by agencies that continue to bury us and our families, forests, waters, farms, and foods in *Toxic Sewage Sludge*, absent any form of comprehensive qualitative analysis or toxicological testing, or do we now draw a line? Knowing where the line should be drawn is a good place to begin.

RACKETEER INFLUENCED AND CORRUPT ORGANIZATIONS ACT

"The Racketeer Influenced and Corrupt Organizations Act (RICO) of 1970 seeks to strengthen the legal tools in evidence gathering by establishing new penal prohibitions and providing enhanced sanctions and new remedies for dealing with the unlawful activities of those engaged in organized crime." The RICO Act is a United States federal law that provides for extended criminal penalties and a civil cause of action for acts performed as part of an ongoing criminal organization.[6]

CLEAN WATER ACT

The Clean Water Act (CWA) is the primary federal law in the United States governing water pollution. Its objective is to restore and maintain the chemical, physical, and biological integrity of the nation's waters; recognizing the responsibilities of the states in addressing pollution and providing assistance to states to do so, including funding for publicly owned treatment works for the improvement of wastewater treatment; and maintaining the integrity of wetlands.

Technically, the name of the law is the Federal Water Pollution Control Act. The first FWPCA was enacted in 1948, but took on its modern form when rewritten in an act entitled the Federal Water Pollution Control Act Amendments of 1972. Major changes have been introduced via amendatory legislation including the Clean Water Act of 1977 and the Water Quality Act of 1987. The Clean Water Act does not directly address groundwater contamination, for the groundwater protection provisions are included in the Safe Drinking Water Act, Resource Conservation and Recovery Act, and the Superfund Act.[7]

POTW BIOSOLIDS MANAGEMENT PROGRAM

The 1987 Water Quality Act (WQA) created a program for management of biosolids *(Toxic Sewage Sludge)* generated by publicly owned treatment works. The Act instructed EPA to develop guidelines for usage and disposal of sewage sludge:

1. Identify uses for sewage sludge, including disposal.
2. Specify factors to be taken into account in determining the measures and practices applicable to each such use or disposal (including publication of information on costs).
3. Identify concentrations of pollutants which interfere with each such use or disposal.

EPA created an Intra-Agency Sludge Task Force to develop comprehensive sludge regulations designed to do the following:

1. Conduct a multimedia examination of sewage sludge management, focusing on sewage sludge generated by POTWs.
2. Develop a cohesive agency policy on sewage sludge management, designed to guide the agency in implementing sewage sludge regulatory and management programs.[8]

ENDANGERED SPECIES ACT

The Endangered Species Act (ESA) provides a program for the conservation of threatened and endangered plants and animals and the habitats in which they are found. The lead federal agencies for implementing ESA are the U.S Fish and Wildlife Service and the U.S. National Oceanic and Atmospheric Administration.

U.S. FISH AND WILDLIFE SERVICE

The U.S Fish and Wildlife Service (FWS) maintains a worldwide list of endangered species that includes birds, insects, fish, reptiles, mammals, crustaceans, flowers, grasses, and trees.[9]

U.S. NATIONAL OCEANIC AND ATMOSPHERIC ADMINISTRATION (NOAA) FISHERIES SERVICE

The law requires federal agencies, in consultation with the U.S. Fish and Wildlife Service and/or the NOAA Fisheries Service, to ensure that actions they authorize, fund or carry out are not likely to jeopardize the continued existence of any listed species or result in the destruction or adverse modification of designated critical habitat of such species. The law prohibits any action that causes a "taking" of any listed species of endangered fish or wildlife.[10]

ENDANGERED SPECIES ACT and EPA

The U.S. EPA Office of Pesticide Programs (OPP) implements key portions of the Endangered Species Act. OPP regulates the use of all pesticides in the United States and establishes maximum levels for pesticide residues in food, thereby safeguarding the nation's food supply.[11]

RESOURCE CONSERVATION AND RECOVERY ACT

"The Resource Conservation and Recovery Act (RCRA) gives EPA authority to control hazardous waste from cradle to grave. This includes the generation, transportation, treatment, storage, and disposal of hazardous waste. RCRA also set forth a framework for the management of non-hazardous solid wastes. The 1986 amendments to RCRA enabled EPA to address environmental problems that could result from underground tanks storing petroleum and other hazardous substances. The Federal Hazardous and Solid Waste Amendments (HSWA) are the 1984 amendments to RCRA that focused on waste minimization and phasing out land disposal of hazardous waste as well as corrective action for releases."[12]

FEDERAL INSECTICIDE, FUNGICIDE, AND RODENTICIDE ACT

One of the most heinous examples of federal government indifference to land-disposed toxic waste is reflected by the Federal

Insecticide, Fungicide and Rodenticide Act (FIFRA) of 1947. The Act states clearly that pesticides can only be refused certification if the risks to health and environment are greater than the benefits to crop yield and quality. Note how that statement leaves a wide opening for government abuse in favor of the agricultural industry, considering that nothing is stated about the criteria for determining such implied risks or benefits, which then are left to subjective definition by the definer alone.[13]

RICO, CWA, ESA, RCRA, and FIFRA are all associated by a common thread of thoughtless acts performed for the planned and intended disposal of *Toxic Sewage Sludge*, which cannot be remediated.

RICO alone informs us there must be prescribed some measure of extended criminal penalties and a civil cause of action for acts performed as part of an ongoing criminal enterprise. We must wonder in amazement why the continuing performance of land-disposed *Toxic Sewage Sludge*, any or all of which constitutes the disposal of multiple and extensive toxic materials into our forests, on our farms, and into our waters, is not considered as being a toxic assault on our total environment, as well as a blatant violation of the many Acts listed above.

The word "Act" as stated in the above referenced sections, refers to a Bill or joint resolution that has passed both the U.S. House of Representatives and Senate in identical form and has been signed into law by the President, or has passed over his veto, thereby becoming a law. If the President signs the Bill, it becomes a law, and is also known as an Act of Congress.

Note well that to my best knowledge, not one elected official, agency manager or employee in Washington State has ever requested a comprehensive qualitative analysis of any *Toxic Sewage Sludge*, nor has any environmental activist or activist organization. And similarly, to my knowledge, no environmental attorney or environmental law firm has ever made such a request or demand, nor has a court of law, despite the obvious and inherent risk, threat, and harm presented by exposure of the public to such toxics. And just as there are no carve-outs, justifications, excuses or exceptions for murder, similarly there should not be any exceptions for exposing humans, wildlife or the environment to known deadly toxic wastes, in any form.

Comprehensive qualitative chemical analysis and safety/toxicology testing of *Toxic Sewage Sludge* are far beyond the technical reach of most agencies and their advisors, other than the EPA, but are well within the expertise of many commercial labs across the U.S. Many of the available commercial labs are certified by the EPA, yet most remain as an unused technical resource, especially for purposes of analysis of *Toxic Sewage Sludge*. The analytical work is needed to clearly define the risk, threat, and harm of exposure to *Toxic Sewage Sludge*. As a result, the toxics and the toxicity contained therein, and the toxic consequences thereof, are wholly unknown and declared not to become known, while the hundreds of thousands of tons of human-originated *Toxic Sewage Sludge* continue to be disposed or dumped on our lands and food crops, thereafter draining, flowing, and migrating into our surface and groundwaters, as oft repeated herein.

Within the state of Washington we see no relief, and further, no interest in such relief or remediation. What may be the risks, threats or other adverse consequences of chronic exposure of our citizens and their air, food, soil, and water to receiving the toxic wastes of others? Who would or should ask these questions ahead of serendipitous land disposal of these toxic wastes? If not you or us, then whom? Definitely it will never be any state or federal agencies or individuals charged with monitoring public health or defining or enforcing the relevant laws, so where do we go for immediate relief? The task then is for those directly affected and other responsible citizens to hold as being accountable those agency persons and the agencies themselves who reveal by their actions such obvious contempt for human and environmental health.

The annual production and release of new and novel chemicals from industry is increasing at a pace that far outstrips our global capacity for assessment and monitoring, especially because they accumulate in our sewage systems, thereafter, to contaminate our forests, farmlands, rangelands, and waters to degrade even further our precious natural resources. The recent cry for an international science-policy body with oversight over the disposal or reuse of chemicals and wastes may provide a forum for informing such actions that are needed to help safeguard the whole Earth system.

Considering the hundreds of POTWs and WWTPs in Washington State alone, much less the tens of thousands nationally, most never even attempting to treat anything beyond a superficial level, the national laws must be reconsidered in light of the decades-long inactivity by municipal, county, and state agencies that have been conferred the responsibility of public health oversight. And we must reflect also on the hundreds, if not thousands, of extant and legacy landfills that continue to discharge their toxic leachate loads into our receiving ground and surface waters, with some leachates being captured and returned, in part, to flow once again through our WWTPs and POTWs to further unload their toxic loads into the resulting *Toxic Sewage Sludge* products, as well as into the next generation of still toxic wastewater effluents. What will it take for us to recover?

May the Great Spirit forgive us for what we have done, or for what we have not done, or for what we continue to allow to happen in silence, and thereby, into perpetuity.

Chapter 22

There Is A Solution!

Intellectuals solve problems; geniuses prevent them.

Albert Einstein, theoretical physicist, (1879 – 1955).[1]

The challenge is clear that we are producing more people, waste, and pollutants than the Earth can accommodate, so something's gotta give. When charted on a graph, we see that fossil fuels, agriculture, consumerism, and toxic wastes, along with cancer and other chronic diseases, are increasing at rates that far exceed the rate of population growth.

Not so long ago, the easiest solution to garbage and other wastes was to simply dump them in wetlands, a river or the sea, or burn it and let it smolder, or unload it into forests or old quarries, or down canyons. The same primitive systems and solutions persist, but are now obscured by fences or forests, or within depressions in the earth, far out of sight and mind. In any case, all of that dumping and hiding of wastes must end, with a mind and budget toward restoration and reconciliation with the land, or our trash and other wastes will become the undoing of us all, including of the entire planet—no exceptions.

For clarity, the nut of the issue centers on our past and current behaviors and attitudes toward our ever-expanding wastes, and those center on our misguided approach to managing wastes by creating yet more wastes, and even more toxic materials, by the unsustainable methods that have been selected. The primary weakness underlying our dysfunctional waste management system is the untenable idea and practice of political commerce. It just does not work.

The actual resolution of these human waste issues is compounded further by emerging new infectious agents and chronic diseases, plus by the accelerating rates of disease incidence and prevalence across global populations, most especially of cancer, and all as a direct result of population growth and resource exploitation, coupled with the unfavorable outcomes of our intended denial or shrouding of the resulting waste products. That's the crux of the problem, so now we must focus on some safe, sustainable, and workable solutions, preferably based on prevention of disease, rather than simply on more untenable or temporary solutions.

Landfilling of municipal, industrial, construction, and other solid wastes must end, globally. Existing landfills, no matter their age or content, must be mined down to bedrock, with the accumulated toxic decomposition products and persistent toxic residuals rendered thereafter as being either recoverable, reusable or repurposable, or otherwise processed to become non-toxic—a big order.

By the same measure, the land disposal of *Toxic Sewage Sludge*, refinery sludge, and other industrial sludges must be banned by law, with such wastes instead being processed to become non-toxic and hopefully repurposed. The point is that we are rapidly approaching the non-sustainability threshold, whereafter each and every discarded item or released gas, substance, fluid or solid waste material simply adds further to our accelerating downfall, and that includes by the likes of *Toxic Sewage Sludge*.

The more difficult challenges will include a transition to the safe and successful, first-time-ever management of wastewater effluents from our wastewater treatment plants and publicly owned treatment works. Since the beginning, we have been conned into believing that such operations actually treat wastewater, which they do not, and that the waste products therefrom, including the dredgings, effluents, and sludges, have also been treated. Thereby, and as we have been told, they are being rendered as "non-toxic, non-infectious or safe," with their ultimate destination being in our food, via our precious lands and surface and groundwaters, or into our once clean atmosphere.

The parallel challenge to Earth's expanding population, with its non-parallel and geometrically expanding waste generation, is the requirement as a global society for more energy, primarily now focused on renewable clean energy. But what must be the working definition of renewable clean energy at this point? Could it be cleaner fossil fuel energy, or cleaner nuclear power, or cleaner wind or solar power that does not require massive resource depletion, or more dams for added clean hydropower? I don't think so. We are now past that, so the focus must be to convert our overly abundant waste resources, such as our tainted waters, including wastewater effluents, and the ever-burgeoning supplies of solid wastes, including sludges and garbage, into renewable clean electrical energy, along with green hydrogen and other forms of renewable green clean energy, but without creating its own resulting toxic waste product residuals. Can it be done? Yes, as is already demonstrated in multiple progressive venues and ventures, most notably by those who think about the future, and who hold the ideals of quality of life, prevention of disease, and curtailing of environmental degradation on the forefront of their minds, now, rather than as an after-the-fact remediation, recovery, repair or cleanup operation. Whoever fosters even the very idea of "cleanup," needs a day on the couch, or in court.

As with nearly any complex technology or engineering challenge, if you immerse the project in enough time, thinking, effort, and cash, you should be able to derive a workable solution, so long as the underlying technology is functional. However, that kind of approach is generally wrought with error, highly difficult, characteristically impractical, and ill-advised, at least from the perspective of agencies and legislators. But, if the overriding objective is to make potable clean water from toxic wastewater effluents derived from wastewater treatment plants, or from captured toxic landfill leachates, or from the orange-red, toxic landfill leachates that ooze from the likes of King County's Cedar Hills Regional Landfill, plus from the dozens of abandoned legacy landfills around the county, for example, it could be done, yes, although the time and cost elements are hardly imaginable. In fact, while such beneficial infrastructure work is absolutely essential for the sake of human and environmental health and survival, it's also substantially impractical in today's political environment. The key challenges rest not with the infrastructure agencies, but with the influential political infrastructure, of and by itself.

If, however, the alternative objective were to determine if the surface and groundwaters that flow to Puget Sound, to the greater Salish Sea, and to the Pacific Ocean from our Washington State wastewater treatment plants, landfills, and *Toxic Sewage Sludge* land-disposal operations, could be restored to their quality of perhaps 200 years ago, the simple answer is, yes, it can be done, but the political will to do so remains far out of touch. Of course it can be done—the engineering exists, but such beneficial work would be deemed too costly, unnecessary, and impractical, and therefore not likely to be contemplated by anyone or any agency. The default objective, then, is that we are destined to live with impaired waters in impaired regions in an ever-more degraded and further-impaired environment, until the day arrives whereby the healthcare costs of such a self-perpetuated, non-beneficial objective are found to exceed the costs for replacing the dysfunctional infrastructure we continue to suffer to this day.

If it is the case, as it so appears, that the Washington State Department of Ecology and King County's Wastewater Treatment Division and Solid Waste Division have yet to identify that there may even be a problem with our failed waste management systems, then no offered or tenable objective will ever be on anyone's table even for discussion, and that then is our fate. The landfills, both legacy and active, still leach and flow to our surface and groundwaters, the wastewater treatment plant effluents all still flow to our surface and groundwaters, and the land-disposed *Toxic Sewage Sludges* still leach, drain, and flow to our surface and groundwaters, all under the contrived fog of, "it's safe," "we're in compliance," "it just goes away," and, "that's the way we do things." If we were to better manage these endlessly growing wastes, one and all, one hundred percent, then public and environmental health would restore and become functional once again. It may be that our agencies' minds are clouded by persistent exposure to our ever-expanding toxic wastes.

Landfill leachates, legacy landfill leachates, toxic wastewater effluents, land-disposed *Toxic Sewage Sludge*, and other land-disposed sludge leachates do not "just go away," they return to us in our air, food, soil, and water—*ad Infinitum.*

In Washington State, as elsewhere, the costs for properly managing toxic wastewater effluents, *Toxic Sewage Sludge*, landfills, landfill leachates, and stormwater runoff would require at least a major investment in physical infrastructure, rather than just more reparative and restorative maintenance exercises. The default failure-to-perform outcome is thus noticed as a steady degradation of our water and waste systems, with stormwater runoff and wastewater effluents declared as being the primary sources of degraded or lost water quality in the region, and with little chance for recovery or restoration. The oft stated theme around Puget Sound in the 1980s was, "Will the last salmon to exit the Strait of Juan de Fuca please turn off the lights." "The salmon are just about gone," was the offered mantra, alright, but with no listeners. Must have been someone else's job or it must simply be "the way we do things."

This is the challenge of calculating the value of commerce versus that of public health, and equally, of calculating the value of commerce versus that of environmental health. We have to change our approach drastically and immediately so that public and environmental health will win for a change, and we can only hope to do that by taking all of the contaminating, polluting, and harmful ingredients out of the out-gassings, runoffs, effluents, leachates, and sludges, plus out of our air, food, soil, and water, as well as out of whatever other waste matrices there may be.

Once upon a time, the global cancer lifetime incidence rate was estimated to be about one in sixteen people who would develop cancer, that is, develop cancer incited primarily by environmental toxics, most commonly by smoking, or so it was thought. Now, and even with the realized curtailment of the said smoking pandemic in many nations, we are at a global cancer incidence rate of about one in five people who will develop cancer, with estimates emerging that within a decade or so we may be approaching a frightening cancer experience of perhaps one in two people globally who will develop cancer, mostly as a result of the carcinogenic or oncogenic mutations incited by toxics in the environment.

The bottom line is that of our approximate $23 trillion U.S. GDP, somewhere between $1 - $2 trillion is spent on cancer, which is bad for citizen victims, but very good for oncologists and for the economy. We are now as a nation in the business of inciting as much cancer as may be possible, so as to assure the long-term success of the cancer business, a nasty statement indeed, but it may be the sad fact of the matter. Cancer is not a general, usual or normal outcome of living, unprovoked, and no, it does not "run in families," as we have heard far too many times—cancer is most often the consequence of exposure to environmental carcinogens, singly or in concert, in current populations or in those before us.

By exhaustive and ongoing research of databases in the U.S. and elsewhere throughout the world, the stark revelation is that the actual cost of cancer to society and to individuals, even just in the U.S., cannot be determined accurately, simply because of the many possible facets of the actual or real costs of cancer to individual cancer victims, as well as to the local and greater economies. The end product of many research approaches to this topic, I will assume, is that the cost of cancer, even just in the U.S., much less worldwide, is indeterminable, and therefore, we must rely upon the best estimates of all agencies to realize simply that environmental contamination and pollution from all sources may be the death of us all, most especially by cancer, both individually and as a nation, and simply from a very stark economic perspective.

The latest reports by the U.S. National Cancer Institute tell us that as of 2019, "the national patient economic burden associated with cancer care was $21.09 billion, made up of patient out-of-pocket costs of $16.22 billion and patient time costs of $4.87 billion."[2] Well published is the statement: "The total economic impact of premature death and disability from cancer worldwide was $895 billion in 2008. This figure represents 1.5 percent of the world's gross domestic product (GDP). This economic toll from cancer is nearly 19 percent higher than heart disease, the second leading cause of economic loss ($895 billion and $753 billion, respectively). This analysis did not include direct medical costs, which would further increase the total economic cost of cancer."[3] These data reveal that the actual costs of cancer to individuals and to society shall remain as being both unknown and likely unknowable, to be forevermore revealed only as a best guess, and without an actionable remedy. And if the externalities of the cancer plague were to be calculated and presented properly, the even greater numbers presented would likely be even more shocking. Indeed, that paltry $895 billion, as presented back then, would likely become dwarfed now before our eyes.

As well, there is another well-ignored calculation that would add to the revealed financial burden of cancer, and that is while deaths from cardiovascular disease and cancer are often in competition as being the leading cause of death, it must be appreciated that many subjects with cardiovascular diseases may expire from their heart disease before they die from their concomitant cancer, with the cause of death then recorded as cardiovascular disease, thereby deflecting the actual cause or cost of death from cancer. Note well that the inciting factors for cardiovascular disease and cancer often result from similar lifestyle and personal behavior patterns, so the assignment of an actual cause of death can be a serious challenge.

Clearly, it's the toxic contaminants and pollutants in *Toxic Sewage Sludge*, plus all that's in toxic wastewater effluents and in landfill leachates, and in stormwater runoff, and in toxic out-gassings, which get to us and our surroundings. Even when we speak just about plastics, including the macroplastics, microplastics, and nanoplastics in the effluents, leachates, and sludges, we get to the point quickly that plastics are in all of it, and much to our detriment. The fecal residues, yard wastes, food wastes, and other actual organics are decomposed by microbes in the wastewater processing system, but it is the toxic metals, and the organic chemicals, including the pharmaceutical and personal care products, and the vast loads of industrial chemicals, plus the plethora of infectious agents, including all of the excipients in the so many functional plastic formulations, that get to us. They all head to the raw sewage processing plants or to landfills to return to us in the air we breathe, the foods we consume, the waters we drink, and the landscapes that are intended to feed and restore us. The agencies deny all of this as a planned and intended cover-up.

There are multiple technologies available for treating raw sewage, *Toxic Sewage Sludge*, and toxic wastewater effluents. In this case we focus on *Toxic Sewage Sludge* and toxic wastewater effluents from wastewater treatment plants. In recent years I've worked with many waste, construction, infrastructure, and process management engineers and engineering firms, including those with specialties in incineration and fluidized bed combustion (FBC), as well as with pyrolysis, biochar, and hydrochar operations. As well, there are many back-end full carbon capture and utilization (FCCU) systems, green hydrogen generation technologies, steam plant electrical generators, and their experienced customers and buyers. To go back to even listening to the mutterings about the usefulness or efficiencies of any of the current systems, I would need a tranquilizer, but here is a brief overview for perspective.

While wind and solar energy are providing short-term cash to developers, in the longer-term they will fail by their present configurations because they are too polluting to develop, operate, maintain, and replace. We put them up, the contractors take them down, and the recyclers make a bundle—they are not at this point sustainable approaches for meeting our long-term, reliable, and resilient power requirements, although a vast array of technology and engineering advancements are in play.

Hydropower has always been beneficial, so long as there may be enough water to turn the turbines, while our precious fish resources decline in their aftermath. Geothermal energy is an excellent approach, as is tidal power, but they both seem to have overly challenging engineering issues, beyond our capabilities to stabilize. Sustainability, reliability, and resilience are the key considerations.

Hydrogen power works well, but has scalability issues, as well as pollution issues associated with generating the electrical power needed to make the said hydrogen in the first place. Generating hydrogen from methane for power is dirty business, for example, it's not green hydrogen.

All fossil fuel power sources are done and over, or so we imagine, and the resulting carbon products can be captured only in part, not yet quite meeting the criteria for sustainability.

Carbon sequestration methods work, for sure, but what do we do with an Earth full of carbon, destined to leak back out? And how much gaseous and liquid CO_2 do we need to put out fires and chill the veggies? In both cases, the CO_2 released reenters the atmosphere, so what did we accomplish?

Coal's deadly byproduct, as one bad example, is carbon dioxide (CO_2). Every million British thermal units (Btu: The amount of heat required to raise the temperature of one pound of water by one degree F) of heat energy generated by burning coal releases about 208 pounds of CO_2, with a ton of coal then having about 20 million Btu, meaning that ton of coal generates about 4,172 pounds of CO_2 when burned. CO_2 emissions are expressed in terms of the energy content of coal as pounds of CO_2 per million Btu. CO_2 forms during the combustion of coal when one atom of carbon (C) unites with two atoms of oxygen (O). The atomic weight of carbon is 12, that of oxygen 16, and CO_2 44. Therefore, one pound of carbon combines with about 2.7 pounds of oxygen to yield about 3.7 pounds of CO_2.

Coal, with a carbon content of around 78% and a heating value of about 14,000 Btu per pound, emits about 204.3 pounds of CO_2 per million Btu when completely burned. This means complete combustion of one short ton (2,000 pounds) of coal will generate about 5,720 pounds (2.86 short tons) of CO_2. According to industry standards, it takes 1,100 pounds of coal to generate one megawatt of electricity. A megawatt is a unit for measuring power that is equivalent to one million watts. One megawatt hour (MWh) then is equal to about

1,000 kilowatts of electricity used continuously for one hour. Therefore, one MWh of coal power creates about 2,100 pounds of CO_2 emissions, so think about that.

Greenhouse gases and other pollutants from the generation of one MWh of electricity per month by combustion of coal include: "2,100 pounds of carbon dioxide [CO_2]; 346 pounds of methane [CH_4]; 185 pounds of coal ash; 3.9 pounds of sulfur dioxide [SO_2]; 1.6 pounds of nitrogen oxides [and nitrogen dioxide (NOx and NO_2)]; 1.7 ounces of radioactive uranium and thorium; 13 mg of mercury (nasty stuff);" requiring also about "16,000 gallons of fresh water," plus about "3.75 square feet of strip-mined land." Now, that's costly and dangerous. "According to the American Coal Ash Association, coal power plants created 102 million tons of coal ash in 2018, making it the second-largest source of waste in the U.S. after trash [i.e., after municipal solid waste]. Coal ash contains trace amounts of various toxic metals, including chromium, arsenic, and cadmium, as well as uranium, and thorium."[4]

As a reminder, then, there's no such thing as "clean coal," just as there's no such thing as "safe" *Toxic Sewage Sludge*, or safe toxic wastewater effluents, or safe landfill leachates, or safe legacy landfill leachates, or safe refinery sludge, or safe radiation waste, etc., old or new. It's a public health versus commerce struggle, and commerce will not win this one in the longer run, because someone, someday, may decide that public health should hold precedence over commerce—how odd!

Waste-to-energy engineering systems provide yet more opportunities to burn something, and they have only partial carbon capture functions, at best. Worst of all are the various and many methane production and capture systems, be they anaerobic digesters from wastewater treatment plants or from concentrated animal feeding operations, biomass processing, landfill gas capture, or earth seepage capture, but what's the point? Combustion of the resulting methane yields CO_2 and water vapor, with the oxygen combining with the carbon to form yet more CO_2 and water vapor, plus some heat and light. The process of burning carbon compounds (combustion), including coal or methane from any of these facilities in air simply yields more CO_2 and water, end of story. Capturing the carbon product efficiently and fully remains a mystery, but is possible.

Having investigated these issues for decades myself, I suspect I may understand the state of the science and engineering, yet I still get the same agency feedback that I "just don't understand." But I do understand that it's often about the cash to be generated in the process, by the process, and for the process, with never an efficient and profitable FCCU outcome at the end. We are a distance from getting there, but we will get there as driven by a life-or-death need, not simply by endless processes designed only to garner more immediate cash for the advocates, promoters, and builders of the generally marginally functional outcomes.

I do believe there is a bright, public health-guided outcome ahead of us, so long as we adopt the core collective mentality that fossil fuel burning must end, landfilling of solid waste will be curtailed, waste-to-renewable clean energy systems will become the standard, including the conversion of *Toxic Sewage Sludge* and other toxic wastes to renewable clean energy, and that wastewater effluents will be processed for the first time ever to meet quality standards reflective of those of the intended receiving waters. All that is required is stout political will to introduce the technology and infrastructure needed, and a firm appreciation of the underlying tenets of public health.

For certain, there are many useful waste processing systems out there, but none to be found in Washington State—we live in a fantasy world of junk food, consumerism, and readily available and packaged healthcare—it's really not here. We compromise our personal health and our living environment by the ceaseless contamination and pollution brought to us by ourselves, and by our timidity, along with our blind faith in the system, and by a failure to demand high-quality infrastructure rather than simply more entertainment, shopping, fast food, and low cost and most convenient electrical power and fuel. It's not sustainable. We're not sustainable under these toxic conditions.

Once the bell tolls,[5] however, I will be ready to pile on, for it is ultimately the collective dysfunction of agencies, environmental law firms, and environmental organizations themselves, and our deference to and the indifference of the pandered commercial interests, referred to as political commerce, that lead us astray. Prevention is a much better idea than recovery, repair, treatment or cleanup. We know well that smokers fully understand their cancer death risk, but are we expected to anticipate death by toxic waste, under the delusion that agencies and their dysfunctional infrastructure will protect us? Whoops! Needs rethinking.

Technology and engineering systems exist that can rescue us from what we have done, what we do, and what we have become since the last ice age, now that we may be nearing the end of the current ice-free era, but we insist on seeing all of the errors of today and of the past to be repeated yet again, only worse each time. Roads and bridges collapse; wastewater treatment plants never really worked or they fail; landfills become overfilled beyond capacity, then leach, seep, and outgas to the adjacent communities, only to be filled more, so as to fail again, as they release their toxic loads on the next set and the next generation of multiple citizen victims, without mercy, sung to the way-too-oft-repeated tune, "We must never let that happen again!" But it did happen again, and it is happening again, *Toxicus ad Infinitum.*

Ice ages? . . . What about ice ages? Based on some reasonably certain science, there have been at least five major ice ages in Earth's history, and although we seem to have been relatively ice-free in our more temperate zones for at least a few or more of the alternating greenhouse-like interglacial warming periods, the cause-and-effect relationships of those cycles are now more clear.[6] The Quaternary Ice Age started about 2.6 million years ago, and since then there have been cycles of glaciations, with ice sheets advancing and retreating on 40,000-year or 100,000-year glacial to interglacial cycles. We are currently in a warmer interglacial cycle, with the last glacial period ending about 10,000 years or so ago, and with the latest interglacial period lasting about 28,000 years.[7] We are on vacation in this present greenhouse period, not a time for abuse.

While the next serious cold glacial period may begin in a few tens of thousands of years or so, this current global warming, intensified by human-forced greenhouse gas emissions, may become the forcing factor that alters our living environment for perhaps the next few hundreds of thousands of years.

Climate forcing is the physical process that affects Earth's climate through forcing factors known as "forcings," because they drive or force the climate to change. In league with the basic laws of thermodynamics, as Earth absorbs energy from the sun, it must eventually emit an equal amount of energy to space. The difference between incoming and outgoing radiation is known as the planet's radiative forcing (RF). When forcings result in incoming energy being greater than outgoing energy, the planet will warm. Conversely, if outgoing energy is greater than incoming energy, the planet will cool. Therefore, Incoming Energy - Outgoing Energy = Radiative Forcing.[8]

In any case, the only forcing factor we may be able to modify is the generation of greenhouse gases, which result in a warmer climate, melting glaciers, and rising oceans, and with reduced habitable land mass, loss of agricultural opportunities, and a really serious headache on our hands. The only way out of this dilemma is to end the cutting of trees, plant new forests, end fossil fuel exploitation, and stop the burying and burning of waste-like things. We caused this knot, now we have to undo it.

Historically, the emergence of earthly land plants generated increased global oxygen levels, along with a substantial parallel reduction in CO_2, as may have been derived from volcanos or elsewhere. Planting trees is the starting place for oxygen generation and for carbon capture, and while younger trees may take up carbon faster from the atmosphere than do older trees, they do not store as much carbon, nor do they store carbon as efficiently, as do older trees. Similarly, soil microbes in soils amended with toxic wastes also may take up carbon faster, albeit temporarily, from the atmosphere than do microbes in natural, healthy soils, but they also will not store as much carbon, nor will they store carbon as efficiently as do the natural microbes in natural soils. The oft told stories to the contrary are simply self-benefiting manipulative myths, not based on science. Why not simply prevent the problem in the first place, instead of having to work to repair the grand mess we have created?

What's worse is that soils amended with artificial nitrogen and other nutrients, most especially from toxic wastes, such as and including from *Toxic Sewage Sludges*, typically suffer a devastating modification of their soil and root microbiomes or microbiota. The beneficial, root-inhabiting microbiota are killed off by *Toxic Sewage Sludge*, for example, as they are sometimes killed by the over-administration of commercial fertilizers, whereby the normal soil and rhizosphere microbiota are destroyed or replaced by microbes introduced by the *Toxic Sewage Sludge* microbiome. The end result is that tree growth, while it may be enhanced temporarily, suffers from the toxic killing of the normal root-inhabiting microbes (and nematodes) that enable and foster nutritional uptake by plants from soils, and that also support the very tree root structural characteristics themselves, which in turn supports the tree growth and stability within an often harsh forest environment.

It is well known that tree seedlings exposed to or treated with *Toxic Sewage Sludge* often succumb to high winds and harsh weather conditions as a result of lost or diminished root function, with Douglas fir seedlings less than three to six feet tall often dying from the toxicity alone, much less from being toppled over by moderate or strong winds. A recent investigation of a local tree farm in the Snoqualmie Forest that had been violated by tens of thousands of tons of fresh *Toxic Sewage Sludge* just months before, revealed large numbers of six- or seven-year-old Douglas fir trees blown down by moderate winds, likely the result of tree root decline in response to *Toxic Sewage Sludge*-incited toxicity to the soil and root microbiomes, as well as to the new roots themselves—nothing new here.

A key driver for reducing carbon from the atmosphere is its natural sequestration by natural means, such as by incorporation into growing plants and soil microbes, and that is the only natural and beneficial resolution to this challenge. The most abundant naturally occurring gases in our atmosphere are nitrogen (78%) and oxygen (21%), with argon being the third most abundant gas, then with traces of carbon dioxide, neon, helium, methane, krypton, hydrogen, nitrous oxide, xenon, ozone, iodine (as sublimated from the solid to the gaseous form), carbon monoxide, and ammonia filling in the rest.[9] The message here is that our delicate atmosphere is comprised primarily of nitrogen, which is essential to plant growth, and as it is sequestered from the atmosphere by root-inhabiting bacterial species, along with essential oxygen, plus a smattering of other gases, and it is not to be trifled with.

Trying to defy the basics of chemistry and physics by such odd measures as direct air capture of carbon dioxide from the atmosphere via carbon sequestration in the earth, or by various forms of pyrolysis to convert one form of organic matter to another, and then to yet another polluting fuel, is simply a fool's errand. Replacing one greenhouse gas generating system with another gets us nowhere, although political commerce may well benefit by the process. However, in the special case of pyrolysis, there are substantial benefits to be gained in agriculture by employing advanced pyrolysis systems to generate clean biochar from organic wastes, with its beneficial use as a soil amendment. In any case, and just as with other products or emissions from combustion or incineration, deep analytical chemistry must be performed on the resulting biochar products to assure that such systems would not be responsible for simply relocating the toxics to a different location.

Biochar, so long as it passes critical comprehensive analytical chemistry testing, is employed well to enhance soil structure, plus the essential soil microbiome, as well as soil physical and chemical properties, such as and including the moisture holding capacity (MHC) of depleted or degraded soils. The strict intent of the biochar analytical chemistry requirement is to assure that toxic chemicals have not survived the pyrolysis process, thereby providing a source of toxics that could be bioassimilated and bioaccumulated by the resulting food or forage crop plants. The critical factors in any pyrolysis system revolve around "Time at Temperature" to assure full thermolytic decomposition of any and all toxic chemicals.

All of the coal and oil reserves in the Earth, plus all that have been recovered, were at one time atmospheric carbon captured by plants and microbes, later to be heated and compressed over millions of years to become carbon-rich coal, peat or crude oil, plus the by-products therefrom. It is trees that sequester and process atmospheric carbon to yield atmospheric oxygen, with carbon-filled solid plant matter and free water left over. Protecting air, food, soil, and water is the way out of this dilemma over the long run, not generating and wasting carbon for the convenience of the wrong kind of energy.

The engineering firms with whom I work have taken on the challenge to identify and design an efficient, safe, and effective integrated facility that aims to convert many or any forms of combustible, carbon-based wastes to renewable clean energy, without residual polluting products. The troublesome waste streams upon which we have focused include *Toxic Sewage Sludges* from wastewater treatment plants, any and all non-recyclable plastics and paper products, plus any other plant derived wastes, including from the lumber industry and from agriculture, as well as from other combustible wastes that afford a high enough Btu value to assure complete and efficient combustion, not incineration.

Combustion and incineration are both high-temperature exothermic redox chemical reactions between reductants (fuels/ feedstocks) and an oxidant (atmospheric oxygen) to produce heat for energy, oxidized gaseous products (flue gas/exhaust gas), and ash. The objective of combustion, and specifically in the case of FBC, is to generate renewable clean energy and eliminate the toxics, case closed. An envisioned fully integrated waste-to-renewable clean energy system of our focus is based on FBC technology. In stark contrast, the focus of incineration is restricted to the reduction of fuel feedstock mass and volume, and to reduce some toxics to a limited extent, but without the generation of renewable clean energy.

The integrated FBC system of our choice is engaged to burn solid fuels, whereby the fuel particles are suspended in a hot fluidized bed of ash and other particulate matter, through which jets of air are blown to provide oxygen for complete combustion or gasification. Incineration cannot meet the renewable clean energy objectives of engineered combustion.

The FBC system includes a fully integrated FCCU system that provides valuable solid and gaseous CO_2 products. No polluting process waters are used or released, and the flexible process is designed to accommodate additional waste streams that may meet the standards for fuel source efficiency, such as selected animal wastes, mixed organic wastes, food processing wastes, plant industry wastes, medical wastes, and selected industrial wastes that will not yield recalcitrant toxic or polluting residues. About 95% of pollutants generated by FBC are captured, with the primary resulting product being high-value renewable clean energy.

The objective of our investigation was to evaluate all existing waste-to-energy systems, ultimately focusing on FBC with FCCU, then to identify the best that could be merged compatibly to meet the end product objective. That objective is to achieve clean heat generation from mixed combustible waste streams with FCCU, combined with the production of high-demand products, plus an absence of residual toxic wastes. This engineering process and technology present the case for achievable decarbonization in the face of an imminent climate crisis.

Our failure to address immediate matters of underprovided critical infrastructure, in an environment of ever-accelerating waste generation, and with a growing demand for autonomous, sustainable, reliable, and renewable clean energy, based on non-fossil fuel-reliant energy, reveals our vulnerability to a changing global environment. Equally, our failure to eliminate from the environment the collected toxic products and byproducts of industry causes an ever-increasing toxic environment for the Earth and its inhabitants, with a bleak outlook for an ever-mounting burden of chronic disease. There is a solution, and the benefits are a function of removing toxics from the environment, capturing all CO_2 and other carbon products from the atmosphere, thereafter, converting the toxic wastes to renewable clean energy and other beneficial, non-polluting products—a climate crisis no longer.

CHAPTER 23

NOTHING BUT ENERGY

In the midst of chaos, there is also opportunity.

Sun Tzu, *The Art of War*, 5th Century BC.[1]

The one-hundred-fifty million dollar question on the table is, "How can hundreds of thousands of tons of *Toxic Sewage Sludge* from five wastewater treatment plants in just one county in Washington State, plus from several smaller wastewater treatment plants in adjacent jurisdictions be collected, transported, and converted from being land-disposed toxic waste into a safer and more functional material for actual beneficial reuse, without resorting to incineration, drying, landfilling or concealment in composts or in other adulterated products, or by land disposal in forests and on farmlands?" The current best estimate is that it would cost about $150,000,000 to bring just one modern fluidized bed combustion facility online within about 20 months, including with full carbon capture and utilization, and with perhaps eight resulting commercially viable products for introduction into commerce.

As discussed at length herein, wastewater treatment plants are major generators of greenhouse gases via the microbial decomposition of raw sewage-borne organic materials alone, and as well with any ensuing microbial decomposition of the sewage products sprayed in forests and on farms. Meanwhile, the toxic chemicals in those *Toxic Sewage Sludges* will remain in, and be continuously disruptive of, the natural soil microbial activities of the soil microbiota in general, as well as that of the plant root microbiome, both of which are essential to soil structural integrity, soil microbiome health, plant growth, carbon sequestration, and nutrient availability. Also, the introduction into soils of massive loads of macroplastics, microplastics, and nanoplastics from *Toxic Sewage Sludges* disrupts the structure and function of once healthy soils. How then, we might ask ourselves, do we produce healthy food in that?

Toxic wastes, such as and including *Toxic Sewage Sludge* and the leachates and runoff therefrom, are sometimes noted as having a transient beneficial effect on plants, similar to that incited by an overdose of high-nitrogen fertilizers on farms or in forests, but thereafter such intrusions become detrimental over the longer run. The combined inherent toxic and high nitrogen shock effect on plant root hairs, and on the root hair-inhabiting nematode, fungal, and other microbial populations, from the toxics in *Toxic Sewage Sludges* presents a direct adverse effect on plant roots and on the greater soil and root microbiomes. Forests most especially do not experience a beneficial long-term outcome from applications of *Toxic Sewage Sludge*, with typical adverse effects noted in root structure failures that may result in falling trees, as well as reduced structural integrity of the resulting wood products.

While there may be transient or short-term sequestration of carbon by adding human waste-originated *Toxic Sewage Sludges* to soils, the recovery times and restoration of the natural soil microbiota-driven sequestration of CO_2 is forever disrupted. Beyond simple carbon sequestration activities, the remaining GHGs are further enhanced by the very nature of disruptions of the soil microbiota when adding the antimicrobial toxic components of *Toxic Sewage Sludge* to soils. The microbiota of the soil and root microbiomes and their biological activities are disrupted by exposure to *Toxic Sewage Sludge*, forevermore.

Wastewater treatment plants can be designed, built, and operated to manage the release of GHGs, from front to back, meaning, from the current process of raw sewage collection and inflows, to settling, to aerobic and anaerobic digestion, to the release of toxic wastewater effluents to our rivers and oceans, and on to the disposal of the remaining residual *Toxic Sewage Sludge* solids.

If that is the case, then why are municipalities, counties, and states allowing and encouraging, by existing code and regulation, the blatant continuation of the extreme contaminating and polluting practices effected by the land disposal of *Toxic Sewage Sludge*? It must be that contamination, pollution, and degradation of our air, food, soil, and water, and of our natural resources, are considered as being common, acceptable, and beneficial practices, as well as providing the lowest cost and most convenient means of waste disposal. Therein lies the definition of political commerce, and that is nuts!

Public health and safety are matters of responsible citizenship and national security, not to be trifled with by elected officials, agencies or their advisors and colleagues in commerce, no matter their personal, social or business interests or affiliations. Reluctance to improve matters of land- and water-disposed toxic wastes, by plan and intent, emphasizes and restates approval, thereby declaring their advocacy of open disposal of toxic waste with consent to undermine the principles of public and environmental health, much to the demise of biodiversity and sustainability. As a result, we are approaching critical environmental and public health tipping points in the form of blatant environmental degradation, defined by the IPCC as "a critical threshold beyond which a system reorganizes, often abruptly and/or irreversibly."[2]

Much to our detriment, we live in an era of planned and intentional greenwashing of *Toxic Sewage Sludge* as a pure marketing ploy by deceptively using green marketing spin to persuade the public that a substance, chemical, material, product, method, policy, regulation, law, code, or action (i.e., the greenwashing entities) is or may be green, safe, or otherwise environmentally friendly. It is the act or practice of making any such greenwashing entity appear to be more environmentally favorable or less damaging than it may in fact be. Greenwashing is in play when an organization dedicates substantial resources to promoting itself and its greenwashing entities as being environmentally favorable, rather than minimizing their actual adverse environmental impacts. The objective of greenwashing is to convey a false impression about such greenwashing entities by offering misleading information about their safety or environmental soundness, generally as weak or unsubstantiated claims with intent to deceive consumers or the public.

Greenwashing is a deceptive marketing practice, often without restraint, and usually with little or no concern about the adverse effects on human or environmental health. Companies or organizations engaged in greenwashing typically fabricate or exaggerate their claims of benefit in an attempt to mislead consumers, otherwise known as citizen victims. Greenwashing is often covered by vague terms such as environmental stewardship, enhances wildlife habitat, restores depleted soils, and the like, but in the case of land-disposed *Toxic*

Sewage Sludge, it is always based on intent to deceive the public by promoting far greater environmental benefit than can be delivered or substantiated. In the specific case of land-disposed *Toxic Sewage Sludge*, the practice of greenwashing is actually harmful to human and environmental health.

Nancy V. Raine, Founding Member of the Sierra Club Grassroots Network, Wastewater Residuals Team, authored an insightful primer on the greenwashing of *Toxic Sewage Sludge* that still holds true today: "A loophole in the 1972 Clean Water Act (CWA) allows large industrial facilities to discharge wastewater that contains hazardous wastes—toxic chemicals, radioactive materials, and infectious wastes—into these same sewers. A loophole in the 1976 Resource Conservation and Recovery Act (RCRA) specifically excludes municipal sewage from being regulated as a hazardous waste. These loopholes set the stage for nearly a half century of recycling complex industrial poisons into our food and water through the use of sewage sludge as a 'fertilizer' and 'soil amendment.'"[3]

Recall that the U.S. EPA tells us that sewage sludge refers to the solids separated during the treatment of municipal wastewater, including during the treatment of domestic septage, yet the term "treatment" remains vague and at best wholly undefined. They tell us that biosolids refers to treated sewage sludge that meets the EPA pollutant and pathogen requirements for land application and surface disposal, but the stated pollutant and pathogen requirements have little or no meaning today. The laboratory test methods used by the EPA to evaluate the safety of *Toxic Sewage Sludge* were, and remain, based on historical methods that are totally irrelevant in today's world of advanced technologies, most especially in the fields of genomics, artificial intelligence, machine learning, toxicology, microbiomics, and analytical chemistry.

When the extant regulations were constructed, the technology did not exist to decipher the kinds and numbers of microbes, including pathogens, at all, or their virulence, or their resistance to treatment, much less the vast array of toxic chemicals in those same matrices. The outcome is that we are all exposed to serious pathogens and toxic chemicals which were in large part unknown back then, or that were considered as being unevaluable, long ago when the regulations and guidelines were prepared. The regulations and test methods have never been improved over the intervening decades so as to be at least partially relevant in today's more toxic and dangerous environment.

To wit, all analytical technologies and methods in use by the agencies are far out of date and thereby considered irrelevant. Just consider this statement by the EPA: "As required by the Clean Water Act Amendments of 1987, the U.S. Environmental Protection Agency (EPA) developed a new regulation to protect public health and the environment from any reasonably anticipated adverse effects of certain pollutants that might be present in any sewage sludge biosolids. This regulation, The Standards for the Use or Disposal of Sewage Sludge (Title 40 of the Code of Federal Regulations [CFR], Part 503), was published in the Federal Register on February 19, 1993, and became effective on March 22, 1993."[4]

The point is that the necessary technology improvements that have been developed since 1993 are well-known and in wide practice by academics and commercial laboratories across the U.S., but are wholly ignored by the U.S. EPA and their state and county colleagues in the management and disposal of *Toxic Sewage Sludge*. The victims of this "failure to perform" are citizens and our fragile living environment. Part 503 has no relevance or value in this modern era.

The good news is that science, technology, and engineering have thoughtful answers awaiting, and in some cases, they are already in action, impeded only by the laggardly approach of seemingly disinterested agencies that have alternate agendas not focused on human and environmental health. Herein we present a brief summary or base of knowledge, explored and revealed over several decades, that spits in the face of agency dogma and their attitudes of indifference that lead to suffering and death. There is a clear reduction-to-practice method and process for reducing *Toxic Sewage Sludge* to valuable products, absent any toxic residuals or effluents, or any climate-changing greenhouse gases, while providing high-value fuels and other renewable clean energy byproducts.

As mentioned, the clear alternative to poorly managed and dangerous composting, landfilling, incineration or land disposal of *Toxic Sewage Sludge* is found in well-designed, field-tested, purpose-engineered facilities that represent the latest generation of industry-proven fluidized bed combustion (FBC) systems. Thermolytic decomposition facilities are designed solely for the safe and efficient processing of *Toxic Sewage Sludge* and other waste materials. These systems can be enhanced to include full carbon capture and utilization (FCCU) technology to assure capture of potential greenhouse gases, and so that there will be no process waters to manage as toxic wastewater effluents.

The resulting FBC FCCU combined facilities provide renewable clean energy for multiple sustainable clean power uses, plus a series of continuously generated products of high value to industry. The single greatest value of the FBC FCCU system and approach is that there are no GHG emissions, and no soil or water pollutants or other residual toxic wastes, for which to seek additional or alternative disposal strategies. Full carbon capture and utilization from any gaseous products within or from these facilities is an essential design criterion to assure that residual carbon from the *Toxic Sewage Sludge* feedstocks is not released to the environment. The age-old *Toxic Sewage Sludge* disposal problem has been resolved!

FBC with FCCU is a fully integrated, reduced-to-practice, mature engineering technology that consumes a vast array of heat-generating carbonaceous feedstocks. Worldwide, there are many manufacturers and constructors of FBC FCCU facilities, with multiple configuration options available as may be needed to address specific requirements of a host client and their specialized or mixed feedstocks.

A typical FBC FCCU facility begins with a fluidized bed burner unit to achieve efficient combustion (note: combustion, not incineration) of the available organic feedstocks, which may be non-recyclable papers and plastics, non-repurposable forest, agricultural or other plant-based feedstocks, and *Toxic Sewage Sludge*, plus animal and specialty wastes that are carbon based and not otherwise recyclable or available for repurposing. Conventional incineration seeks only to achieve reduction of mass, volume, and toxicity, plus the generation of heat for energy, while the resulting gas and ash products typically require containment or land disposal.

In the case of FBC FCCU, the intent is to achieve clean and efficient combustion of the selected feedstocks to produce renewable clean heat energy for generation of electrical energy for the grid, plus for the production of renewable green hydrogen by electrolysis in both gaseous and solid forms. The resulting green hydrogen product is of high value for use as transportation fuel and for other useful products. The FBC FCCU system also provides autonomous renewable clean energy to support its own base power requirement.

There are high-value commercial products to be realized from a purpose-designed FBC FCCU facility, all absent escaped greenhouse gases or toxic wastewater effluents. Those products include: Heat for the generation of renewable clean electrical energy for the grid and to support the FBC FCCU facility; renewable clean energy for the hydrolysis of water to produce solid and gaseous forms of hydrogen for the clean transportation industry; an ammonia product that is further processed into high-value nitrogen fertilizer; solid and gaseous carbon dioxide for the food processing and other industries; a hydrochar product that is valuable to the graphene industry and other specialty carbon monolayer materials, and for the production of carbon filters to achieve FCCU; and a clean ash product for use in sustainable-structured, enhanced concrete production.

The FBC FCCU system offers a solution and an alternative to incineration, composting or land disposal of *Toxic Sewage Sludge*, with autonomous, sustainable, and renewable power generation opportunities for purposes of microgrid security as perhaps the single greatest outtake. Sustainable energy is hugely important in today's world. It means you can keep critical systems going without causing adverse effects to the environment or to people.

Another core technology entrained within the FBC FCCU facility includes the production of a novel form of pure hydrochar for use in a gas filtering system as an efficient adsorbent for the carbon content of the resulting carbonaceous gases, which then recycle the adsorbed carbon back through the combustion system to further convert the captured carbon to additional renewable clean heat energy. There are also several additional, highly sought-after commercial uses for the resulting pure form of hydrochar generated by the system, other than just as an adsorbent reactant in the carbon filtering system.

FBC FCCU facilities can receive managed amounts of municipal solid waste (MSW) or refuse-derived fuel (RDF, i.e., fuel produced from the combustible components of MSW) in the case of a stressed solid waste disposal situation, whereby a landfill may be rendered as being unavailable or inaccessible to receive the combustible organic components of MSW. This opportunity presents a critical backup system for landfills that may be subject to the adverse effects of weather conditions or other unexpected restrictions or disasters, such as earthquake, road closure, tsunami or the attainment of full landfill capacity.

It should be appreciated that conventional landfills present risks of reliable availability, and that conventional wastewater treatment plants may also experience leaks, spills, overflows, blockages, uncontrollable emissions, power failures, and system operating breakdowns that may result in the release of unprocessed or poorly processed effluents and sludges. FBC FCCU facilities can receive non-processed, partially processed or fully processed *Toxic Sewage Sludge*, no matter the stage of interrupted processing or post-dewatering. Similarly, FBC FCCU facilities can also accommodate prescribed amounts of concentrated animal feeding operation wastes or food wastes under conditions of operational stress. To underscore this key point, Chelsea Wald, author of *Pipe Dreams: The Urgent Global Quest to Transform the Toilet*, tells us that:

> . . . large, centralized systems can benefit from economies of scale, but that scale may itself be a liability in the face of climate change and other threats, such as terrorism. Centralized treatment plants tend to be located in low-lying areas to take advantage of gravity, and near bodies of water to release their effluents. So storm surges and floods can easily bring them down, and sometimes it takes weeks to get them working again. Drought also affects their operations: Too little water can cause a sewage flow to stagnate, and thirsty tree roots can penetrate underground pipes in search of water. What's more, if the power goes out, pumping stations and wastewater treatment plants can go offline. And since centralized systems are practically invisible to users, people just keep showering and flushing, even when the wastewater they're creating will just flow directly to rivers, lakes, and the sea.[5]

Renewable clean energy produced by a modern FBC FCCU facility can avoid such headaches and be an asset to any region or community. It can provide redundant electrical power grid security as an "autonomous energy source" that can contribute to a greater power grid system in a time of emergency. An FBC FCCU facility can provide power wherever it may be located. A key benefit of an autonomous energy source is that it is based on the concept of redundancy and autonomy, i.e., it is served by itself.

Purpose-designed renewable clean energy, as is provided by FBC FCCU facilities, are configured with multiple standalone engineering "islands." Each island has a discrete function that is isolated to assure safety, efficiency, and reliability, and to serve its separate role as part of a fully integrated system. The objective of the FBC FCCU facility is to reduce *Toxic Sewage Sludge* and other high Btu, organic, non-recyclable waste materials to renewable clean energy, along with commercially valuable products, and without generating toxic exhaust gases, toxic process waters, toxic ash or toxic residual byproducts.

The dedicated functional islands within the system focus individually on the core burn unit island that houses the FBC system, a well-tested engineering approach that is reduced to practice globally to address multiple forms of waste feedstocks. A novel design feature is that the core burn unit island yields heat, exhaust gases, and ash, absent any process water.

Heat from the core burn unit island is directed to generate steam via the steam-driven electricity generator island, with the resulting electricity provided to the grid and to the hydrolysis island. The hydrolysis island generates clean green hydrogen for use in the production of solid and gaseous hydrogen products for use as fuels, as well as hydrogen for ammonia-originated, nitrogen-based fertilizers.

The exhaust gases captured in the FCCU island also generate solid CO_2 and gaseous CO_2 products for use in the food processing industry, plus a hydrochar pure carbon product that captures additional residual carbon from the exhaust gas stream by a novel carbon capture process. The hydrochar product is useful in commerce for making graphene and other high-value, pure carbon products for the next-generation electronics industry.

The conversion of toxic wastes to renewable clean energy and to the resulting products derived therefrom addresses squarely the matter of a conflict between commerce and the greater interests of public and environmental health. Sustainability must apply to the entire living environment, thereby intending to mitigate environmental degradation. This type of purpose-designed facility will transform toxic wastes into something useful, clean, and beneficial to society.

Obsolete agency ideas and practices, and their outdated methods and collaborations for the disposal of *Toxic Sewage Sludge,* must be laid by the wayside in favor of systems that seek to achieve the better interests of the citizens and of the regions served. The land disposal of *Toxic Sewage Sludge* is a dangerous and outdated practice that must be replaced immediately by engineered systems that seek no more than to render such toxic wastes safe from the perspectives of human and environmental health. There is urgency in this need, as the human toll is high and is accelerating as greater land areas in our forests and farms are being subjected to arcane practices, such as and including for the land disposal of *Toxic Sewage Sludge*, only for the financial betterment of the field of players, bound by questionable contracting practices between agencies and their collaborators and participants in environmentally detrimental forms of commerce.

Ultimately, the objective of this work is to focus on matters of public health, specifically on the prevention of cancer and other chronic diseases. With an enhanced *Toxic Sewage Sludge* FBC FCCU processing system, the collected mutagens and carcinogens will go into these facilities and thereby be rendered non-toxic. Nothing will go into the air, food, soil or water. It's simple, safe, resilient, cost-efficient, beneficial, and clean.

We get in our own way with toxic waste generation and management, and with all of the subsequent issues we have covered herein. Climates do change, as we have said, but the climate challenges we are now facing are, in part, of our own doing; it didn't just happen, we are making it happen, and happen faster, and unless change is brought about, we will make it happen even faster to the suffering of us all.

We can return the Earth's air, soil, and water back to their historical and sacred forms. As Rachel Carson stated so beautifully in her seminal work, *Silent Spring*, ". . . those who contemplate the beauty of the Earth find reserves of strength that will endure as long as life lasts."[6] There is opportunity for renewal of the Earth with every Spring, and all it takes is a dream, a strong will, and a lot of hard work.

The motivations of Washington State and its counties are most likely heavily influenced by matters of cost, politics, and commerce, as we have stated, and not on matters of human and environmental health. The costs incurred to improve the quality of *Toxic Sewage Sludge* from Class B to Class A, or even to Exceptional Quality Class A sludge, for example, are remarkably without merit, because: i) Their class system has no health or ecological relevance; and, ii) In light of the perhaps $1 trillion annual cost of cancer in the United States alone, the costs to treat or manage carcinogenic *Toxic Sewage Sludge* components to reduce cancer incidence, prevalence, death, and the associated societal costs, and to enhance human and environmental health, are less than trivial, at best. It may be that each dollar spent on enhanced waste management, could generate more than a dollar in realized health-cost relief, by precluding many chronic diseases. The cost of lost workdays alone could easily balance the expenditures for prevention.

Globally, human populations, fossil fuel exploitation, agriculture, consumerism, toxic waste pollution, cancer, and nearly all other chronic diseases, and the causes thereof, are increasing, while there have been few mitigations of our environmentally degradative behaviors or in our methods used to render our wastes as being less toxic or even marginally safe.

Agriculture now represents an $8 trillion industry globally, yet presents great risk by its polluting excesses in order to assure the crop production requirements for an expanding global population.[7] When combined with the production of fossil fuels to meet the needs of transportation, consumerism, and industry, including the food and plastics industries, we must be approaching a self-realization that we have reached, or are about to reach, the limits of our overly exploitative nature, as the inhabitants of a finite planet. Change is upon us, by default, unless we take stock of our self-indulgent driving forces, for the good of us all, and it begins with us, and our agencies.

Lest we forget who warned us, and who worked also to protect us from ourselves, there have been many scientists over many decades who projected that an ever-polluting world would impose irreversible limits on the capacity of the Earth to support us, unless we move forward to protect our natural resources and we have not. The contamination and pollution of our sacred lands, including our forests and farmlands, leaves us on the brink of global famine and the consequences thereof. No matter the advances in air, land, and water conservation, the exploitative nature of industry casts nearly all progress asunder.

Of the many who have strived to protect us from ourselves, Norman Borlaug is one brilliant example who foresaw how the multitude of advances in agronomy and healthcare would likely be lost to the rampant and excessive greed of commerce. Despite his observations of our errant behaviors, Dr. Borlaug continued to invent new agronomic technologies designed to feed the world. As an American agronomist, he led initiatives worldwide that contributed to increases in agricultural production and was awarded multiple honors for his work, including the Nobel Peace Prize, the Presidential Medal of Freedom, and the Congressional Gold Medal. Dr. Borlaug has been called "the father of the Green Revolution," and is credited with protecting the lives of more than a billion people worldwide from starvation. As well, and as a means of reducing deforestation, Dr. Borlaug advocated for increasing crop yields with his methods subsequently named the "Borlaug hypothesis." His theory states that increased agricultural productivity on farmland helps to control deforestation due to a lower demand for new farmland.[8] By this work we celebrate the better works of Dr. Borlaug who shined a bright light on technologies and methods for increasing crop production in a near-starving world, among the never-dimming antics of the fossil fuel and other contaminating industries.

One wonders what Dr. Borlaug would have to say today about the vast wheatlands of Central and Eastern Washington and other areas of the United States, buried regularly in *Toxic Sewage Sludge* year after year, awaiting infiltration into the aquifers below with each rain and snow melt. Does wheat actually grow better, or more productively, after a good dousing of *Toxic Sewage Sludge*? Is this staple food thereby made healthier for the world to consume? Are forests better off having had their soil microbiome, and most especially their root mycobiome, degraded by the likes of *Toxic Sewage Sludge*, rather than left alone? We're more than likely to find out, so long as we refuse to investigate, evaluate, and change for the better.

CHAPTER 24

A PRECAUTIONARY TALE

What we do makes a difference, and you have to decide what kind of difference you want to make.

Dr. Jane Goodall, Ethologist and Conservationist, (1934-present).[1]

Recently, we convened a meeting of six in an old building in Central Western Washington State. The agenda pivoted around the fate of a former pyrolysis facility that reduced expired tires and other carbonous wastes to carbon, destined thereafter for other uses, such as for crude oil that could be refined into new diesel fuel. The owner stated clearly that he plans to sell the facility, while I declared that I wanted to demolish it all and restore the contaminated grounds to their former pristine condition, all as friendly discourse, of course.

Five decades or more before, the owner had been a local schoolteacher, barely a way to make a living. By his own research he discovered that the best way to make a useful living was to become the local garbageman, for garbage is a never-ending resource, with few restrictions on its fate back then. He entered into contracts with homeowners, farmers, communities, dairies, businesses, and other generators of solid waste, and that fulfilled his need for a greater source of reliable income, for garbage and other forms of solid waste will always be with us. The local population was growing, and so was its solid waste supply—and so was his new business.

The dilemma presented next was just where would one dispose of this new and ever-accumulating mass of high-value municipal, agricultural, and industrial solid waste, along with some of the *Toxic Sewage Sludges* from local wastewater treatment plants, plus some of the ever-expanding fecal wastes from animal agricultural operations? That was all back in "the old days," so it was safe to have this discussion for historical perspective.

The answer presented itself without much challenge; all local solid wastes in any form and from any source were simply dumped down any convenient culvert or canyon just out of town, along the many winding roads that led to the prairies, foothills, and mountains to the east. Disposal was easy, the owner recounted for the rest of us, for there were and still exist today nearly limitless culverts and canyons along these rural roads of Washington State, paved or not, and with all dumped wastes quickly disappearing from sight, without a disposal fee, tipping fee, or other charge or inspection, and without any requirement for a permit. That was the way they did things back then, with plenty of fresh evidence to declare that the practice continues in similar and less-than-formal fashion to this day.

The next interesting revelation presented by the former solid waste entrepreneur was that all of his newly minted solid waste dump sites were immediately colonized by massive packs of rats. Family fun on weekends then became trips to the many and ever-growing solid waste dump sites around the countryside to shoot the rats with .22 caliber rifles, without fear of missing, for there were so many rats that you couldn't miss, even if you missed a targeted rat. There were no regulations at all for the first decades of the garbage and rat population explosions, as engendered by the wanton and free disposal of solid wastes, garbage, and sludges dumped throughout the adjoining foothills, so everyone made an unrestrained living from the practice, and the rat-shooting opportunities were without end—oh joy!

Decades later, someone made the decision to end open dumping of garbage on any available land or in any waters, and they opted as well to cover the historical, do-it-yourself commercial garbage dumps with soil, so as to hide the crimes and limit the ever-increasing rat and scavenger bird populations. To this day, thereby, we still notice the landfill leachates that flow without end to our ground and surface waters, often indicated by iron-oxidizing bacterial trails. The toxic waters are thereafter noticed as orange, reddish-brown colored streams and rivulets that carry away the toxic wastes from the former unrestricted dump sites into the wetlands and ground and surface waters below, to the demise of salmon, orcas, and other natural wildlife populations.

Many of our schools, parks, dog parks, and sports fields are constructed over these legacy landfills, as a catch-me-if-you-can remedy, yet the practice continues. As a more than peculiar habit or hobby, I continue to hike or bike to and around suspected former do-it-yourself legacy landfills, imaging and making videos of the colorful leachates that originate therefrom, wondering if the toxic flows may ever cease—I doubt it.

The ex-schoolteacher turned garbage entrepreneur's revealing and actual stories of the practices and events of days-gone-by were factual and amusing to the rest of us, but we all knew well, between chuckles, that the open dumping practices and locations of some decades-before landfills are still active and flourishing, although now under the casual or permissive eyes of county and state agencies. The municipal solid wastes, *Toxic Sewage Sludges*, sewage sludge grit, and other forms of domestic and industrial waste, plus the plastics and many other recyclables included along with the masses of solid waste, are still dumped down canyons, although now with massive, decomposable plastic tarps below and above the garbage, as a "cover" story that depicts some form of solid waste management—not!

Today these solid waste dump sites are referred to as landfills, but in fact are no more than open pit, canyon-dumped or former quarry-dumped, solid waste landfills, all leaking and flushing from below, but now with some meager efforts at soil coverage, tarp or liner containment, gas emissions monitoring, methane collection, and landfill leachate management. Landfills are still no more than just open garbage pits, albeit under another name, and they all still out-gas, leak, leach, and stink, and they all still attract airborne and ground-borne scavengers, and they're all still public nuisances, awaiting to be filled to over-capacity, then to leak, leach, and stink forevermore. Once I began locating, identifying, and recording the colorful and odoriferous outflows just from the many legacy landfills within and about King County and a few other counties in Washington State, it became apparent that we will all be living with these holocaust-like deposits for decades or centuries to come.

As an aside, when I worked as a technology analyst for an investment banking firm a few years ago, at one memorable meeting with an investment banking executive, he professed that his firm loved to invest in quarries, because you made money by selling the rock, gravel, and sand on the front end, and then you made even more money afterwards as the former quarry was thereafter filled with solid waste, and what's more, you also made a bundle later by the sale of the surface real estate for development, so how could you lose on that one?

The adverse human and environmental health consequences of residing near a landfill are clear to all, yet remain poorly defined—a topic ignored by agencies. And while monitoring the cause-and-effect public health consequences appears to be limited to methane sampling, the toxics in the runoff, leachates, and emitted gases are obscured as being no more than the annoying complaints of nearby neighbors. If you may wonder what infectious agents or toxics may be held within those landfill-sourced matrices, you may as well just pack up and leave, for the agencies will not generate or reveal such negative data, or they just do not know, and they certainly do not want to find out, and then have to reveal the findings, or actually have to do something about it. It'll just go away, is the prevailing philosophy.

I inquired recently of a landfill management engineer the usual and standard questions of: "What's in the emitted gases?" "What's in the landfill leachates?" "Where do the gases and leachates go?" "What flows to the adjacent and receiving soils and waters?" "What are the downstream present and long-term consequences?" "How are the toxic consequences managed?" "What are the adverse consequences to human health and to local and distant environments?" "Who is directly liable for the long-term consequences of acute and chronic diseases resulting from exposure to the emitted gases and to the outpouring of toxic waters over the next decades or more?" "What are the workable alternatives to land disposal?" The response was at first simply a startled blank stare, followed by the offhanded comment (not since repeated), that if the local citizens are bothered, why don't they just move? That was not the first time I had heard that remark—it seems to be an agency-wide snark, likely uttered only at closed agency meetings.

That obvious and implied response, as revealed by the initial blank stare, is that no one really knows or cares exactly what's in it, or its fate in the environment, or the immediate or long-term consequences of such actions. Therefore, and by the underlying tenants of the precautionary principle, it should be on the mind of any agency, firm or person involved in the land disposal of municipal solid waste, aka, garbage, that there is something seriously wrong with the practice of landfilling, of and by itself. Such primitive practices must be terminated, post haste! Perhaps at least a fleeting consideration of employing a modern waste-to-energy facility, or two, or more, may solve this toxic dilemma.

As an additional tale for emphasis, we recently evaluated yet another former pyrolysis facility whereupon rested the rusted remnants of the structure and it's still leaking toxic wastes, this one focused on agricultural and forest-based organic materials as the intended fuel feedstock source. We investigated the site as a possible location for a hoped-for new waste-to-renewable clean energy facility but discovered that it is located directly atop and within a major flowing, perennial wetland, as had been approved by the "responsible" county and state agencies, along with the concerned federal agency. Once constructed, permitted, and operating, the facility was cited for the regular release of toxic waste residuals from its pyrolysis operations directly into the underlying wetlands that flow directly into Puget Sound.

How could that be? Aren't wetlands and Puget Sound protected? And, once contaminated and polluted, aren't they expected to be remediated, i.e., restored by reversing or stopping environmental damage? Never happened here—rarely ever happens! Of course our life-sustaining wetlands are now more protected in the name of environmental health, yet if wetlands are in fact the cradle of life, then that cradle needs some immediate cleanup and restoration, if at all possible.

The rusting hulk now resides as a monument to the obvious flaws in local and state agency planning, siting, and remediation systems. Once cited, the owners simply walked away, with the residual monstrosity still dripping and leaching contaminants and pollutants into the flowing wetlands that still percolate down and directly into Puget Sound. A casual inspection revealed the typical "Red Goo," as the environmental community calls it, meaning the indicator red, iron-oxidizing bacterial masses that form in oxygen-deprived anaerobic waters and conditions beneath such rusted facilities that will continue to ooze for many decades or centuries to come. The associated flowing toxics were never analyzed, evaluated or monitored, and presumably never will be.

So there it sits, in the midst of thickly forested wetlands, never to be restored, a relic of the state's past permitting behaviors and practices. How many more such neglected sites are there in Washington State and elsewhere across the country? And what will be the adverse impacts and troublesome human and environmental health consequences as yet to be made known? And would it have been a good idea to have the investors, builders, and operators provide some form of a security bond beforehand as a condition of permitting that would cover the fully anticipated remediation costs, once the venture was found to be a contaminating and polluting violation and failure? Was it the hoped-for rewards in economic development that drove the project from the beginning, environmental consequences be damned?

A casual visit to any currently active or legacy landfill, as I continue to do, reveals the same brown, orange or red leachates that flow freely, wholly unencumbered, untreated, and unattended, into any and all convenient wetlands, groundwaters, aquifers, streams, and rivers, and into the nearest sound and ocean for dilution, to the demise of any and all wildlife. One of my favorite examples of a non-stop Red Goo, spring-fed, stream flow winds its way through several culverts under major roadways beneath signs that read: "This stream is in your care. Within the greater Lake Washington Watershed." Oh, my Gosh!

Solid waste disposal is still a national disgrace, managed in large part by agencies and major corporations that make claims of: "The volatile organic compounds in the emitted gases are a fully captured source of renewable clean energy," and that, "They capture all of the highly toxic landfill leachates that flow uncaptured to all adjacent ground and surface waters;" or "Every single drop of leachate is accounted for"—yet another unlikely and near-comical story.

Our failed attempts at managing solid and liquid wastes, municipal, industrial or otherwise, require serious, revised management consideration. Just how would one capture or collect land-disposed solid wastes, land-disposed *Toxic Sewage Sludges*, landfill leachates, and VOCs from landfills and surface-disposed sludges, once dumped in forests, on prairies, or on our farms and food crops? It can't be done after the fact and so it becomes clear that states, counties, and municipalities across the U.S. cannot now and have never stepped-up to the task of managing our wastes properly. As it stands, nearly all our wastes are burned, buried, hidden or disposed of inappropriately, all as part of our greater Toxic Waste Relocation Program.

For the sake of our well-being and for the health of our land, and as suggested repeatedly herein, we need to abolish the land disposal of *Toxic Sewage Sludge,* and the land and ocean dumping of other waste products, but where will it go? The challenge and the solutions are clear: *Toxic Sewage Sludge* and other toxic wastes cannot be repurposed or reused somewhere else; it all remains as toxic waste until thermolytic decomposition by high heat renders such toxic wastes as being markedly less toxic or non-toxic, far beyond what may be provided by incineration.

Toxic Sewage Sludge can be processed safely by enhanced thermal decomposition to yield heat for energy, clean exhaust gas equivalent to ambient air, and nearly inert ash that can be used for the production of industrial products. As it stands, *Toxic Sewage Sludge* is disposed in our forests and on our farms and rangelands, simply because we fail to embrace existing thermolytic decomposition technologies that easily reduce *Toxic Sewage Sludge* and its toxic chemicals, metals, and pathogens to basic elements that can be repurposed safely.

The land disposal of *Toxic Sewage Sludge* faces significant controversy, and a resistance movement to the practice is rightfully building against this misguided policy. Science doesn't support the disposal of *Toxic Sewage Sludge* across the landscape, in that the supposed benefits are more than offset by the risks to human and environmental health. As stated, a vast number of chemical and biological contaminants exist in *Toxic Sewage Sludge*, they persist after land disposal, and they cannot be cleaned up later, no matter the source of that bad idea. Many investigations have identified only a tiny fraction of the total contaminant load, and we cannot say with any degree of confidence what the true range of contaminant risk may be—it's unknowable. We then begin to realize that effectively all current and legacy industrial chemicals end up in our sewage, and during ineffective treatment processes they move into the *Toxic Sewage Sludge*. If you apply the resulting *Toxic Sewage Sludge* product to the land, and if you have filled our canyons and wetlands with trash, then you have transferred our cumulative toxic wastes into the living and once life-giving landscape.

Having walked the wastes of China, Korea, India, Russia, Georgia (Republic of Georgia), Latin America, and the Antarctic, and having observed some remarkable improvements firsthand in those countries, I still wonder why we are unable to confess that disposal practices for our solid wastes and sewage products in the U.S. are not improved. The definitions of developed and lesser-developed nations may need revision.

While there have been many attempts to reveal the science surrounding the risks, threat, and harm from even casual exposure to *Toxic Sewage Sludge* and other toxic wastes, these revelations are generally lacking in toxicological data. Regardless, and for all of the reasons pointed out in this tome, we can and must end these destructive practices. As Native American culture and my Tribal friends have so often stated: "The quality of our future and of our children's future, seven generations and beyond, depends on what we do today."

Thankfully, I have seen a gradual awakening in recent times. Not so long ago, I received a call from a young woman in distress. Until recently, she said, she had been an employee of the county wastewater treatment division, assigned to work in the local forests east of Seattle. After listening to her story, it became clear that she had assisted with the land disposal of *Toxic Sewage Sludge* in the local Snoqualmie forests. She really wanted to talk, specifically about her not-so-thrilling, real-life, in-the-forest experiences. The troubled young woman had observed firsthand the blatant and irresponsible environmental destruction caused by the disposal of *Toxic Sewage Sludge* in the forests, and was bewildered as to why the county would elect to dump their toxic loads down steep hillsides above the many waterways below.

She insisted on total anonymity in her revelations, repeating various episodes whereby the stated guidelines were not followed, and that the sludge loads were simply dumped as guided by convenience, and for no reason other than that they were always behind schedule. After complaining to her supervisor that she objected to allowing the sludge to flow downhill into the wetlands and streams, she was strongly reprimanded. They had work to do, she was chided, and a tight schedule to follow.

The young woman continued to unload her distress over the phone about what she had witnessed, repeatedly. After several more such inappropriate sludge-dumping experiences, she told me, she complained again, but at a higher level at the county. "I was instantly transferred to another job in the forest," as she put it, "and was no longer involved with the sewage sludge dumping." While working on her new assignment, she noticed more instances of thoughtless sludge disposal in the forests, including on steep hillsides, along streams, and adjacent to wetlands, with little regard for contamination of the forest waters. Upon voicing her concerns again, and with the hope of encouraging greater care in their methods as the sludge disposal work continued, she was abruptly terminated by the county without further explanation.

Her final remarks were along the lines that she had gone to college to study forestry with a mind to a career in wildlife conservation, and that the destructive behaviors she had observed were in direct opposition to all she had studied. She could not believe that what she was witnessing was county and state policy.

She had heard about me from a similarly concerned co-worker, she stated, found my number, and wanted to reveal her personal experiences, and that was it. Others at the county held similar concerns, she added, but valued their positions. A while later, at an environmental science conference at the University of Washington, a young woman approached me following my presentation and stated simply, "Thank you for what you are doing," and walked away—same voice!

We are the authors of our own fate. The sages of gloom and doom may foretell the end of an Earth that supports human life, and sometimes of an Earth that supports any form of life at all, but we're not those people! We must study the literature, both science and popular press, and strive to bring the conversation down to an intelligible level for everyone. The messages to be told from our time here will likely reveal the scenarios of prophesied outcomes, sometimes great, sometimes dismal, almost always of decline, but also always holding promise. The evidence is bold. What we must do is state the problems with clarity and courage, just like this young woman, because they are real, while in the end we must construct a workable scenario that will provide a way out—a really plausible way out. We have to uplift our world. People need to know there are actual solutions to the most critical challenges we face today, whether political or commercial interests engage in the good fight or not.

Resources

PRAEFATIO: *BON APPÉTIT!*

1. Seeger P. (2019). Quite Early Morning. The Smithsonian Folkways Collection. *Smithsonian Folkways Recordings.* https://unearthed.greenpeace.org/2020/02/04/sewage-sludge-landspreading-environment-agency-report/
2. Sung H., Ferlay J., Siegel R. L., Laversanne M., Soerjomataram I., Jemal A., and Bray F. (2021, May). Global Cancer Statistics 2020: Estimates of Incidence and Mortality Worldwide for 36 Cancers in 185 Countries. *National Library of Medicine.* https://pubmed.ncbi.nlm.nih.gov/33538338/
3. USDA. (2022). Organic Regulations. *US Department of Agriculture: Agricultural Marketing Service.* https://www.ams.usda.gov/rules-regulations/organic
4. USDA. (2022, April 27). Animal Products: Animal Policy & Regulatory Issues. *U.S. Department of Agriculture Economic Research Service.* https://www.ers.usda.gov/topics/animal-products/
5. Smith-Heavenrich S. (2003). Sludge By Any Name Will Never Be "Organic." *Maine Organic Farmers and Gardeners.* https://www.mofga.org/advocacy/toxics/sludge/

INTRODUCTION: IT'S NOT WHAT YOU THINK

1. Lewis D. L. (2014). *Science for Sale: How The US Government Uses Powerful Corporations and Leading Universities to Support Government Policies, Silence Top Scientists, Jeopardize Our Health, and Protect Corporate Profits.* Skyhorse Publishing, New York, N.Y.
2. Dowler C. and Boren Z. (2020, April 2). Revealed: salmonella, toxic chemicals and plastic found in sewage spread on farmland. *Unearthed.*
3. Socolow R. (2022, March 7). Global Collapse Is In View. *Bulletin of the Atomic Scientists.* https://thebulletin.org/2022/03/global-collapse-is-in-view/
4. DOE. (2022). Learn About Biosolids. *State of Washington Department of Ecology.* https://ecology.wa.gov/Waste-Toxics/Reducing-recycling-waste/Biosolids/Learn-about
5. King County. (2021, October 13). Loop Biosolids. King County, Seattle, WA. https://kingcounty.gov/services/environment/wastewater/resource-recovery/loop-biosolids/biosolids.aspx

PROLOGUE: NOT ON MY STREET!

1. Carson R. L. (1962). *Silent Spring.* Houghton Mifflin Company, New York, N.Y.
2. Carson R. L. (1962).
3. Knight JZ. (2022). Introduction to RSE by Ramtha. *Ramtha's School of Enlightenment*, Yelm, WA. www.ramtha.com.
4. DOE. (2022). Critical Aquifer Recharge Areas. *State of Washington Department of Ecology.* https://ecology.wa.gov/Water-Shorelines/Water-quality/Groundwater/Protecting-aquifers/Critical-aquifer-recharge-areas
5. PTC. (2019). Sludge Is Not Coming to Yelm! *Preserve the Commons*, Yelm, WA. https://www.preservethecommons.org/

6. Rosane E. (2019, February 15). Citizen Organization Say They'll Take Legal Action If Biosolids Site Approved. *Nisqually Valley News*. https://www.yelmonline.com/stories/citizen-organization-say-theyll-take-legal-action-if-biosolids-site-approved,101216 https://www.preservethecommons.org/biosolids-near-yelm.html
7. Rosane E. (2019, March 29). Thurston County Staff Examining What Authority They Have on Biosolids Application. *Nisqually Valley News*. https://www.yelmonline.com/stories/thurston-county-staff-examining-what-authority-they-have-on-biosolids-application,101080
8. Lyon P. (2018, December 17). Department of Ecology: State Environmental Policy Act Determination of Nonsignificance (DNS) – Fire Mountain Farms (FMF, Inc).
9. Northwest Toxic Communities Coalition. (2022). Sequim, WA. https://nwtoxiccommunities.org/
10. Horkings-Brigham M. (2019). *Cannabis the Conundrum: Medical Mystery, Environmental Enigma, Neighborhood Nightmare*. Horus Publishing, Inc., WA.
11. Rosane E. (2019, June 27). Biosolids Opposition Group Holds Meeting With Microbiologist. *Nisqually Valley News*. https://www.yelmonline.com/stories/biosolids-opposition-group-holds-public-meeting-with-microbiologist,99667
12. Rosane E. (2019, October 31). Application for Biosolids Site on 128th Avenue Closed Out by Department of Ecology. *Nisqually Valley News*. https://www.yelmonline.com/stories/application-for-biosolids-site-on-128th-avenue-closed-out-by-department-of-ecology,96360

CHAPTER 1: DR. HONOUR

1. Boyden A. L. (2008, June 2). *Echoes from Hospital and White House: A Record of Mrs. Rebecca R. Pomroy's Experience in Wartimes (1884).* Kessinger's Rare Reprints.
2. Rotondi J. P. (2019, December 5). Why Laos Has Been Bombed More Than Any Other Country. History. https://www.history.com/news/laos-most-bombed-country-vietnam-war
3. Augustyn A. (2022). Bay of Pigs invasion. The Editors of *Encyclopaedia Britanica*. https://www.britannica.com/event/Bay-of-Pigs-invasion
4. Carson R. L. (1962). *Silent Spring.* Houghton Mifflin Company, New York, N.Y.
5. Merino N., Aronson H.S., Bojanova D. P., Feyhl-Buska J., Wong M. L., Zhang S., and Giovannelli D.. (2019, April 15). Living at the Extremes: Extremophiles and the Limits of Life in a Planetary Context. *Frontiers in Microbiology.* https://www.frontiersin.org/articles/10.3389/fmicb.2019.00780/full
6. Pauling E. C. (1937 – 1997). The Pauling Blog. https://paulingblog.wordpress.com/2012/06/04/edward-crellin-pauling-1937-1997
7. Honour R. C. (1973, August). *Lysis of Phytophthora Parasitica Mycelium and Oospores in Soil.* University of California, Riverside.
8. Petsko E. (2019, April 5). Where Does All the Human Poop in Antarctica End Up? *Mental Floss: Environment.* https://www.mentalfloss.com/article/575512/what-happens-to-human-waste-in-antarctica

9. USC Norris Comprehensive Cancer Center. (2021, August 20). USC Norris Comprehensive Cancer Center Receives Renewal of Prestigious Designation from the National Cancer Institute. https://uscnorriscancer.usc.edu/2021/08/20/usc-norris-comprehensive-cancer-center-receives-renewal-of-prestigious-designation-from-the-national-cancer-institute/
10. Seebach J. and Feinberg J. (2021, December 6). Bill to Protect Grand Canyon Area From Mining and Other Threats Gains Support. *Pew*. https://www.pewtrusts.org/en/research-and-analysis/articles/2021/12/06/bill-to-protect-grand-canyon-area-from-mining-and-other-threats-gains-support

CHAPTER 2: A BRIEF HISTORY OF SLUDGE

1. Wilson D. (2001). *Fateful Harvest: The True Story of a Small Town, a Global Industry, and a Toxic Secret.* Harper Collins Publishers Inc., New York, NY.
2. Donovan A. (2013, March 19). Early Sanitation in Boston: A Brief History of Sewers in Boston. *UMass Boston*. https://blogs.umb.edu/pumpingstation/2013/03/19/a-history-of-sewers-in-boston/
3. Omnia Mechanical Group. (2020, October 1). The History of Plumbing in New York City. https://sanitary.nyc/the-history-of-plumbing-in-new-york-city/ https://www.nytimes.com/2020/08/14/nyregion/freshkills-garbage-dump-nyc.html
4. City of Linden. (2021, November 19). Aries Linden Biosolids Gasification Plan Achieves Mechanical Completion. https://linden-nj.gov/aries-linden-biosolids-gasification-plan-achieves-mechanical-completion/
5. Sullivan R. (2020, August 14). How the World's Largest Garbage Dump Evolved Into a Green Oasis. *The New York Times*. https://www.nytimes.com/2020/08/14/nyregion/freshkills-garbage-dump-nyc.html
6. Morningstar, S. (2022). Sand Point Naval Air Station Landmark District. *Seattle Department of Neighborhoods*. City of Seattle, WA. https://www.seattle.gov/neighborhoods/programs-and-services/historic-preservation/historic-districts/sand-point-naval-air-station-landmark-district
7. Caldbrick J. (2013, January 2). Union Bay Natural Area (Seattle). HistoryLink.org. https://www.historylink.org/file/10182
8. Mapes L. V. (2019, February 25). Decline of salmon adds to the struggle of Puget Sound's orcas. *The Seattle Times*. https://www.adn.com/nation-world/2019/02/25/decline-of-salmon-adds-to-the-struggle-of-puget-sounds-orcas/
9. Abramo A. (2020, December 8). Outdated sewage treatment is suffocating fish in Puget Sound. *Crosscut*. https://crosscut.com/environment/2020/12/outdated-sewage-treatment-suffocating-fish-puget-sound
10. UIA (2022). Environmental degradation. *The Encyclopedia of World Problems and Human Potential*. http://encyclopedia.uia.org/en/problem/135569
11. Borlaug N. (1970). The Nobel Peace Prize 1970 - Norman Borlaug Biographical. *The Nobel Prize*. https://www.nobelprize.org/prizes/peace/1970/borlaug/biographical/
12. Carson R. L. (1962). *Silent Spring*. Houghton Mifflin Company, New York, N.Y.

CHAPTER 3: DON'T YOU KNOW CANCER IS OVER?
1. Schweitzer A. (n.d.). Albert Schweitzer Quotes. Retrieved August 24, 2022, from BrainyQuote.com. https://www.brainyquote.com/quotes/albert_schweitzer_104993 2. WHO. (2022, February 3). Cancer: What causes cancer? *World Health Organization*. https://www.who.int/news-room/fact-sheets/detail/cancer
3. OSHA. (2022). Hexavalent Chromium. *Occupational Safety and Health Administration. United States Department of Labor*. https://www.osha.gov/hexavalent-chromium
4. Simoneit B. R. T., Medeiros P. M. and Didyk B. M. (2005, October). Combustion Products of Plastics as Indicators for Refuse Burning in the Atmosphere. *National Library of Medicine*. Pubmed.gov https://pubmed.ncbi.nlm.nih.gov/16201617/
5. EPA. (2022, March 24). CADDIS Volume 2: Unspecified Toxic Chemicals. *U.S. Environmental Protection Agency*. https://www.epa.gov/caddis-vol2/unspecified-toxic-chemicals
6. Ferley J., Colombet M., Soerjomataram I., Parkin D. M., Piñeros M., Znaor A., and Bray F.. (2021, April 5). Cancer statistics for the year 2020: An overview. *National Library of Medicine*. PubMed.gov https://pubmed.ncbi.nlm.nih.gov/33818764/#affiliation-1
7. Eastwood C. (1965). *For A Few Dollars More*. Film directed by Sergio Leone.
8. Carroll L. (1865). Alice's Adventures in Wonderland. Macmillan Publishers Ltd. London, UK.
9. AG Ferguson B. (2022). Mission, Vision & Values. *Washington State Office of the Attorney General*. https://www.atg.wa.gov/mission-vision-values
10. Veflen Olson N. and Motarjemi Y. (2014). Food Safety Assurance System: Food Safety and Ethics. *Encyclopedia of Food Safety*, Vol. 4. ScienceDirect. https://www.sciencedirect.com/science/article/pii/B9780123786128004376
11. Loop®. (2022). Loop is a resource, sourced by you. https://www.loopforyoursoil.com/what-is-loop/

CHAPTER 4: SLUDGE-TRACKING 101
1. Eliot G. (1872). *Middlemarch, A Study of Provincial Life*. William Blackwood and Sons, Scotland, UK.
2. Brabcová V., Nováková M., Davidová A., and Baldrian P. (2016, February 2). Dead fungal mycelium in forest soil represents a decomposition hotspot and a habitat for a specific microbial community. *New Phytologist Foundation*. https://nph.onlinelibrary.wiley.com/doi/10.1111/nph.13849
3. Kantavichai R., Briggs D. G., and Turnblom E. C. (2010). Effect of thinning, fertilization with biosolids, and weather on interannual ring specific gravity and carbon accumulation of a 55-year-old Douglas-fir stand in western Washington. *ResearchGate*. https://www.researchgate.net/publication/233580183_Effect_of_thinning_fertilization_with_biosolids_and_weather_on_interannual_ring_specific_gravity_and_carbon_accumulation_of_a_55-year-old_Douglas-fir_stand_in_western_Washington
4. SEI. (2022). Legal Definition of Ecocide. *Stop Ecocide International*. https://www.stopecocide.earth/legal-definition-of-ecocide

5. EPA. (2022, March 23). Toxic Substances Control Act (TSCA) and Federal Facilities. *United States Environmental Protection Agency.* https://www.epa.gov/enforcement/toxic-substances-control-act-tsca-and-federal-facilities
6. Mountains to Sound Greenway Trust. (2022). https://mtsgreenway.org/

CHAPTER 5: COMMERCE VS. PUBLIC HEALTH
1. Taylor D. E. (2014). Toxic Communities: Environmental Racism, Industrial Pollution, and Residential Mobility. *New York University Press*. New York and London.
2. Saltman J. (2019, April 3). Metro Vancouver, other communities, sending tonnes of trach to U.S. *Vancouver Sun*, BC, Canada. https://vancouversun.com/news/local-news/metro-vancouver-other-communities-sending-tonnes-of-trash-to-u-s
3. Taylor D. E. (2014).
4. Evers G. (2022, June 2). Great Pacific Garbage Patch. *National Geographic Society.* https://education.nationalgeographic.org/resource/great-pacific-garbage-patch
5. Franz H. (2022). Commissioner of Public Lands Hilary Franz. Washington State Department of Natural Resources. https://www.dnr.wa.gov/commissioner
6. DNR. (2022). About the Washington Department of Natural Resources. https://www.dnr.wa.gov/about-washington-department-natural-resources
7. DOE. (2022). About Us. *Washington Department of Ecology.* https://ecology.wa.gov/About-us
8. WDFW. (2022). About WDFW. *Washington Department of Fish and Wildlife.* https://wdfw.wa.gov/about
9. DOH. (2022). About Us. Washington State Department of Health. https://doh.wa.gov/about-us
10. DOC. (2022). About Us. Washington State Department of Commerce. https://www.commerce.wa.gov/
11. USDA. (2022). About the U.S. Department of Agriculture. https://www.usda.gov/our-agency/about-usda

CHAPTER 6: AS DOGS WILL DO
1. Swift J. (1726). Gulliver's Travels. *British Library.*

CHAPTER 7: POLLUTION AS POLICY
1. Auden W. H. (1957, March 1). First Things First. *The New Yorker*, Pg. 38.
2. CHRL. (2018, September 17). Cedar Hills Regional Landfill. King County, WA. https://kingcounty.gov/depts/dnrp/solid-waste/facilities/landfills/cedar-hills.aspx
3. King County. (2022). Biosolids: Turn your dirt around. King County Wastewater Services, WA. https://kingcounty.gov/services/environment/wastewater/resource-recovery/loop-biosolids.aspx

CHAPTER 8: TOXIC INSULT
1. Hemingway E. (1935, October 1). Monologue to the Maestro: A High Seas Letter. *Esquire.*

2. National Park Service. (2022). Manhatten Project. Welcome to Hanford, Washington. National Park Service U.S. Department of the Interior. https://www.nps.gov/mapr/hanford-wa.htm
3. BYJU's. (2022). Organometallic Compounds. https://byjus.com/chemistry/organometallic-compounds/
4. Lewis D. L. (2014). *Science for Sale: How The US Government Uses Powerful Corporations and Leading Universities to Support Government Policies, Silence Top Scientists, Jeopardize Our Health, and Protect Corporate Profits*. Skyhorse Publishing, New York, N.Y.
5. EPA. (2021, February). EPA Biennial Report – No. 8 (Reporting Period 2018-2019). United States Environmental Protection Agency. https://www.epa.gov/sites/default/files/2021-02/documents/biennial-review8-factsheet-2021.pdf
6. Perkins T. (2021, May 28). 'Forever chemicals' found in home fertilizer made from sewage sludge. *The Guardian*. https://www.theguardian.com/environment/2021/may/28/home-fertilizer-toxic-pfas-forever-chemicals-sewage-sludge
7. EPA. (1994, September). A Plain English Guide to the EPA Part 503 Biosolids Rule. United States Environmental Protection Agency Office of Wastewater Management.
8. Lewis D. L. (2014).
9. Lewis D. L. (2014).

CHAPTER 9: THE MICROBIOME OF SLUDGE

1. Carrington D. (2020, August 17). Microplastic particles now discoverable in human organs. *The Guardian*. https://www.theguardian.com/environment/2020/aug/17/microplastic-particles-discovered-in-human-organs
2. Major N., Schierstaedt J., Jechalke S., Nesme J., Ban S. G., Cerne M., Sorensen S. J., Ban D., and Schikora A.. (2020, July 9). Composted Sewage Sludge Influences the Microbiome and Persistence of Human Pathogens in Soil. *National Library of Medicine*. https://www.ncbi.nlm.nih.gov/pmc/articles/PMC7409118/
3. Mantilla-Calderon D. and Hong P-Y. (2017, July 1). Fate and Persistence of a Pathogenic NDM-1-Positive Escherichia coli Strain in Anaerobic and Aerobic Sludge Microcosms. PubMed.gov. https://pubmed.ncbi.nlm.nih.gov/28411227/
4. Li M., Song G., Liu R., Huang X., and Liu H. (2021, September 25). Inactivation and risk control of pathogenic microorganisms in municipal sludge treatment: A review. *National Library of Medicine*. https://www.ncbi.nlm.nih.gov/
5. Ubukata K., Nonoguchi R., Matsuhashi M., and Konno M. (1989, May). Expression and inducibility in Staphylococcus aureus of the mecA gene, which encodes a methicillin-resistant S. aureus-specific penicillin-binding protein. *American Society for Microbiology Journal of Bacteriology*. https://www.ncbi.nlm.nih.gov/pmc/articles/PMC209980/
6. DOH. (2022). Valley Fever (Coccidioidomycosis). *Washington State Department of Health*. https://doh.wa.gov/you-and-your-family/illness-and-disease-z/valley-fever-coccidioidomycosis

7. Arnaud C. H. (2020, November 21). Monitoring COVID-19 in sewage. *c&en Chemical and Engineering News*. https://cen.acs.org/environment/water/Monitoring-COVID-19-sewage/98/i45
8. Venkatasan A. K., Done H. Y., and Halden R. U. (2014, May). United States National Sewage Sludge Depository at Arizona State University – A New Resource and Research Tool for Environmental Scientists, Engineers, and Epidemiologists. ResearchGate. https://www.researchgate.net/publication/262305017_United_States_National_Sewage_Sludge_Repository_at_Arizona_State_University_-_A_New_Resource_and_Research_Tool_for_Environmental_Scientists_Engineers_and_Epidemiologists

CHAPTER 10: THAT'S THE WAY WE DO THINGS
1. Roberts P. (2008) *The End of Food.* Houghton Mifflin Harcourt Publishing Company,
2. SEI. (2021, June). Independent Expert Panel for the Legal Definition of Ecocide *EcocideLaw*. https://ecocidelaw.com/independent-expert-drafting-panel/
3. Ingham E. R. (2022). Soil Nematodes. *United States Department of Agriculture Natural Resources Conservation Service*. https://www.nrcs.usda.gov/wps/portal/nrcs/detailfull/soils/health/biology/?cid=nrcs142p2_053866
4. Sung H., Ferlay J., Siegel R. L., Laversanne M., Soerjamataram I., Jemal F., Bray F. (2020). Global Cancer Statistics 2020: GLOBOCAN Estimates of Incidence and Mortality Worldwide for 36 Cancers in 185 Countries. *National Library of Medicine*. https://pubmed.ncbi.nlm.nih.gov/33538338/

CHAPTER 11: IS SCIENCE FOR SALE?
1. Ghosh A. (2017, July 24). *The Great Derangement: Climate Change and the Unthinkable*. University of Chicago Press.
2. Lewis D. L. (2014). *Science for Sale: How The US Government Uses Powerful Corporations and Leading Universities to Support Government Policies, Silence Top Scientists, Jeopardize Our Health, and Protect Corporate Profits*. Skyhorse Publishing, New York, N.Y.

CHAPTER 12: WHAT'S REALLY IN IT?
1. Mathney J. M. J. (2011, March 16). A Critical Review of the U. S. EPA's Risk Assessment for the Land Application of Sewage Sludge. *New Solutions: A Journal of Environmental and Occupational Health Policy*. https://www.semanticscholar.org/paper/A-Critical-Review-of-the-U.S.-EPA%27s-Risk-Assessment-Mathney/4213f4b41573aa6981783081d656dfbbe3442642
2. Paltrow G. (1998). *Shakespeare in Love*. Directed by Jon Madden. The Bedford Falls Co.
3. EPA. (2022, April 29). Basic Information about Biosolids – Classes of Biosolids. *U.S. Environmental Protection Agency*. https://www.epa.gov/biosolids/basic-information-about-biosolids
4. Daley R. (1976). *The Outlaw Josey Wales*. The Malpaso Company.

5. Lacitus E. (2018). 1 bicyclist dead, 1 hurt in cougar attack near Snoqualmie. *The Seattle Times*. https://www.seattletimes.com/seattle-news/eastside/1-dead-1-injured-in-cougar-attack-on-eastside/
6. Fesler S. (2018). Nearly Six Million Residents in Central Puget Sound by 2050. The Urbanist. https://www.theurbanist.org/2018/01/26/nearly-six-million-residents-puget-sound-area-2050-psrc-says/
7. Apple J., Wold R., Stark K., Bos J., Williams P., Hamel N., Yang S., Selleck J., Moore S. K., Rice J., Kantor S., Krembs C., Hannach G, and Newton J. (Eds). (2020). Puget Sound marine waters: 2019 overview. *PSEMP Marine Waters Workgroup*. https://swinomish-nsn.gov/media/110779/2019_marine_waters_overview.pdf
8. Weiss M. (2021, April 21). In the Tales Told by Sewage, Public Health and Privacy Collide. *Undark*. https://undark.org/2021/04/21/covid-19-data-down-the-drain/?ct=t
9. Wells-Barnett I. B. (2014 edition). *The Light of Truth: Writings of an Anti-Lynching Crusader*. Penguin Classics.

CHAPTER 13: SMOKING GUN

1. Wells Hemingway E. (1940, October 21). *For Whom the Bell Tolls*. Charles Scribner's Sons.

CHAPTER 14: A WORLD SMOTHERED IN PLASTIC

1. UN Environment Programme. (2018, May 25). Our planet is drowning in plastic pollution. United Nations. https://www.unep.org/news-and-stories/story/our-planet-drowning-plastic-pollution
2. Vuleta B. (2022, March 31). 51+ Disturbing Plastic Waste Statistics [2022]. *SeedScientific*. https://seedscientific.com/plastic-waste-statistics/
3. Volcovici V. (2022, May 4). U.S. plastic recycling rate drops close to 5% - report. Reuters. https://finance.yahoo.com/news/u-plastic-recycling-rate-drops-120951448.html
4. Vuleta B. (2022, March 31).
5. Phillips E. K. (2020, October 17). Al Capone convicted on this day in 1931 after boasting 'They can't collect legal taxes from illegal money.' *Forbes*. https://www.forbes.com/sites/kellyphillipserb/2020/10/17/al-capone-convicted-on-this-day-in-1931-after-boasting-they-cant-collect-legal-taxes-from-illegal-money/?sh=67aeaca91435
6. Law K. L., Starr N., Siegler T. R., Jenna R. Jambeck J. R., Mallos N. J., George H. and Leonard G. H. (2020, October 30). The United States' contribution of plastic waste to land and ocean. ScienceAdvances. https://www.science.org/doi/10.1126/sciadv.abd0288
7. Vuleta B. (2022, March 31).

CHAPTER 15: PARTICULATE PLASTICS IN POO

1. Vuleta B. (2022, March 31). 51+ Disturbing Plastic Waste Statistics [2022]. *SeedScientific*. https://seedscientific.com/plastic-waste-statistics/
2. Thompson R C., Olsen Y., Mitchell R. P., Davis A., Rowland S. J., John A. W. G., McGonigle D., and Russell A. E. (2004, May 7) Lost at Sea: Where Is All the

Plastic? *Science*. Vol 304. Issue 5672. https://www.science.org/doi/10.1126/science.1094559
3. UN. (2022). Our planet is choking on plastic. *United Nations Environment Programme*. https://www.unep.org/interactives/beat-plastic-pollution/
4. Dell J. (2019, April 30). Six Times More Plastic Waste is Burned in U.S. than is Recycled. *Plastic Pollution Coalition*. https://www.plasticpollutioncoalition.org/blog/2019/4/29/six-times-more-plastic-waste-is-burned-in-us-than-is-recycled
5. McGrath T. J., Poma G., Matsukami H., Malarvannan G., Kajiwara N. and Covaci A. (2021, January 6). Short- and Medium-chain Chlorinated Paraffins in Polyvinylchloride and Rubber Consumer Products and Toys Purchased on the Belgian Market. *National Library of Medicine*. https://www.ncbi.nlm.nih.gov/pmc/articles/PMC7908593/
6. Rizzuto S., Baho D. L., Jones K. C., Zhang H., Leu E., and Nizzetto L. (2021, August 25). Binding of waterborne pharmaceutical and personal care products to natural dissolved organic matter. *National Library of Medicine*. https://pubmed.ncbi.nlm.nih.gov/34088045/
7. Law K. L., Starr N., Siegler T. R., Jambeck J. R., and Leonard G. H. (2020, October 30). The United States' contribution of plastic waste to land and ocean. *ScienceAdvances* Vol 6, Issue 44. https://www.science.org/doi/10.1126/sciadv.abd0288
8. Simon M. (2021, September 22). Baby Poop Is Loaded With Microplastics. *Beyond Plastics*. https://www.beyondplastics.org/news-stories/baby-poop-is-loaded-with-microplastics
9. Machado A., Lau C. W., Kloas W., and Bergmann J. (2019, April). Microplastics Can Change Soil Properties and Affect Plant Performance. ResearchGate. https://www.researchgate.net/publication/332666674_Microplastics_Can_Change_Soil_Properties_and_Affect_Plant_Performance

CHAPTER 16: PERSISTENCE PAYS

1. Linklater R., Krizan K. (2004). *Before Sunset*. Castle Rock Entertainment.
2. NOAA Fisheries. (2021, July 6). Listing of Southern Resident Killer Whale Under the ESA. *NOAA Office of Protected Resources*. https://www.fisheries.noaa.gov/action/listing-southern-resident-killer-whale-under-esa

CHAPTER 17: SPEAKING TRUTH TO POWER

1. Wells-Barnett I. B. (2014 edition). *The Light of Truth: Writings of an Anti-Lynching Crusader*. Penguin Classics.
2. American Friends Service Committee. (1955, January 1). Speak Truth to Power: A Quaker Search for an Alternative to Violence. American Friends Service Committee.

CHAPTER 18: JUST LIKE MY SHOES

1. Gregoire C. (2012). Environment & Energy. *Washington State Office of the Governor*. https://www.digitalarchives.wa.gov/governorgregoire/priorities/environment/default.asp

2. U.S. Army. (2022). USAMRIID Special Pathogens Laboratory – Resource for all DOD. *Army Medical Research Institute of Infectious Diseases*. https://usamriid.health.mil/SPL/index.htm
3. CDC. (2021, July 1). Zoonotic Diseases. *Centers for Disease Control and Prevention*. https://www.cdc.gov/onehealth/basics/zoonotic-diseases.html
4. FAOSTAT. (2022). Crops and livestock products. *Food and Agriculture Organization of the United Nations*. https://www.fao.org/faostat/en/#data/QCL
5. Melton-Celsa A. R. (2014, July 31). Shiga Toxin (Stx) Classification, Structure, and Function. *National Library of Medicine*. https://www.ncbi.nlm.nih.gov/pmc/articles/PMC4270005/

CHAPTER 19: THE DOCUMENTARIAN

1. Carson R. L. (1962). *Silent Spring*. Houghton Mifflin Company, New York, N.Y.
2. U.S. DOE. (2022, August 3). About Hanford Cleanup. *United States Department of Energy Office of River Protection: Richland Operations Office*. https://www.hanford.gov/page.cfm/AboutHanfordCleanup
3. Jackson S. L. (1994). *Pulp Fiction* – Samuel L. Jackson Speech. Directed by Quentin Tarantino. A Band Apart and Jersey Films.

CHAPTER 20: NEXUS

1. Linden E. (2006). *The Winds of Change: Climate, Weather, and the Destruction of Civilizations*. Simon and Shuster, New York, NY.
2. United Nations. (2022, July 28). UN General Assembly declares access to clean and healthy environment is a universal human right. *UN News*. https://news.un.org/en/story/2022/07/1123482
3. Bar-on Y. M., Phillips R., Milo R. (2018, May 21). The biomass distribution on Earth. *Proceedings of the National Academy of Sciences, PNAS* https://doi.org/10.1073/pnas.1711842115
4. Census of Marine Life. (2011, August 24). How many species on Earth? About 8.7 million, new estimate says. *ScienceDaily*. https://www.sciencedaily.com/releases/2011/08/110823180459.htm
5. IUCN. (2022). The IUCN Red List of Threatened Species™. *The International Union for Conservation of Nature*. https://www.iucnredlist.org/
6. Wood D. (2021, November 14). How Earth's Orbit & Tilt Affects Climate Change. Study.com. https://study.com/academy/lesson/how-earths-orbit-tilt-impacts-climate-change.html
7. Maurer P. (2020, September 17). Communities facing conflict, climate change and environmental degradation walk a tightrope for survival. *International Committee of the Red Cross ICRC*. https://reliefweb.int/report/world/communities-facing-conflict-climate-change-and-environmental-degradation-walk-tightrope
8. Yaffa J. (2022, January 10). The Great Siberian Thaw. *The New Yorker*. https://www.newyorker.com/magazine/2022/01/17/the-great-siberian-thaw
9. IPCC. (2018). Global Warming of 1.5°C. *The Intergovernmental Panel on Climate Change SR1.5*. https://www.ipcc.ch/sr15
10. UNEP. (2022). SDGs will address 'three planetary crises' harming life on earth. *United Nations UN News: Global perspective Human stories*. https://news.un.org/en/story/2021/04/1090762

11. Greshko M. and National Geographic staff. (2019, September 26). What are mass extinctions, and what causes them? National Geographic. https://www.nationalgeographic. com/science/article/mass-extinction
12. Pierce F. (2010, February 24). From ocean to ozone: Earth's nine life-support systems. Center for Biological Diversity. https://www.biologicaldiversity.org/news/center/articles/2010/new-scientist-02-24-2010.html
13. University of Oxford. (2022, January 20). An Estimated 1.2 million people died in 2019 from antibiotic-resistant infections. University of Oxford News and Events. https://www.ox.ac.uk/news/2022-01-20-estimated-12-million-people-died-2019-antibiotic-resistant-bacterial-infections
14. University of Oxford. (2022, January 20).
15. IPCC. (2018).
16. IPCC. (2018).
17. DOI. (2022). Tackling the Climate Crisis. *U.S. Department of the Interior.* https://www.doi.gov/priorities/tackling-climate-crisis

CHAPTER 21: KNOW THE RULES

1. Rice S. A. (1957). *Green Planet: how plants keep the Earth alive.* Rutgers University Press 2009, New Brunswick, N.J.
2. Lewis D. L. (2014). *Science for Sale: How The US Government Uses Powerful Corporations and Leading Universities to Support Government Policies, Silence Top Scientists, Jeopardize Our Health, and Protect Corporate Profits*. Skyhorse Publishing, New York, N.Y.
3. Lewis D. L. (2014).
4. Wilson D. (2001). *Fateful Harvest: The True Story of a Small Town, a Global Industry, and a Toxic Secret.* Harper Collins Publishers Inc., New York, NY.
5. Hoffman L. (2022, February 28). Biosolids may be coming to Toledo. *Willapa Harbor Herald.*
6. Roukis G. S. and Charnov B. H. (1985, May). RICO (Racketeer Influenced and Corrupt Organizations) Statute – Implications for Organized Labor. *U.S. Department of Justice Office of Justice Programs.* https://uscode.house.gov/view.xhtml?path=/prelim@title18/part1/chapter96&edition=prelim
7. BOEM. (2022). Clean Water Act (CWA). *U.S. Department of the Interior Bureau of Ocean Energy Management.* https://www.boem.gov/environment/environmental-assessment/clean-water-act-cwa
8. EPA. (2022, July 6). *U.S. Environmental Protection Agency.* Biosolids Laws and Regulations. https://www.epa.gov/biosolids/biosolids-laws-and-regulations
9. FWS. (1973, December 28). Endangered Species Act. *U.S. Fish & Wildlife Services.* https://www.fws.gov/law/endangered-species-act
10. NOAA. (1972-1973). Laws and Policies: Marine Mammal Protection Act: Endangered Species Act. *NOAA Fisheries.* https://www.fisheries.noaa.gov/topic/laws-policies/endangered-species-act
11. EPA. (2021, November 10). Office of Pesticide Programs: Contacts by Division and Topic. *U.S. Environmental Protection Agency.* https://www.epa.gov/pesticide-contacts/office-pesticide-programs-contacts-division-and-topic

12. EPA. (2021, September 28). Summary of the Resource Conservation and Recovery Act. *U.S. Environmental Protection Agency.* https://www.epa.gov/laws-regulations/summary-resource-conservation-and-recovery-act
13. EPA. (2021, September 28). Summary of the Federal Insecticide, Fungicide, and Rodenticide Act. United States Environmental Protection Agency. https://www.epa.gov/laws-regulations/summary-federal-insecticide-fungicide-and-rodenticide-act

CHAPTER 22: THERE IS A SOLUTION!
1. Einstein A. (1879-1955). Quote of Albert Einstein – Intellectuals solve problems, geniuses create them. *Bubo Quote.* https://www.buboquote.com/en/quote/1189-einstein-intellectuals-solve-problems-geniuses-prevent-them
2. NCI. (2021, October 26). Annual Report to the Nation Part 2: Patient economic burden of cancer care more than $21 billion in the United States in 2019. *National Cancer Institute.* https://www.cancer.gov/news-events/press-releases/2021/annual-report-nation-part-2-economic-burden
3. American Cancer Society. (2010). The Global Economic Cost of Cancer. *American Cancer Society, Inc.* No. 005444. http://phrma-docs.phrma.org/sites/default/files/pdf/08-17-2010_economic_impact_study.pdf
4. Nussy B. (2020, June 5). Straight Facts on the Environmental Impact of Coal: CO_2 Emissions, Pollution, Land and Water. *FreeingEnergy.* https://www.freeingenergy.com/environmental-impact-coal-water-co2-so2-mercury-pollution/
5. Donne J. (1624). *Devotions upon Emergent Occasions, Meditation XVII.*
6. NOAA. (2021, October). Glacial-Interglacial Cycles. *U.S. Department of Commerce National Oceanic and Atmospheric Administration.* https://www.ncei.noaa.gov/sites/default/files/2021-11/1%20Glacial-Interglacial%20Cycles-Final-OCT%202021.pdf
7. Blum M. D. (2013). Glacial-Interglacial Scale Fluvial Responses. *Encyclopedia of Quaternary Science* – Second Edition. https://www.sciencedirect.com/topics/earth-and-planetary-sciences/glacial-interglacial-cycle
8. NOAA. (2022). Climate Forcing. *National Oceanic and Atmospheric Administration* https://www.climate.gov/maps-data/climate-data-primer/predicting-climate/climate-forcing
9. Sharp T., and Dobrijevic D. (2021, December 22). Earth's atmosphere: Facts about our planet's protective blanket. *Space.com.* https://www.space.com/17683-earth-atmosphere.html

CHAPTER 23: NOTHING BUT ENERGY
1. Sun Tzu. (5th Century BC). *The Art of War.*
2. IPCC. (2021). Summary for Policymakers. In: Climate Change 2021: The Physical Science Basis. *Contribution of Working Group I to the Sixth Assessment Report of the Intergovernmental Panel on Climate Change* [Masson-Delmotte V., Zhai P., Pirani A., Connors S. L., Péan C., Berger S., Caud N., Chen Y., Goldfarb L., Gomis M. I., Huang M., Leitzell K., Lonnoy E., Matthews J. B. R., Maycock T. K., Waterfield T., Yelekçi O., Yu R, and Zhou

B. (eds.)]. Cambridge University Press, Cambridge, United Kingdom and New York, NY, USA, pp. 3–32, doi:10.1017/9781009157896.001.
3. Raine N. V. (2022). A Primer on the Greenwashing of Sewage Wastes Pollution. Email.
4. EPA. (1994, September). A Plain English Guide to the EPA Part 503 Biosolids Rule. United States Environmental Protection Agency. https://www.epa.gov/sites/default/files/2018-12/documents/plain-english-guide-part503-biosolids-rule.pdf
5. Wald C. (2022, March 22). *Pipe Dreams: The Urgent Global Quest to Transform the Toilet.* Avid Reader Press/Simon and Shuster.
6. Carson R. L. (1962). *Silent Spring.* Houghton Mifflin Company, New York, N.Y.
7. Van Nieuwkoop M. (2019, June 17). Do the costs of the global food system outweigh its monetary value? *World Bank Blogs.* https://blogs.worldbank.org/voices/do-costs-global-food-system-outweigh-its-monetary-value
8. Borlaug N. (1970). The Nobel Peace Prize 1970 - Norman Borlaug Biographical. *The Nobel Prize.* https://www.nobelprize.org/prizes/peace/1970/borlaug/biographical/

CHAPTER 24: A PECAUTIONARY TALE

1. Goodall J. (2021, August 16). Decide What Kind of Difference You Want to Make: Leadership Quote of the Week. *Laidlaw Scholars.* https://laidlawscholars.network/posts/decide-what-kind-of-difference-you-want-to-make

Acknowledgments

Our deepest gratitude to family, friends, and colleagues who have supported us throughout this effort, and to everyone who reads this tome and is thereby inspired to create a healthier and more sustainable world for all to enjoy now and far into the future.

RCH & MH-B

HORUS PUBLISHING, INC.
P.O Box 1127
McKenna, WA 98558, U.S.A.
https://www.sludgetracker.com
info@sludgetracker.com

Made in United States
Orlando, FL
03 January 2023

28003674R00232